Brock/Springer Series in Contemporary Bioscience

Hypersaline Environments

Brock/Springer Series in Contemporary Bioscience

Series Editor: Thomas D. Brock
University of Wisconsin–Madison

Tom Fenchel
ECOLOGY OF PROTOZOA: The Biology of Free-living Phagotrophic Protists

Johanna Döbereiner and Fábio O. Pedrosa
NITROGEN-FIXING BACTERIA IN NONLEGUMINOUS CROP PLANTS

Tsutomu Hattori
THE VIABLE COUNT: Quantitative and Environmental Aspects

Roman Saliwanchik
PROTECTING BIOTECHNOLOGY INVENTIONS: A Guide for Scientists

Hans G. Schlegel and Botho Bowien (Editors)
AUTOTROPHIC BACTERIA

Barbara Javor
HYPERSALINE ENVIRONMENTS: Microbiology and Biogeochemistry

Barbara Javor

Hypersaline Environments

Microbiology and Biogeochemistry

With 40 Figures and 55 Tables

Springer-Verlag
Berlin Heidelberg New York
London Paris Tokyo

Library of Congress Cataloging-in-Publication Data

Javor, Barbara.
Hypersaline environments.

(Brock/Springer series in contemporary bioscience)
Includes bibliographies and index.
1. Halophilic Microorganisms. 2. Salinity.
3. Microbial Ecology. 4. Biogeochemistry
I. Title. II. Series.
QR97.S3J38 1989 576 89-4230

Cover photograph: *Aphanothece halophytica* in mucilaginous packets from a microbial mat. From the Exportadora de Sal saltern, Mexico. Phase contrast.

ISBN 3-540-50666-7 Springer-Verlag Berlin Heidelberg New York

ISBN 0-387-50666-7 Springer-Verlag New York Berlin Heidelberg

Production and editorial supervision: Science Tech Publishers

Printed in the United States of America

10 9 8 7 6 5 4 3 2 1

Contents

Preface

From the original idea of describing halophilic bacteria and their natural habitats, this volume grew to become a cross-disciplinary examination of the variety of halophilic microorganisms that thrive in extremely hypersaline environments and also a presentation of their roles in modifying the geochemistry of these milieux. Halophilic microorganisms may occur in almost any hypersaline environment: tropical to polar, terrestrial to submarine, acidic to alkaline, or aerobic to anaerobic. The organic and inorganic by-products of halophilic microorganisms affect the morphology, the kinetics of precipitation, and the actual occurrence of many evaporite minerals. Biological activity can also be traced through stable isotopic signatures and the composition of trapped fluids in evaporite minerals.

There is no doubt that an understanding of the geochemical environment of halophilic microoganisms will lead to both new questions and new insights into their biology. Likewise, the exploration and analysis of both recent and ancient evaporite deposits should consider both general microbial processes that can help shape the environment as well as specific microorganisms or types of microoganisms that leave distinct biomarkers. It is hoped that this volume will help bridge the gap for the many microbiologists, geologists, chemists, and limnologists who share an interest in hypersalinity and evaporites.

For the purpose of describing microorganisms and biogeochemical processes in hypersaline environments, the descriptions in this book deal basically with salinities >7–10%. Many organisms that inhabit normal seawater (3.5% salinity) are able to tolerate slightly elevated salinities, but only truly halophilic or halotolerant species can thrive in habitats characterized by extreme hypersalinity and the precipitation of evaporite minerals.

I wish to acknowledge the library facilities of the University of California at San Diego and the computer search service of Chemical Abstracts. I also wish to thank Claude ZoBell for making available his reprint collection of many of the older works cited. I want to thank Preston E. Cloud and the late Mel Webster for inspiring me to view the world from a biogeochemical point of view. Lastly, I am grateful to Marina, who taught me that microbial mats are "algae gardens" and beautiful sunsets are "halobacterial skies."

Introduction

Hypersaline environments are not mere scientific curiosities nor dead-ends of geochemical processes. They have been known to exist since Precambrian times, and ancient evaporites are a rich source of information concerning the development of oceanic and continental basins. A growing body of biogeochemical knowledge suggests that microorganisms and their products have played major roles in modifying both the precipitation and the diagenesis of chemical sediments in hypersaline environments.

Some evaporite minerals and their associated secondary minerals (e.g., native sulfur) are economically important. Ancient halite deposits are the primary source of the world's sodium chloride while potash deposits are mined for use in agriculture and industry. Native sulfur associated with gypsum is believed to be a biological end-product of the activities of sulfate-reducing and sulfide-oxidizing bacteria.

Evaporite minerals are also of indirect economic importance. In the petroleum industry, it is recognized that not only do many evaporite deposits contain reservoirs of petroleum, but they were probably the source rocks for petroleum genesis. While evaporite minerals may be poor in heavy metals, the brines from which the minerals precipitate are not. Evaporites are sometimes associated with stratiform, metal-bearing, sedimentary rocks. Geological reconstruction of the sedimentary basins containing such deposits shows that the metalliferous deposits were probably a result of the biogeochemical interaction between the activities of anaerobic bacteria, metal-enriched brines, and porewater migration.

Evidence indicates that different classes of microbially-produced dissolved organic carbon (DOC) can modify, inhibit, or enhance the precipitation of various evaporite minerals and affect their crystal morphology. Models that predict the order of precipitation of minerals during an evaporation event have generally ignored the possible role of organic matter and microbial activity in modifying ion activity and the solubility of chemical precipitates. It is unknown to what extent DOC may have modified the distribution of ancient evaporites.

Using the tenet that the present is the key to the past, the study of modern evaporite deposits has been applied to the understanding of major trends in oceanic and continental evolution. The study of stable isotope fractionation in sedimentary sulfates and carbonates throughout Phanerozoic history has provided a unique window to observe changes in the distribution and organic productivity of ancient oceans. Studies in paleontology and plate tectonics have provided evidence to reconstruct the relative positions of the continents and world's oceans during the last 600 million years. Stable isotope analyses of ancient evaporites support the model of a dynamic crust marked by changing ocean basins and continental margins. These analyses indicate periodic changes in sea level, the relative extent of shallow seas (which may indicate high organic productivity), and the relative amount of sulfur and carbon locked up in rocks during different geological periods.

Stable isotope studies are useful because CO_2-fixing organisms discriminate against ^{13}C and sulfate-reducing bacteria discriminate against ^{34}S. Although these two major biological activities have been recorded in ancient evaporite minerals, it remains uncertain to what extent stable isotope fractionation has occurred *in situ* under hypersaline conditions or under normal marine conditions, to be reflected in the chemical precipitates. Investigations of modern evaporite basins may provide the answers.

Many hypersaline environments are inimical to macroscopic life, but are actually the preferred habitats of a wide variety of microorganisms. Both primary productivity and degradation occur in evaporite environments at high salinities. Among both eucaryotes and procaryotes, a broad variety of biochemical adaptations have evolved to allow the organisms to cope with osmotic and ionic stress, temperature changes, various light environments, and variations in Eh (particularly oxygen and sulfide concentrations). Yet to be addressed are questions concerning pressure tolerance and the chemical mechanisms that prevent the most salt-tolerant microorganisms from thriving in certain brines.

Investigations of metabolic activities of isolated microorganisms from hypersaline habitats provide clues to their potential activities in nature. However, there remain significant gaps between laboratory studies, field investigations, and interpretations from ancient evaporite environments. Some unanswered questions are given below.

Although osmotic pressure due to dissolved salts is a major factor influencing biological activity, all hypersaline environments of equal osmotic pressure do not show the same extent of biological development. The chemical composition of the brine obviously plays a role. Why are some hypersaline environments inimical or nearly inimical to life? Such environments include marine-derived bitterns brines, undiluted Dead Sea water, and Don Juan Pond in Antarctica (a $CaCl_2$ brine). Crystalline NaCl and mirabilite ($Na_2SO_4 \cdot 10\ H_2O$) may contain viable bacteria, but anhydrous Na_2SO_4 (thenardite) does not.

Are carbon, nitrogen, and sulfur mineral cycles complete in hypersaline environments, or are they attenuated or interrupted? Photoautotrophy, methanogenesis, and anaerobic decomposition (fermentation) appear to occur in high salinities, but sulfate reduction may be attenuated by extremely hypersaline conditions more so than these other processes. Attempts to demonstrate N_2 fixation in extremely hypersaline environments have been unsuccessful, but such habitats may not be N-limited and therefore nitrogen fixation may not be expected.

How important is anoxygenic photosynthesis in driving the carbon cycle in certain hypersaline environments? In some modern, extremely saline habitats, the only significant primary producers are cyanobacteria (e.g., marine-derived systems) or phototrophic bacteria (e.g., alkaline systems). Some cyanobacteria can thrive by anoxygenic photosynthesis. Environments dominated by anoxygenic cyanobacteria or phototrophic bacteria would not only be distinguished by totally or predominantly anoxic conditions, but the hydrocarbons of the buried autochthanous organic matter would largely or entirely be of procaryotic origin. Knowledge of the ecology and biogeochemistry of anoxygenic photosynthesis may help in understanding the origin of the important process of oxygenic photosynthesis in the Precambrian.

What limits the development of microbial mats in hypersaline environments? Organic-rich sediments associated with microbial mats do not appear to be important once brines have evaporated past the point of gypsum saturation and precipitation. High salinities alone may inhibit the growth of the microbial communities thriving in microbial mats, or the precipitation of gypsum may prevent nutrient recycling from the sediments below, a process that is necessary to maintain a microbial mat. Since microbial mats are potential stromatolite deposits, an understanding of the effect of salinity on mat formation may help in understanding the geomicrobiology of ancient and modern stromatolites.

Why do certain microorganisms dominate in hypersaline environments? Certain species of the green alga *Dunaliella* are common in hypersaline brines. Other eucaryotic algae, including diatoms, are usually present but they never appear to dominate. *Dunaliella* can grow in a broad range of salinities, yet cyanobacteria are often the dominant primary producers in salinities between about 5 and 15%. Yeasts and fungi as a whole are osmotolerant, yet they have never been described as important agents of biodegradation in hypersaline environments. In culture medium, most species of halobacteria thrive best in 20–25% salinity. They grow poorly in marine-derived brines of 25% or greater salinities, the brines in which they are most abundant in solar salterns. Are they merely passively concentrated by brine evaporation, or are there significant differences between growth observed in the laboratory and actual growth under natural conditions?

Through cross-disciplinary studies of a variety of marine and non-marine hypersaline environments, of the different halotolerant and halophilic microorganisms isolated from these habitats, and of the organic and inorganic geochemical markers associated with the sediments, it should be possible to reconstruct more accurately ancient evaporite basins and to recognize the contribution of microorganisms and their by-products to the mineralogy of the sediments. In turn, the analyses of ancient evaporites and observations and measurements of biogeochemical processes in modern hypersaline environments may provide a basis for improvement and innovation in chemical technology and biotechnology. The role of microorganisms in solar salt production is well established, and the use of organic agents in potash processing is standard in the industry. Biotechnology using halophilic microorganisms is only just beginning, with β-carotene and glycerol production from *Dunaliella* the pioneer. Halophilic microorganisms may prove to be rich sources of fine chemicals and novel pharmaceuticals, as well as important keys to exploration in the petroleum industry.

1

Geology and Chemistry

The terms *brines* and *evaporites* are used in a generic way to describe concentrated solutions of ions and the chemical precipitates from those solutions. Based on the order of precipitation of evaporite minerals from brines of known composition, geochemists have deduced the chemistry of brines from which ancient evaporites have been deposited. Modern evaporite basins and their brines also serve as models to reconstruct the physical and temporal development of ancient evaporite deposits. The following discussion highlights some of the major geological aspects of hypersaline environments as well as chemical and physical aspects of brines. The discussion of brine chemistry emphasizes the factors that are important in biological productivity. The modern evaporite areas discussed in this book are shown in Figure 1.1.

1.1 Brine and evaporite evolution

Evaporites, defined as minerals precipitated from solution as a result of the evaporation of water, occur in both marine-derived (thalassic) and non-marine (athalassic) systems. Evaporites may be deposited in polar regions (in arid zones where cold temperatures promote the freezing-out of salts), continental lakes, subterranean aquifers, and subtropical marine environments. In marine environments, strong brines and evaporite formation can occur in coastal intertidal zones, supratidal zones (sabkhas), lagoons fed by direct flow from the sea, and sea-marginal lakes and ponds fed by percolation through a natural barrier. A summary of chemical composition of a variety of hypersaline environments is given in Table 1.1. Evaporites are also recorded in larger, deeper basins and sub-sealevel basins with inflow from the sea. For the pur-

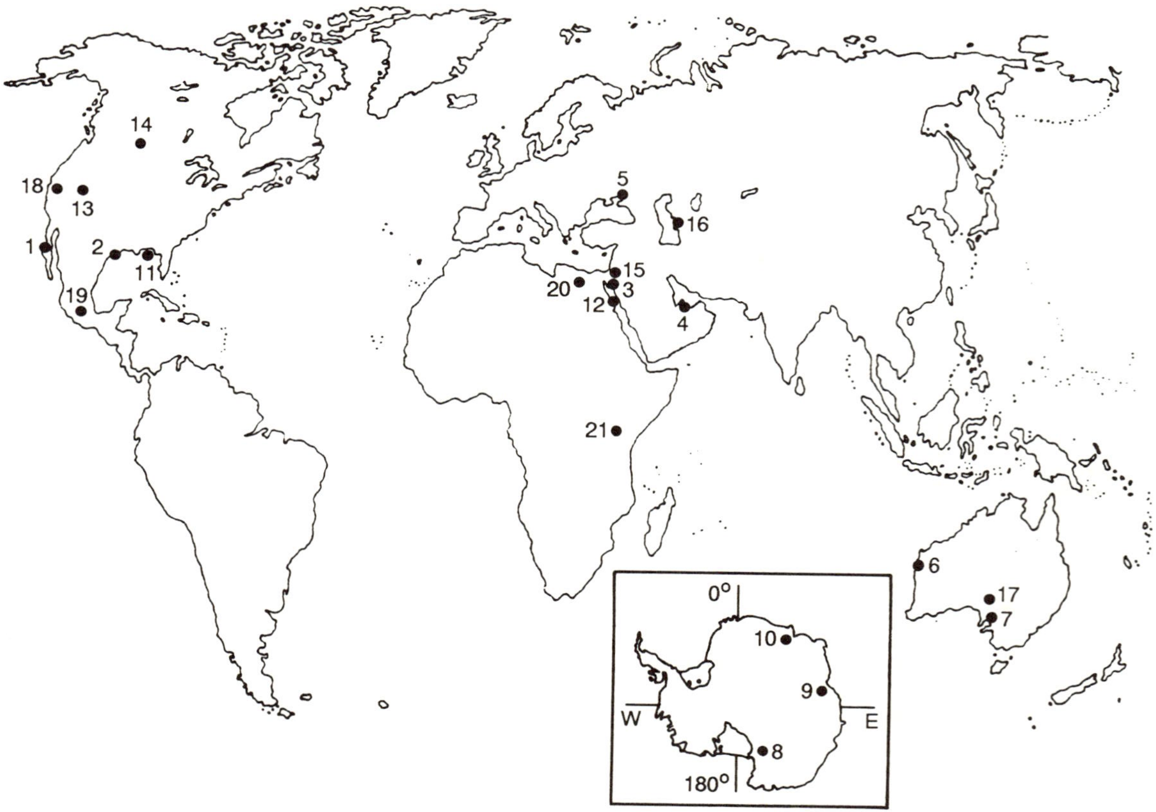

Figure 1.1 Modern evaporite basins discussed in this volume. Thalassic basins: 1, Laguna Ojo de Liebre, Mexico; 2) Laguna Madre, Texas and Mexico; 3, Solar Lake, Gavish Sabkha, and Ras Mohammad pool, Sinai; 4) Trucial coast, Persian Gulf; 5, Putrid Sea, Sea of Azov; 6, Shark Bay, Western Australia; 7, Spencer Gulf, South Australia. Antarctic lakes (inset): 8, Dry Valley lakes (Lake Bonney, Lake Vanda, and Don Juan pond); 9, Deep Lake, Vestfold Hills; 10, Syowa Oasis lakes (Lake Hunazoko and Lake Suribati). Submarine basins: 11, Gulf of Mexico (Orca Basin and East Flower Garden); 12, Red Sea. Neutral athalassic basins: 13, Great Salt Lake, Utah; 14, Saskatchewan lakes, Canada; 15, Dead Sea, Israel-Jordan; 16, Kara Bogaz Gol, Caspian Sea; 17, Lake Eyre, South Australia. Alkaline athalassic basins: 18, western Great Basin lakes, U.S.A. (Mono Lake, Owens Lake, Searles Lake, and Big Soda Lake); 19, Lake Texcoco, Mexico; 20, Wadi Natrun, Egypt; 21, Lake Magadi, Kenya.

Table 1.1 Major ion concentrations of extremely hypersaline, non-alkaline brines ($g \cdot l^{-1}$)

Ion	Seawater	Seawater at onset of gypsum saturation	Seawater at onset of NaCl saturation	Seawater at onset of potash saturation	Great Salt Lake, North Arm	Dead Sea, lower water mass before overturn
Na^+	10.8	49.5	98.4	61.4	105	39.7
Mg^{2+}	1.3	6.8	14.5	39.3	11.1	42.4
Ca^{2+}	0.4	1.7	0.4	0.2	0.3	17.2
K^+	0.4	2.0	4.9	12.8	6.7	7.6
Cl^-	19.4	91.5	187	189	181	219
SO_4^{2-}	2.7	12.5	19.3	51.2	27.0	0.4
TDS, %	3.5	16.4	32.4	35.4	33.3	32.7
References	Holser, 1979a	Javor, 1983a, 1983b	Javor, 1983a, 1983b	Javor, 1983a, 1983b	Post, 1977	Nissenbaum, 1975

Table 1.1 continued

	Saskatchewan lakes		Antarctic lakes			
Ions	Lake Mushiki	Little Manitou Lake	Lake Suribati	Lake Vanda (64.6 m)	Lake Bonney, East lobe (32.5 m)	Don Juan Pond
Na^+	62.2	12.3	69.2	6.1	56.9	11.5
Mg^{2+}	29.0	9.5	8.5	7.4	21.7	1.2
Ca^{2+}	low	low	1.2	24.4	1.2	114
K^+	low	low	2.5	0.6	2.3	0.2
Cl^-	114	18.0	130	74.3	162	212
SO_4^{2-}	237	39.6	8.1	0.6	2.9	0.01
TDS, %	34.2	8.1	22.0	11.4	24.8	33.9
Reference	Hammer, 1978	Hammer, 1978	Tominaga & Fukui, 1981	Torii et al., 1975	Torii et al., 1975	Meyer et al., 1962

TDS, total dissolved solids.

poses of definition, athalassic brines and evaporites refer to systems that are not associated with marine environments, although the salts and brines may have been associated with a marine environment at an earlier time in geologic history. For that reason, the ion content of the Great Salt Lake, Utah (U.S.A.), is similar to that of evaporated seawater (Table 1.1).

Evaporites constitute about 2% of all the sediments on the platforms of continents (Ronov, 1968). Evaporite deposits have been recorded from the Precambrian (with the oldest dating from 3.5 b.y. [Button, 1982]) through Phanerozoic time. Some of the greatest salt deposits known are associated with the Paleozoic, particularly the Permian (Figure 1.2). Permian evaporite basins probably covered more than 4.5 million km^2 of the earth's surface and produced nearly 1.3 million km^3 of rock salt (Zharkov, 1981, 1984). The distribution of modern evaporites is very localized and rather insignificant in terms of geologic time and sediment volume. Analyses of sedimentary sulfates through geologic time have been used to interpret major trends in crustal and atmospheric history (Claypool et al., 1980; Veizer et al., 1980) (see Chapter 4). Studies of the distribution and composition of evaporites through geologic time have also demonstrated the intimate and dynamic interactions between the cycles of sulfide/sulfate and organic carbon/carbonate that control global O_2 and CO_2 production and consumption (Garrels and Perry, 1974; Garrels and Lerman, 1984).

Chemical, physical, and geological aspects of the evolution of brines and evaporites have been thoroughly discussed in a number of reviews, including those largely on marine systems (Borchert and Muir, 1964; Braitsch, 1971) and those largely on non-marine systems (Hardie and Eugster, 1970; Eugster and Hardie, 1978). Hardie (1984) identified some of the major features that distinguish marine from non-marine evaporites, including kinds of primary and secondary evaporite minerals, associated non-evaporite rocks and sediments, kinds of fossils, trace elements, stable isotope signatures, and fluid inclusion geochemistry. In some cases, the organic geochemical "fingerprints" of strictly marine or strictly non-marine, halophilic organisms should be detectable in the dissolved organic carbon (DOC) or particulate organic carbon (POC) associated with the evaporites.

When normal seawater (3.5% salinity) undergoes evaporation, minerals precipitate in order of their solubility. In a simple batch evaporation sequence, the parent brines become concentrated (Figure 1.3) and the major minerals precipitate in the order: calcium carbonate (calcite or aragonite), gypsum, halite, and certain K-Mg minerals (potash) (Table 1.2). The theoretical succession of minerals and the observed succession are often different, especially in the potash phases (Dean, 1978; Harvie et al., 1980). The solubility of various minerals is strongly affected by the common ion effect and ion complexing (Holser, 1979a) as well as DOC in some cases (Chapter 3). The effects of DOC have not been included in any models of evaporite precipitation sequences.

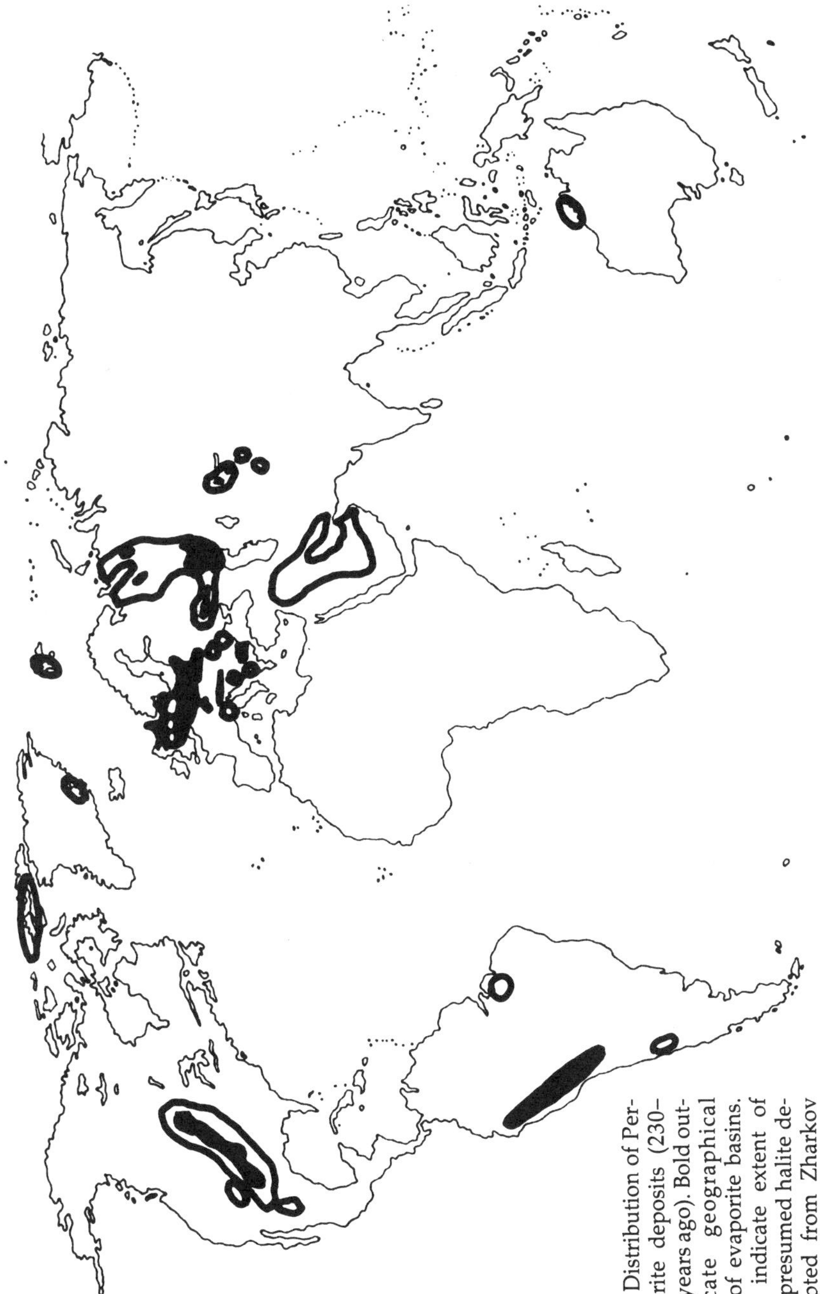

Figure 1.2 Distribution of Permian evaporite deposits (230–280 million years ago). Bold outlines indicate geographical boundaries of evaporite basins. Black areas indicate extent of known and presumed halite deposits. Adapted from Zharkov (1981, 1984).

Table 1.2 Evaporite minerals discussed in the text

Name	Composition	Usual occurrence[a]
Calcite	$CaCO_3$	1°, 2°, *
Aragonite	$CaCO_3$	1°, 2°, *
Dolomite	$CaMg(CO_3)_2$	2°, *
Gypsum	$CaSO_4 \cdot 2\ H_2O$	1°, *
Anhydrite	$CaSO_4$	1°, 2°, *?
Halite, rock salt	$NaCl$	1°, *
Hydrohalite	$NaCl \cdot 2\ H_2O$	1°
Mirabilite	$Na_2SO_4 \cdot 10\ H_2O$	1°, *
Thenardite	Na_2SO_4	2°
Antarctite	$CaCl_2 \cdot 6\ H_2O$	1°
Tachyhydrite	$CaCl_2 \cdot (MgCl_2)_2 \cdot 12\ H_2O$	1°, 2°
Trona	$NaHCO_3 \cdot Na_2CO_3 \cdot 12\ H_2O$	1°, 2°, *
Natron	$Na_2CO_3 \cdot 10\ H_2O$	1°
Nahcolite	$NaHCO_3$	1°, 2°, *
Sylvite	KCl	1°, 2°, *?
Carnallite	$KCl \cdot MgCl_2 \cdot 6\ H_2O$	1°, *
Celestite	$SrSO_4$	1°, 2°, *

[a]1° = as a primary precipitate, 2° = as a secondary precipitate, * = demonstrated influence of biological activity or dissolved organic carbon on occurrence, solubility, or crystal morphology. (From Sonnenfeld, 1984).

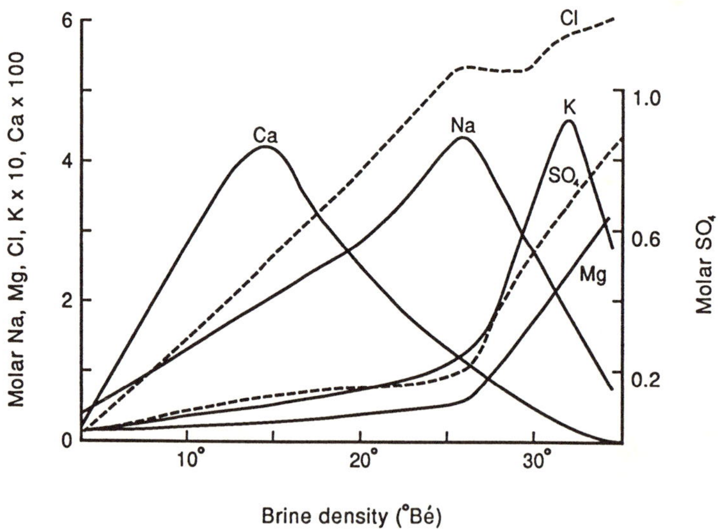

Figure 1.3 The composition of seawater brines from the Exportadora de Sal saltern during the course of evaporation (from Javor, 1983a).

Evaporite models are further complicated when the brine composition deviates from that predicted by batch evaporation. Factors that can complicate the prediction of mineral sequences include: influx or contact with less saline water or with brines of different composition; contact with soluble rocks and sediments; reflux of denser brines from the evaporation basin; biological activity or DOC; and non-steady state conditions (such as periodic brine freshening or freezing-out of salts). If primary precipitates are altered during diagenesis to produce secondary minerals, the composition of the interstitial brines may also change. This factor is important for understanding microbial activities in hypersaline sediments. Although a number of porewater and mineral analyses of evaporite sediments have been conducted, relatively few have been directly interfaced with measurements of biological activity (see Chapters 14, 15, 18, and 19).

Non-marine evaporites and brines are often of markedly different ion composition than those found in marine environments (Table 1.1 and Chapters 17–20). The ion composition depends on the rock types associated with the aquifers feeding the system. The solute composition of the inflowing waters into some alkaline lakes (Chapter 20), such as those of the African Rift valley and the Great Basin of North America, is typically a result of associations with volcanic rocks. Inflowing springs are sometimes recharged from hot groundwater reservoirs causing the selective dissolution of subterranean minerals. Potassium ions are often screened out by ion exchange or adsorption. Ca^{2+} (and Mg^{2+}) may be low in HCO_3^--rich systems because their carbonate salts are less soluble in freshwater than in seawater, an effect caused by ion complexing (Berner, 1971).

The dominant anions of Na^+-rich, non-marine brines consist of HCO_3^-/CO_3^{2-}, Cl^-, and SO_4^{2-} (present in varying concentrations). Sulfate reduction and the concomitant increase in HCO_3^- concentration has been documented as the mechanism of alkalinization of the inflowing springs into the Wadi Natrun lakes (Abd-el-Malek and Rizk, 1963). High concentrations of fluoride and borate may also accumulate in some systems and eventually precipitate, especially as sodium salts. A detailed study of the evolution of such brines and evaporites has been made in several lake systems described in Chapter 20. All of these alkaline systems are associated with either high organic composition or microbial decomposition in the sediments.

In hypersaline lakes subjected to extremely cold conditions, hydrated minerals such as mirabilite often precipitate by freezing out. The Great Salt Lake and some of the Saskatchewan lakes experience such chemical precipitation during the winter but redissolution occurs in warmer weather. Many of the Antarctic hypersaline lakes are nearly constantly subjected to extremely cold conditions in which the hydrated minerals remain stable (Chapter 16).

1.2 Chemical and physical aspects of concentrated brines

Total ion composition Salinity or total ion composition can be measured by a variety of techniques. Unfortunately, different investigators employ different methods and there often is no means of comparison between studies.

The following discussion outlines techniques that are used by both geochemists and biologists.

Salinity or total dissolved salt content is best measured by chemical analyses and summation for all specific dissolved ions. Although the analytical methods may be time-consuming and expensive, the problems of analytical error due to fluctuations in brine temperature are therefore avoided. In addition, enrichments or depletions of certain ions are detected. Salinity is then expressed as the sum of the weight of the ions per kg water, per kg brine, or per liter brine. The molar or molal concentration of contributing species can be calculated to determine charge balance and ionic strength.

In seawater-derived brines, the salinity can be expressed as the degree of evaporation based on the concentration of a conservative ion such as Br^-, Mg^{2+} (if the brines have not evaporated so far as to precipitate potash salts), or Cl^- (if the brines have not precipitated NaCl). Other measurements listed by Holser (1979a) include measurement of volume or mass ratios of salts or brines relative to the original seawater.

Measurement of salt content by dry weight determination (total dissolved solids, or TDS) is not particularly accurate in strong brines because minerals lose their water of hydration at different temperatures and caking in the evaporation vessels may not permit proper dehydration of all the hygroscopic salts. Gypsum loses all its hydration water at 57°C (Hardie, 1967). TDS measurements in the halite crystallizing basins of salterns by this technique led to the reports of salinities greater than 40% (Rodriguez-Valera et al., 1981, 1985), whereas salinity measurements by specific ion content show brine salinity to be ca. 32–35% in crystallizers (Table 1.1).

Hydrometry is a convenient method for measurement of brine density as well as salinity if the ionic composition is known. Hydrometers based on specific gravity (sp. gr.) or on a Baumé scale (°Bé) are commonly used. At a temperature of 20°C, 0° Bé is equivalent to a sp. gr. of 1.000. °Bé at 20°C can be converted to specific gravity by the equation

$$\text{Sp. gr.} = 145/(145 - {}^\circ\text{Bé})$$

Hydrometer measurements at temperatures other than 20°C require compensation calculations.

Like hydrometry, refractometry is suitable for salinity measurements in seawater-derived systems (based on batch evaporation) or in other saline bodies where ion composition can be measured or predicted. Refractometry, like hydrometry, is temperature-sensitive. An advantage of refractometry is that only a small volume of brine (<0.1 ml) is needed for measurement, making refractometry a convenient and inexpensive method for determining salinity of sediment porewater.

The specific conductance of hypersaline brines is a measurement often employed by limnologists and oceanographers who use this method to compare relative salinities of less saline lakes or normal seawater. Although the

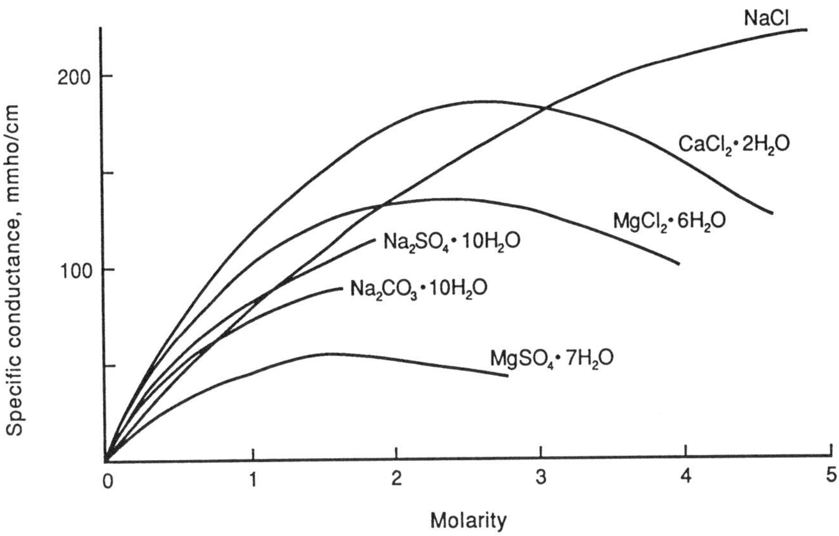

Figure 1.4 The specific conductance of different salt solutions at 20°C (calculated from *CRC Handbook of Chemistry and Physics* [Weast, 1984]).

specific conductance gives some measure of salinity, it is not particularly informative for strong brines because different ions affect the electrical conductivity in different ways (Figure 1.4.) NaCl solutions have a lower conductivity than other common chloride salts below 2–3 M concentration, but conductivity increases as NaCl approaches saturation. The specific conductance of $CaCl_2$ and $MgCl_2$ solutions actually decrease in concentrated brines. Carbonate and sulfate solutions are less electrically conductive than chloride solutions. Conductivity measurements, which are expressed as mmhos·cm^{-1} (a mmho is equivalent to a millisiemen, mS), are temperature-sensitive.

As brines increase in concentration, a variety of other physical and chemical attributes are affected (Table 1.3). The vapor pressure decreases as does the water activity. The succession of evaporite precipitation is strongly de-

Table 1.3 Some chemical and physical effects associated with increasing concentration of brines

Higher TDS and specific gravity	Lower specific heat
Change in ion content and ratios	Increased surface tension
Increased ionic strength	Increased viscosity
Change in activity coefficients	Lower gas solubility
Change in pH	Lower gas diffusion coefficients
Lower water content	Increased electrical conductance (usually)
Lower water activity	

TDS, total dissolved solids.

pendent on the relative humidity of the brines and of the atmosphere. Using the measurements of brine vapor pressures of Rothbaum (1958) and converting them to water activity, a_w (defined as the vapor pressure of the brine divided by the vapor pressure of pure water), it is possible to determine the stability of brines and chemical precipitates from seawater in any evaporation regime at 20°–30°C (Figure 1.5). Evaporation will not proceed unless the relative humidity ($= 100 \times a_w$) of the atmosphere is less than that of the brine. The data are expressed here in terms of a_w rather than vapor pressure to emphasize the functional water content, a factor essential for the success of organisms in hypersaline brines. While halobacteria have been noted to thrive in NaCl-saturated marine brines ($a_w = \leq 0.77$ at 25°C), they are absent and unable to grow in bitterns brines ($a_w = \leq 0.63$ at 25°C) (Javor, 1983b, 1984). The lower a_w limit for halobacteria is not as low as that of xerophilic fungi growing in concentrated sugar solutions ($a_w = 0.61$) (Horowitz, 1979).

Brine concentration can also be expressed as osmolality (Os), defined as the freezing-point depression (°C) of a solution relative to that of distilled water, divided by 1.86. It is expressed as $Os \cdot kg^{-1}$ water. Like water activity, osmolality describes the interaction of water and solutes in brines. Unlike a_w, osmolality is unaffected by temperature. Both types of measurements are useful for comparing biological activities in strong ionic and non-ionic solutions.

pH and alkalinity In dilute freshwater, pH measurements are indicative of the alkalinity. However, the pH typically gives little indication of the total

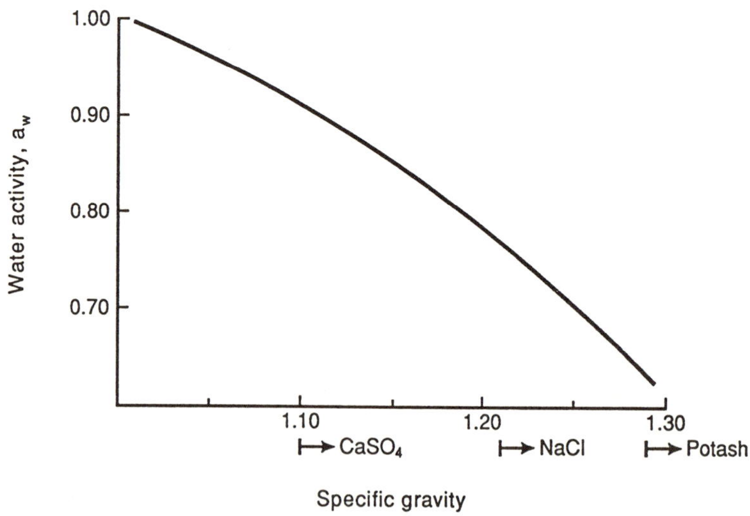

Figure 1.5 Water activity of concentrated seawater brines at 25°C (calculated from Rothbaum, 1958).

alkalinity of brines. The pH of seawater increases with salinity until it is concentrated to about 5% (about pH 9), and then the pH drops with increasing salinity (Copeland, 1967). Amit and Bentor (1971) showed that the pH of Dead Sea and salt spring brines as well as seawater increased upon dilution. The effect was found only in solutions containing HCO_3^- salts. They concluded that the dissociation of HCO_3^- is depressed in strong brines. As dilution causes dissociation and the formation of OH^-, pH increases. A detailed review of the effects of salts on the pH of brines was presented by Krumgalz (1980).

When measuring pH, the use of a liquid junction in the reference electrode may lead to significant error in the estimation of brine pH. Ben-Yaakov and Sass (1977) estimated the pH of artificial Dead Sea brine in an electrochemical cell without a liquid junction and found that the pH of the artificial brine (5.86) was lower than that measured employing a liquid junction reference electrode (6.22). In addition to electrode uncertainties, the use of dilute standard buffers for electrode calibration may be inappropriate for correct pH determination of brines.

Changes in both ion pairing and in activity coefficients of dissolved ions contribute to the pH (or a_H) and ionic strength of brines. As a result of ion complexing, the true ionic strength may be much lower than that calculated from the molalities of each ion (Berner, 1971). Seawater has an ionic strength of 0.7. Between a calculated ionic strength of 1.0 and 4.0 in NaCl solutions with HCO_3^- and CO_3^{2-}, the activity coefficients of Na^+, Mg^{2+}, and Ca^{2+} increase, but those of HCO_3^- and CO_3^{2-} decrease (Garrels, 1967). At ionic strengths >3, over 90% of the free carbonate is complexed in ion pairs. Because HCO_3^- complexes even more strongly with Mg^{2+} than with Na^+ in artificial seawater (Garrels et al., 1961), it is reasonable to expect that even less free HCO_3^- and CO_3^{2-} are found in concentrated seawater than in NaCl solutions.

Because much of the dissolved $HCO_3^- + CO_3^{2-}$ is bound in complexes, the contribution of these ions to the salinity and to the total CO_2 must be measured by alkalinity titration (total alkalinity). In seawater-derived brines which are devoid of sulfide and are low in ammonia, total alkalinity is conservative with brine concentration (Figure 1.6). Evaporation of seawater under laboratory conditions results in a distinct deflection in total alkalinity where $CaCO_3$ precipitates (between 2.5 $\times$ and 4 $\times$ seawater concentration) (Lazar et al., 1983). The total alkalinity can increase significantly in hypersaline sediments due to bacterial sulfate reduction (Chapter 4). The mechanism of seawater buffering in anaerobic porewaters was described by Ben-Yaakov (1973).

In concentrated seawater, borate may contribute significantly to the total alkalinity since it apparently concentrates conservatively until the potash phase of brine evolution. Borate accessory minerals are sometimes noted with potash salts (Holser, 1979a). Boron concentration in normal seawater is 0.4 mM. In

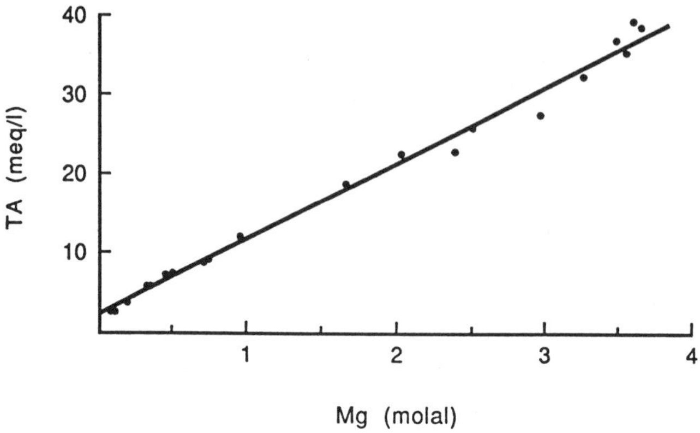

Figure 1.6 Total alkalinity (TA) of seawater brines from a solar saltern at different concentration levels. (expressed as magnesium ion molality).

some alkaline deposits, borate salts may be major precipitates. It is significant that the growth of some halophilic bacteria isolated from alkaline lake brines and sediments was unaffected by 1–5% sodium borate (Nehrkorn and Schwartz, 1961).

Although activity coefficients, ion complexing, and total alkalinity have been measured in brines, the apparent dissociation constants of the carbonate system

$$CO_2 + H_2O \longrightarrow H_2CO_3 \xrightarrow{K'_1} H^+ + HCO_3^-$$

$$HCO_3^- \xrightarrow{K'_2} H^+ + CO_3^{2-}$$

have been determined accurately only in Dead Sea brines (Sass and Ben-Yaakov, 1977). The investigators found that as salinity increased, the pK'_1 and pK'_2 decreased while the activity coefficient of H^+ increased. Because the composition of Dead Sea brines is so markedly different from that of concentrated marine seawater, the apparent dissociation constants of the carbonate system of thalassic brines must be determined independently.

The total alkalinity ($HCO_3^- + CO_3^{2-} + BO_4^{3-} + OH^-$) of a brine can be easily determined by acid titration because this measurement does not require the knowledge of the first and second dissociation constants of the carbonate system. However, measurements of primary productivity employing $^{14}CO_3^{2-}$ cannot be accurately performed in marine-derived brines since such methods employ the carbonate alkalinity in the determination of the final specific activity of the radiotracer. The calculation of carbonate activity requires the knowldege of the apparent dissociation constants of the carbonate system of each salinity in question. Errors may be especially introduced when primary productivity rates are determined at different depths in a salinity-stratified water column or at different dates when salinities are subject to seasonal changes. The resulting $^{14}CO_2$ fixation rates are affected by both the biological composition and activity as well as the carbonate chemistry of the brines of different salinities. All the productivity measurements in Solar Lake by Cohen and co-workers (Chapter 14) were performed with ^{14}C radiotracers without calculations of the carbonate activity and they should therefore be regarded with caution.

Specific heat Specific heat or the thermal capacity of brines (the heat required to raise the temperature of a given mass or volume of brine compared to that of pure water) decreases with increasing salinity. This means that for a given volume of brine, less heat is required to raise the temperature a given number of degrees than is necessary for an equivalent volume of pure water. The relatively low heat capacity of brines contributes to the ability of density-stratified brines to remain hot in heliothermal lakes such as Solar Lake.

It is believed that saltern crystallizer brines colored red by halobacteria retain more heat due to lower reflectance of the brine, but this effect has never been reported in the literature. However, an appropriate experiment was performed at Exportadora de Sal saltern in Baja California (H. Estrada, personal communication). Flasks containing crystallizer brines with halobacterial enrichments of different bacterial densities (achieved by including 0, 0.1, 0.5 or 1.0% peptone in the brine) were placed in the sun and the temperatures were recorded over a 6.5-hr period (Figure 1.7). The redder, more turbid brines not only heated faster but attained higher temperatures than the clearer brines, even though all the brines were heated significantly over the maximum ambient temperature (23.3°C). Garrett (1965) calculated that a salt pond with 15 cm of brine and a perfectly reflective bottom should retain >96% of the incoming solar radiation. The heat retention is even greater when the brines are turbid.

Dissolved gases The diffusivity of gases at 25°C and 1 atm pressure decreases with increasing TDS or viscosity of a solution. For example, in 3 M NaCl, the diffusivity of CO_2 (defined as $10^5 \times cm^2 \cdot sec^{-1}$) is only 75% of that rate measured in pure water (Ratcliff and Holdcroft, 1963). Not only do gases diffuse more slowly as brine density increases, but the capacity to hold dissolved gases decreases. Kinsman et al. (1974) showed that the O_2 solubility at 22°C decreased in marine brines from about 7 $mg \cdot kg^{-1}$ water in normal seawater, to about 4 $mg \cdot kg^{-1}$ water at the onset of gypsum saturation, and to about 2 $mg \cdot kg^{-1}$ water at the onset of halite precipitation. In spite of their large surface-to-volume ratios, saltern crystallizers are typically anaerobic due to bacterial consumption of the O_2 that diffuses from the atmosphere or is produced by *Dunaliella* (Javor, 1983b).

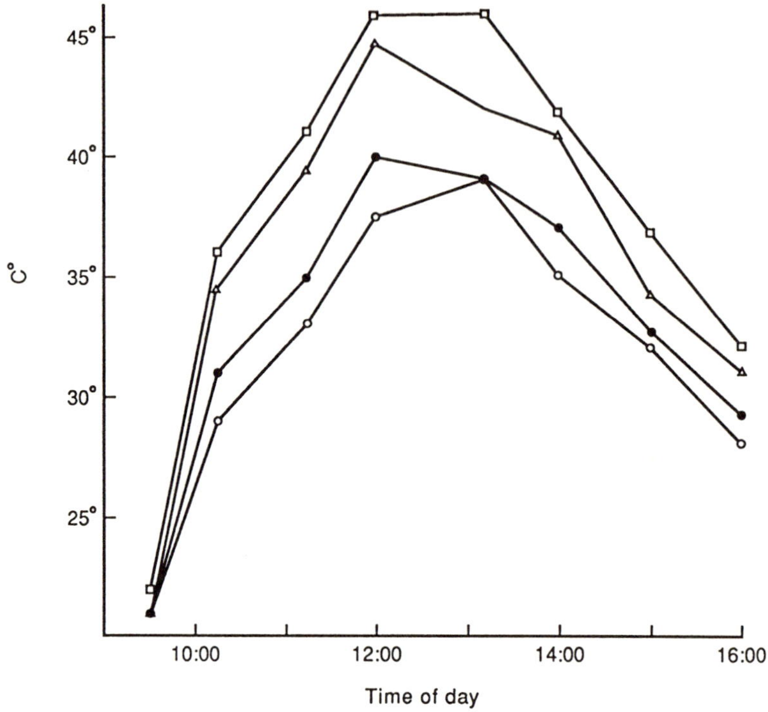

Figure 1.7 Temperatures attained during the course of a day by flasks of saltern crystallizer brines of the same specific gravity but enriched with different densities of halobacteria. Open circles: O.D.$_{580}$ = 0.034 (control, no bacterial enrichment). Closed circles: O.D.$_{580}$ = 0.180. Triangles: O.D.$_{580}$ = 0.740. Squares: O.D.$_{580}$ = 0.850. (Data of Hilario Estrada, Exportadora de Sal, Baja California).

1.3 Trace elements and nutrients

Trace elements Evaporite salts rarely precipitate as pure minerals but rather they typically include trace amounts of heavy metals. Heavy metals also accumulate in the brines associated with evaporite salts. Stratiform, metalliferous deposits overlain by evaporite deposits account for about 30% of the copper produced in the world (Renfro, 1974). Some of the best known metalliferous deposits associated with evaporites are the Kupferschiefer beds of the Zechstein evaporites in Europe. Some deposits associated with evaporites may also contain significant amounts of gold, lead, zinc, and cobalt. For example, in the Elk Point evaporite sequence (Saskatchewan, Canada), the basal shale and anhydrite zones are particularly enriched in Cu, Pb, and Zn, although these metals are found in much higher concentrations in the shale (Thiede and Cameron, 1978). Heavy metals also accumulate in halite and potash minerals where they not only affect the crystal morphology and color of the salts, but they may catalyze crystallization.

Trace metal accumulation in most hypersaline environments appears to involve both biological and purely chemical mechanisms. Metal-bearing brines in contact with low-redox, high-sulfide groundwater can precipitate metal sulfides. The trace metal content of some brines correlates with salinity and low Eh. Metals such as Cu and Zn are quite soluble in brines because they are largely complexed as chlorides or as organo-chlorides (Rose, 1976; Hallberg et al., 1980). In addition, trace metals can be selectively removed and concentrated by microorganisms. Trace metals are believed to accumulate in stratiform deposits by the migration of metal-enriched brines to zones of mixing with terrestrial groundwaters characterized by high Eh and low pH (Renfro, 1974). Alternatively, migrating brines may leach metals of igneous origin and redeposit them where chemical and structural conditions are favorable (Davidson, 1965).

Trace metal concentrations have been analyzed in many modern evaporite brines and sediments, and they all show essentially the same enrichment patterns: heavy metal concentrations are significantly higher in the stronger brines of stratified water columns, and the highest concentrations are often associated with the accumulation of dissolved sulfide and organic matter.

Besides heavy metals, other trace elements that occur in halite and potash salts include Br, I, Li, B, Rb, and others. Sr accumulates in gypsum and more soluble evaporites. A thorough discussion of the mechanisms of these trace element enrichments has been given by Holser (1979b).

Nutrients Concentrations of the major nutrients of biological importance (NH_4^+, NO_3^-, and PO_4^{3-}) in most of the hypersaline environments presented in this book are summarized in Table 1.4. Moderately oligotrophic conditions promote microbial mat development, while under eutrophic conditions planktonic populations prevail. Under anaerobic conditions in hard-water brines of high organic activity, NH_4^+ can reach millimolar concentrations. In hard-water brines, PO_4^{3-} concentrations are typically low while they can be several millimolar in hypersaline soda lakes. In sabkha environments, sedimentary phosphate can be rather high (up to nearly 1 mg per g sediment) relative to overlying brines which contain up to several micromolar dissolved PO_4^{3-} (Lyons et al., 1984). In the Wadi Natrun soda lakes there is a high positive correlation between PO_4^{3-}, CO_3^{2-}, and NO_3^- concentrations, while there is a large negative correlation between PO_4^{3-} and NH_4^+ concentrations (see Table 20.2).

Although phosphates are rare in evaporite sequences, ammonia commonly accumulates in evaporite salts. Inorganic nitrogen in evaporite deposits is almost entirely biogenic. Carnallite may contain 0.06% NH_4Cl while sylvites accumulate up to 0.014% (Sonnenfeld, 1984). In these salts NH_4Cl can sub-

Table 1.4 General limnology of some extremely hypersaline lakes and brines

Name, location	Depth	Meromixis[a]	TDS %	pH	Nutrients	References
Non-marine, neutral pH lakes						
GREAT SALT LAKE, Utah						
N. Arm	11 m	–	27.5–33.0	7.7	NO_3, 0 NH_4, 0–54 μM Org N, 6.7 mg·l^{-1} PO_4, 3–16 μM DOC, 43 mg·l^{-1}	Post, 1977, 1981
S. Arm	11 m	–	12.0–25.9	8.2	Same as N. Arm	
DEAD SEA, Israel/Jordan	400 m	±	33.3	6.4	NO_3, 0–0.3 μM NH_4, 0.11–0.42 mM PO_4, 1 μM DOC, 4.2–8.3 mg·l^{-1}	Neev and Emery, 1967 Nissenbaum, 1975 Oren, 1981
LAKE EYRE, Australia	3.4 m	–	30.0		NH_4, 1 mM PO_4, 3.8 mM	Baas-Becking and Kaplan, 1956
PINK LAKE, Australia	0.7–1 m	–	up to 34.0	7.6–8.6	NO_3, 0 PO_4, <0.2 μM	Hammer, 1981 Timms, 1983
LITTLE MANITOU LAKE, Saskatchewan	5.2 m	–	6.7–18.2 8.14 avg	8.2	NO_3, 17 μM NH_4, 3 mM PO_4, 4 μM P total, 21 μM	Hammer, 1978

Marine lakes						
SOLAR SALT PONDS, Calif., Mexico	up to 1 m	–	up to saturation	7–9	NO_3, 0–37 μM NH_4, 0–110 μM PO_4, 0–5 μM DOC, up to 100 $mg \cdot l^{-1}$	Javor, 1983a, 1983b, unpublished data
SOLAR LAKE, Sinai	5 m	–	6.8–18.0	6.8–8.8	nd	Cohen et al., 1977a
GAVISH SABKHA, Sinai	0.7 m	–	12.0–36.0	nd	NO_3, 1.6–59 μM PO_4, 1.2–8.9 μM	Gerdes et al., 1985
DEEP LAKE, Antarctica	36 m	–	28.0	7.4	NO_3, 1.2 μM PO_4, 0.6 μM	Kerry et al., 1977 Campbell, 1978
LAKE SURIBATI, Antarctica	31.2 m	+	11.3–22.0	7–8	NO_3, 0 NH_4, 0–575 μM PO_4, 0.4–50 μM DOC, 103–186 $mg \cdot l^{-1}$	Tominaga and Fukui, 1981 Wright and Burton, 1981
Non-marine, alkaline lakes						
MONO LAKE, Calif.	26 m	±	9.0+	9.7	NO_3, 84–1065 μM NH_4, 56–140 μM PO_4, 500–700 μM DOC, 68–85 $mg \cdot l^{-1}$	Mason, 1967 Winkler, 1977 R. Oremland, pers. comm. J. Melack, pers. comm.
BIG SODA LAKE, Nevada	34.5 m 65 m (max)	+	2.6 (top) 8.8 (bot)	9.7 9.7	NH_4, <5 μM (top), 2.5 mM (bot) DOC, 20 $mg \cdot l^{-1}$ (top) 60 $mg \cdot l^{-1}$ (bot)	Cloern et al., 1983b
LAKE MAGADI, Kenya	shallow	–	up to 31.3	9.5–11.2	PO_4, 0.25–1.01 mM DOC, 165 $mg \cdot kg^{-1}$	Jones et al., 1977
WADI NATRUN, Egypt (6 lakes)	shallow	–	9.2–39.4	11	NO_3, 53–237 μM NH_4, 2–461 μM PO_4, 133–683 μM DOC, 136–1552 $mg \cdot ^{-1}$	Imhoff et al., 1979

[a] + = meromictic, – = monomictic, ± = occasionally meromictic or in transition.
TDS total dissolved solids, DOC, dissolved organic carbon.

stitute in the crystal lattices. NH_4Cl has also been reported from various evaporite minerals including dolomite, clay intercalations and brines from German Zechstein deposits, and other potash salts. NH_4^+ increases gypsum solubility but decreases that of KCl. NH_4^+ can also catalyze the conversion of gypsum or anhydrite to other sulfate minerals.

Because the most highly productive hypersaline environments would be expected to be rich in nitrogen and phosphate nutrients, it may be useful to correlate nutrient content and bitumen content of ancient evaporites as a possible key for petroleum exploration. Like organic waste products from microorganisms (see Chapter 3), the inorganic waste product, ammonia, can modify the precipitation of evaporites. These observations stress the fact that many processes in hypersaline environments are under biogeochemical control.

References

Abd-el-Malek, Y. and **Rizk, S.G.** 1963. Bacterial sulphate reduction and the development of alkalinity. III. Experiments under natural conditions in the Wadi Natrun. *Journal of Applied Bacteriology* 26: 20–26.

Amit, O. and **Bentor, Y.K.** 1971. pH dilution curves of saline waters. *Chemical Geology* 7: 307–313.

Baas-Becking, L.G.M. and **Kaplan, I.R.** 1956. The microbiological origin of the sulphur nodules of Lake Eyre. *Transactions of the Royal Society of South Australia* 79: 52–65.

Ben-Yaakov, S. 1973. pH buffering of porewater of recent anoxic marine sediments. *Limnology and Oceanography* 18: 86–94.

Ben-Yaakov, S. and **Sass, E.** 1977. Independent estimate of the pH of Dead Sea brine. *Limnology and Oceanography* 22: 374–376.

Berner, R.A. 1971. *Principles of Chemical Sedimentology*. McGraw-Hill, New York. 240 pp.

Borchert, H. and **Muir, R.O.** 1964. *Salt Deposits. The Origin, Metamorphism and Deformation of Evaporites*. Van Nostrand-Rheinhold, Princeton, 338 pp.

Braitsch, O. 1971. *Salt Deposits. Their Origin and Composition*. Springer-Verlag, Berlin, 297 pp.

Button, A. 1982. Sedimentary iron deposits, evaporites and phosphorites. State of the art report, pp. 259–273 in Holland, H.D. and Schidlowski, M. (editors), *Mineral Deposits and the Evolution of the Biosphere*, Dahlem Konferenzen, Springer-Verlag, New York.

Campbell, P.J. 1978. Primary productivity of a hypersaline Antarctic lake. *Australian Journal of Marine and Freshwater Research* 29: 717–724.

Claypool, G.E., Holser, W.T., Kaplan, I.R., Sakai, H., and Zak, I. 1980. The age curves of sulfur and oxygen isotopes in marine sulfate and their mutual interpretation. *Chemical Geology* 28: 199–260.

Cloern, J.E., Cole, B.E., and **Oremland, R.S.** 1983. Seasonal changes in the chemistry and biology of a meromictic lake (Big Soda Lake, Nevada, U.S.A.). *Hydrobiologia* 105: 195–206.

Cohen, Y., Goldberg, M., Krumbein, W.E., and **Shilo, M.** 1977. Solar Lake (Sinai). 1. Physical and chemical limnology. *Limnology and Oceanography* 22: 597–608.

Copeland, B.J. 1967. Environmental characteristics of hypersaline lagoons. *Texas University Contributions to Marine Science* 12: 207–218.

Davidson, C.F. 1965. A possible mode of origin of strata-bound copper ores. *Economic Geology* 60: 942–954.

Dean, W.E. 1978. Theoretical versus observed successions from evaporation of seawater, pp. 74–85 in Dean, W.E. and Schreiber, B.C. (editors), *Marine Evaporites*, Society of Economic Paleontologists and Mineralogists Short Course, no. 4, Tulsa.

Eugster, H.P. and **Hardie, L.A.** 1978. Saline lakes, pp. 237–293 in Lerman, A. (editor), *Lakes. Chemistry Geology Physics*, Springer-Verlag, New York.

Garrels, R.M. 1967. Ion-sensitive electrodes and individual ion activity coefficients, pp. 344–361 in Eisenman, G. (editor), *Glass Electrodes for Hydrogen and Other Cations. Principles and Practice.* Marcel Decker, New York.

Garrels, R.M. and **Perry, E.A.** 1974. Cycling of carbon, oxygen, and sulfur through geologic time, pp. 303–336 in Goldberg, E. (editor), *The Sea*, vol. 3, John Wiley and Sons, New York.

Garrels, R.M. and **Lerman, A.** 1984. Coupling of the sedimentary sulfur and carbon cycles—an improved model. *American Journal of Science* 284: 989–1007.

Garrels, R.M., Thompson, M.E., and **Siever, R.** 1961. Control of carbonate solubility by carbonate complexes. *American Journal of Science* 259: 24–45.

Garrett, D. 1965. Factors in the design of solar salt plants. Part II. Optimum operation of solar ponds, pp. 176–187 in Rau, J.L. (editor), *Second Symposium on Salt*, vol. 2, Northern Ohio Geological Society, Cleveland.

Gerdes, G., Krumbein, W.E., and **Holtkamp, E.** 1985. Salinity and water activity related zonation of microbial communities and potential stromatolites of the Gavish Sabkha, pp. 238–266 in Friedman, G.M. and Krumbein, W.E. (editors), *Hypersaline Ecosystems. The Gavish Sabkha*, Ecological Studies 53, Springer-Verlag, New York.

Hallberg, R.O., Bubela, B., and **Ferguson, J.** 1980. Metal chelation in sedimentary systems. *Geomicrobiology Journal* 2: 99–114.

Hammer, U.T. 1978. The saline lakes of Saskatchewan. III. Chemical characterization. *Internationale Revue der gesamten Hydrobiologie* 63: 311–335.

Hammer, U.T. 1981. A comparative study of primary production and related factors in four saline lakes in Victoria, Australia. *Internationale Revue der gesamten Hydrobiologie* 66: 701–743.

Hammer, U.T. and **Haynes, R.C.** 1978. The saline lakes of Saskatchewan. II. Locale, hydrography and other physical aspects. *Internationale Revue der gesamten Hydrobiologie* 63: 179–203.

Hardie, L.A. 1967. The gypsum-anhydrite equilibrium at one atmosphere pressure. *American Mineralogist* 52: 172–200.

Hardie, L.A. 1984. Evaporites: marine or non-marine. *American Journal of Science* 284: 193–240.

Hardie, L.A. and **Eugster, H.P.** 1970. The evolution of closed-basin brines. *Mineralogical Society of America Special Paper* 3: 273–290.

Harvie, C.E., Weare, J.H., Hardie, L.A., and **Eugster, H.P.** 1980. Evaporation of seawater: calculated mineral sequences. *Science* 208: 498–500.

Holser, W.T. 1979a. Mineralogy of evaporites, pp. 211–294 in Burns, R.G. (editor), *Marine Minerals*, Mineralogical Society of America Short Course Notes, vol. 6, Washington, D.C.

Holser, W.T. 1979b. Trace elements and isotopes in evaporites, pp. 295–346 in Burns, R.G. (editor), *Marine Minerals*, Mineralogical Society of America Short Course Notes, vol. 6, Washington, D.C.

Horowitz, N.N. 1979. Biological water requirements, pp. 15–27 in Shilo, M. (editor), *Strategies of Microbial Life in Extreme Environments*, Dahlem Konferenzen, Berlin.

Imhoff, J.F., Sahl, H.G., Soliman, G.S.H., and **Trüper, H.G.** 1979. The Wadi Natrun:

chemical composition and microbial mass developments in alkaline brines of eutrophic desert lakes. *Geomicrobiology Journal* 1: 219–234.

Javor, B.J. 1983a. Nutrients and ecology of the Western Salt and Exportadora de Sal saltern brines, pp. 195–205 in Schreiber, B.C. and Harner, H.L. (editors), *Sixth International Symposium on Salt*, vol. 1. The Salt Institute. Toronto.

Javor, B.J. 1983b. Planktonic standing crop and nutrients in a saltern ecosystem. *Limnology and Oceanography* 28: 153–159.

Javor, B.J. 1984. Growth potential of halophilic bacteria isolated from solar salt environments: carbon sources and salt requirements. *Applied and Environmental Microbiology* 48: 352–360.

Jones, B.F., Eugster, H.P., and **Rettig, S.L.** 1977. Hydrochemistry of the Lake Magadi basin, Kenya. *Geochimica et Cosmochimica Acta* 41: 53–72.

Kerry, K.R., Grace, D.R., Williams, R., and **Burton, H.R.** 1977. Studies on some saline lakes of the Vestfold Hills, Antarctica, pp. 839–858 in Llano, G.A. (editor), *Adaptations Within Antarctic Ecosystems*, Smithsonian Institution, Washington, D.C.

Kinsman, D.J.J., Boardman, M., and **Borcsik, M.** 1974. An experimental determination of the solubility of oxygen in marine brines, pp. 325–327 in Coogan, A.J. (editor), *Fourth Symposium on Salt*, vol. 1, The Northern Ohio Geological Society, Cleveland.

Krumgalz, B. 1980. Salt effect on pH of hypersaline solutions, pp. 73–83 in Nissenbaum, A. (editor), *Hypersaline Brines and Evaporitic Environments*, Developments in Sedimentology 26, Elsevier Scientific, New York.

Lazar, B., Starinsky, A., Katz, A., Sass, E., and **Ben-Yaakov, S.** 1983. The carbonate system in hypersaline solutions: alkalinity and $CaCO_3$ solubility of evaporated seawater. *Limnology and Oceanography* 28: 978–986.

Lyons, W.B., Hines, M.E., and **Gaudette, H.E.** 1984. Major and minor element pore water geochemistry of modern marine sabkhas: the influence of cyanobacterial mats, pp. 411–423 in Cohen, Y., Castenholz, R.W. and Halvorson, H.O. (editor), *Microbial Mats: Stromatolites*, MBL Lectures in Biology, vol. 3, Alan R. Liss, New York.

Mason, D.T. 1967. Limnology of Mono Lake. *University of California Publications in Zoology* 83: 1–102.

Neev, D. and **Emery, K.O.** 1967. The Dead Sea. Depositional processes and environments of evaporites. *Bulletin of the Israel Geological Survey* 41, 147 pp.

Nehrkorn, A. and **Schwartz, W.** 1961. Untersuchungen über Lebensgemeinschaften halophiler Mikroorganismen. I. Mikroorganismen aus Salzeen der californischen Wustengebeite und aus einer Natriumchlorid-Sole. *Zeitschrift für Allgemeine Mikrobiologie* 1: 121–141.

Nissenbaum, A. 1975. The microbiology and biogeochemistry of the Dead Sea. *Microbial Ecology* 2: 139–161.

Oren, A. 1981. Approaches to the microbial ecology of the Dead Sea. *Kieler Meeresforschungen* 5: 416–424.

Post, F.J. 1977. The microbial ecology of the Great Salt Lake. *Microbial Ecology* 3: 143–165.

Post, F.J. 1981. Microbiology of the Great Salt Lake North Arm. *Hydrobiologia* 81: 59–69.

Ratcliff, G.A. and **Holdcroft, J.G.** 1963. Diffusivities of gases in aqueous electrolyte solutions. *Transactions of the Institution of Chemical Engineers* 41: 315–319.

Renfro, A.R. 1974. Genesis of evaporite-associated stratiform metalliferous deposits—a sabkha process. *Economic Geology* 69: 33–45.

Rodriguez-Valera, F., Ruiz-Berraquero, R., and **Ramos-Cormenzana, A.** 1981. Characteristics of the heterotrophic bacterial populations in hypersaline environments of different salt concentrations. *Microbial Ecology* 7: 235–243.

Rodriguez-Valera, F., Ventosa, A., Juez, G., and **Imhoff, J.F.** 1985. Variation of environmental features and microbial populations with salt concentration in a multi-pond saltern. *Microbial Ecology* 11: 107–115.

Ronov, A.B. 1968. Probable changes in the composition of sea water during the course of geological time. *Sedimentology* 10: 25–43.

Rose, A.W. 1976. The effect of cuprous chloride complexes on the origin of red-bed copper and related deposits. *Economic Geology* 71: 1036–1048.

Rothbaum, H.P. 1958. Vapor pressure of seawater concentrates. *Journal of Chemical Engineering Data* 3: 50–52.

Sass, E. and **Ben-Yaakov, S.** 1977. The carbonate system in hypersaline solutions: Dead Sea brines. *Marine Chemistry* 5: 183–199.

Sonnenfeld, P. 1984. *Brines and Evaporites.* Academic Press. New York. 613 pp.

Thiede, D.S. and **Cameron, E.N.** 1978. Concentration of heavy metals in the Elk Point evaporitic sequence, Saskatchewan. *Economic Geology* 73: 405–415.

Timms, B.V. 1983. A study of benthic communities in some shallow saline lakes of western Victoria, Australia. *Hydrobiologia* 105: 165–177.

Tominaga, H. and **Fukui, F.** 1981. Saline lakes at Syowa Oasis, Antarctica. *Hydrobiologia* 82: 375–389.

Veizer, J., Holser, W.T., and **Wilgus, C.K.** 1980. Correlation of $^{13}C/^{12}C$ and $^{34}S/^{32}S$ secular variations. *Geochimica et Cosmochimica Acta* 44: 579–587.

Weast, R.C. (editor). 1984. *CRC Handbook of Chemistry and Physics.* CRC Press. Boca Raton.

Winkler, D.W. (editor). 1977. An ecological study of Mono Lake, California. *Institute of Ecology Publication* 12, University of California, Davis, 190 pp.

Wright, S.W. and **Burton, H.R.** 1981. The biology of Antarctic saline lakes. *Hydrobiologia* 82: 319–338.

Zharkov, M.A. 1981. *History of Paleozoic Salt Accumulation.* Springer-Verlag. New York. 308 pp.

Zharkov, M.A. 1984. *Paleozoic Salt Bearing Formations of the World.* Springer-Verlag. New York. 427 pp.

2

Biology

Two themes of this volume are 1) the comparative biology of different hypersaline systems and 2) the diversity of microorganisms in the various hypersaline habitats, topics that cannot be adequately addressed in a single chapter. From the tabulation of the different organisms known to inhabit various hypersaline environments, it is clear that macroscopic diversity is lacking but that microscopic and metabolic diversity are not. Comparisons of community metabolism or potential metabolic activity in different hypersaline habitats or in waters of different salinities in a single habitat do not necessarily indicate that productivity and decomposition processes decrease as salinity increases. Brines much more concentrated than seawater are often highly productive. The following discussion and tables outline the taxonomy, salinity tolerances, and biological activities of organisms from a variety of hypersaline environments. The summary presented here should serve as a starting point for the more detailed analyses of how halophilic microorganisms cope with and modify their environments, and how their activities can be recognized in evaporite sediments.

2.1 Distribution of organisms

Depending on a host of chemical and physical conditions, extremely hypersaline lakes may have rather diverse assemblages of organisms, particularly microorganisms. Table 2.1 lists the biota of lakes described in Table 1.4. Diversity and productivity or standing crop appear to be largely unrelated in many hypersaline lakes (Table 2.2). Some of the highest biomasses and pro-

Table 2.1 Biota of extremely hypersaline lakes

Lake and references	Organisms and salinities (%)
Marine origin	
SOLAR SALT PONDS worldwide Davis, 1978 Golubic, 1980 Hof, 1935 Javor, unpubl. Klug et al., 1985 Mathrani and Boone, 1985 Mitchell and Geddes, 1977 Post et al., 1983 Rodriguez-Valera et al., 1981	Cyanobacteria: 13 spp. including *Aphanothece halophytica* (or other similar forms) are common or abundant in 5.5–21.3%; Green algae (5.5–10.2%): *Batophora oerstedii, Cladophora* sp.; *Dunaliella salina* and *D. viridis* also at higher salinities; Dinoflagellates: 4 species in 5.5–10.2%; Cryptomonad *Cryptomonas* sp. common in 5.5–10.2%; Diatoms common in 5.5–10.2%: *Amphora* sp., *Navicula* spp., *Nitzschia* spp., *Amphiprora paludosa* (up to 21.3%); Protozoa in >15.0%: 14 ciliates, 10 zooflagellates, 4 sarcodines; Bacteria, including moderate (up to ca. 25%) and extreme halophiles (mostly above 25%), phototrophic bacteria (up to ca. 15%), sulfur oxidizers (up to 30%), methanogens (moderate halophiles), sulfate reducers (up to NaCl saturation); Invertebrates: *Artemia salina, Ephydra* sp.
LIMANS Rubentschik, 1926a, 1926b, 1929, 1933, 1946 Saslowsky, 1928	Various aerobic and anaerobic bacteria including urea oxidizers (up to 13%), sulfate reducers (up to 30%), denitrifiers (up to 30%), nitrifiers (up to 15%), methanogens (up to 15%).
SOLAR LAKE Cohen et al., 1977b, 1977c Gerdes et al., 1985b Giani et al., 1984 Hirsch, 1980 Krumbein et al., 1977 Por, 1972 Potts, 1980 Walsby et al., 1983 Wilbert and Kahan, 1981	Cyanobacteria: 24 species, dominated by *Oscillatoria limnetica, O. salina, Microcoleus* sp., *Aphanothece halophytica, Aphanocapsa littoralis, Dactylococcopsis salina*; Diatoms: *Amphora coffeaeformis, Nitzschia thermalis, Nitzschia* sp., *Navicula* sp.; Phototrophic bacteria: *Chromatium violescens*, green *Prosthecochloris* sp., 121 morphologically distinct bacteria including *Caulobacter, Hyphomicrobium, Desulfovibrio*, sulfur oxidizers (*Achromatium volutans, Beggiatoa* sp.), fermenters, methanogens, flexibacteria; 28 eukaryotic protists including 16 spp. of ciliates; Invertebrates: acoel flatworm *Macrostomum* sp., *Artemia salina* (surface water only), copepod *Robertsonia* sp., insects *Octhebius* sp. and *Paraberosus* sp. (larvae).
GAVISH SABKHA (>9.0% salinity) Erlich and Dor, 1985 Gerdes et al., 1985a, 1985b	Cyanobacteria: 21 spp. of which 14 tolerated up to 20%, *Aphanothece halophytica* and *Oscillatoria arenaria* tolerated over 25%; Phototrophic bacteria: *Ectothiorhodospira, Chromatium, Thiocapsa*, Chlorobiaceae, Chloroflexiaceae; other bacteria include sulfur oxidizers, sulfate reducers, budding bacteria, halobacteria; Green alga *Dunaliella*; Diatoms (up to 20%): *Amphora coffeaeformis, Navicula* spp., *Nitzschia* spp.; Invertebrates: 1 nematode, 1 turbellarian worm, 2 copepods (*Robertsonia salsa* and *Nitocra lacustris*), 2 ostracods (*Cyprideis torosa* and *Paracyprideinae* sp.), rotifers, fly larvae (*Bezza* sp., *Atylotus agrestis, Hecamede grisescens*), beetle larvae (2 spp.), no *Artemia salina*.
DEEP LAKE Campbell, 1978 Hand, 1980 Kerry et al., 1977	Green alga *Chlamydomonas* (*Dunaliella*?); centric and pennate diatoms, rare silicoflagellates; rod-shaped bacteria (not viable in lake water medium); no invertebrates reported.
LAKE SURIBATI Wright and Burton, 1981	Green alga *Dunaliella*; Diatom *Tropidoneis*; no phototrophic bacteria; no invertebrates recorded.

(continued next page)

Table 2.1 Continued

Lake and references	Organisms and salinities (%)
Non-marine origin	
GREAT SALT LAKE Cronin and Post, 1977 Post, 1977, 1981 Felix and Rushforth, 1979 Pack, 1919	Cyanobacteria: *Aphanothece halophytica* (common), *Microcoleus lyngbyaceus* (rare, S. Arm); Green algae *Dunaliella salina* (N. Arm only), *D. viridis*; Diatoms in S. Arm: *Amphora coffeaeformis* (most common), *Biddulphia levis, Navicula* spp., *Entomoneis pulchra, A. delicatissima, Rhopalodia musculus, Nitzschia palea, Nitzschia epithemiodes, Surirella striatula*; Fungus *Chladosporium*; Bacteria: halobacteria; sulfate reducers, methanogens, fermenters; no nitrifiers or N_2 fixers; Protozoa include 3 flagellates, ciliates *Uroleptus packii* and *Prorodon utahensis*, 2 amoebae; Invertebrates: *Artemia salina, Ephydra hians, E. gracilis, Ephydra.* sp.
DEAD SEA Elazari-Volcani, 1940, 1943a, 1943b Nissenbaum, 1975 Oren, 1983, 1986 Oren and Shilo, 1982	Green alga *Dunaliella parva*; Bacteria: *Halobacterium* spp., various aerobic and anaerobic eubacteria including fermenters, denitrifiers, sulfur oxidizers, possibly sulfate reducers; no nitrifiers or nitrogen fixers; one protozoan (amoeba) but not active in Dead Sea salinities; no higher organisms.
LAKE EYRE Baas-Becking and Kaplan, 1956	Green algae: *Dunaliella* spp.; Diatom: *Amphora coffeaeformis*; Cyanobacteria include *Nodularia spumigena, Lyngbya* sp; Bacteria include sulfate reducers, sulfur oxidizers, fermenters, denitrifiers, methanogens; Protozoa: colorless ciliates and flagellates; Fungus: unidentified Chytid; Invertebrate: *Parartemia zietziana.*
PINK LAKE Hammer, 1981 Timms, 1983	Green alga: *Dunaliella salina*; Invertebrates: *Parartemia zietziana*, ostracods *Diacypris compacta* and *Reticypris herbsti.*
LITTLE MANITOU LAKE Hammer et al., 1983	Green algae: *Ctenocladus circinnatus, Rhizoclonium hieroglyphicum, Enteromorpha prolifera*; Diatoms (common): *Chaetoceros elmorei, Melosira granulata, Fragilaria crotonensis, Navicula cincta, Amphora coffeaeformis, Hantzschia amphioxys, Nitzschia* spp.; Cyanobacteria: *Microcystis aeroginosa, Lyngbya birgei, Nodularia spumigena*; Others: *Euglena* spp., rare occurrences of other species including dinoflagellates; Invertebrate: *Artemia salina.*
MONO LAKE Blinn, 1971 Mason, 1967 Winkler, 1977 Melack, 1983	Green algae: *Nannochloris* (dominant), *Chlamydomonas* sp. or *Dunaliella, Ctenocladus circinnatus*; Diatoms: *Nitzschia communis* (dominant), *Amphora coffeaeformis* and others; Cyanobacteria: over 6 spp. including 2 spp. of *Dactylococcopsis*; Fungi include *Chytridomycetes*; Rotifers: *Brachionus plicatus, Hexarthra jenkinae*; protozoa; Invertebrates: *Artemia monica, Ephydra hians*, Ceratopogonidae (biting midge), unidentified oligochaete.
BIG SODA LAKE Oremland et al., 1982 Priscu et al., 1982 Cloern et al., 1983a, 1983b	Mixolimnion phytoplankton: Diatoms *Nitzschia palea* and *Chaetoceros* sp.; Phototrophic bacteria *Ectothiorhodospira vacuolata* and/or *Thiocapsa* sp.; Chemoautotrophic bacteria. Monimolimnion: Bacteria (sulfide producers, methanogens).

(continued next page)

Table 2.1 Continued

Lake and references	Organisms and salinities (%)
LAKE MAGADI Jones et al., 1977 Eugster, 1980 Tindall et al., 1980 Imhoff et al., 1981 De Rosa et al., 1983	Phototrophic bacteria *Ectothiorhodospira vacuolata*; alkaliphilic halobacteria; "considerable biological activity" including sulfate reduction.
WADI NATRUN Imhoff and Trüper, 1977 Imhoff et al., 1978, 1979 Tindall et al., 1984 Weisser and Trüper, 1985	Cyanobacteria: *Spirulina* sp. (up to 9.2%), *Synechococcus* sp. (up to 24%); Green alga: *Dunaliella salina* (up to 37.4%); Phototrophic bacteria: *Ectothiorhodospira* spp., *Chromatium* sp., green bacteria; Other bacteria: alkaliphilic halobacteria, chemoautotrophs, sulfate reducers, halophilic *Bacillus*; various unidentified bacteria; flagellated protozoa (up to 37.4%); no invertebrates.

ductivity rates known in non-polluted systems are found in extremely hypersaline lakes and ponds.

Most marine-derived, hypersaline environments can be divided into four classes based on salinity and organism distribution (Por, 1980). The first class (ca. 6–7 to 10% salinity) contains a diverse biota of marine origin. Between about 10% and 14% salinity, the second class contains a biota largely of particularly adapted, halophilic or halotolerant organisms related to freshwater species.

The third class includes the salinity range of ca. 14% to 30%. These salinities are marked by the disappearance of cyanobacteria and most crustaceans besides *Artemia* (and *Parartemia* in Australia). Besides certain phototrophic bacteria, both moderately halophilic and extremely halophilic, non-phototrophic bacteria thrive in these salinities. The fourth class (>30% salinity) contains microscopic organisms only. Table 2.3 lists the upper limits of salinity, temperature, and pH of a variety of microorganisms and invertebrates found in extremely hypersaline lakes.

Most freshwater or terrestrial bacteria do not thrive or even tolerate exposure to ≥7–10% salt (Hill and White, 1929; Hof, 1935; ZoBell et al., 1937). Moderately halophilic bacteria isolated from hypersaline soils belong to genera typical of normal soils and are more resistant to low salt concentrations than bacteria from hypersaline water (Quesada et al., 1983; Rodriguez-Valera, 1986). Many invertebrates have been documented from rather hypersaline brines (Table 2.3). Salt-tolerant marine fish (*Cyprinodon variegatus* and *Menidia beryllina*) were found in hypersaline lagoons in 11.0% salinity but they were excluded once the salinity rose to 14.0% (Copeland, 1967).

Depending on the total salinity, specific ion content, nutrients, and other chemical and physical factors, *Dunaliella* and halophilic bacteria may be absent from the most concentrated brines. Strong brines that appear to be completely sterile have been documented in the $CaCl_2$ waters of Don Juan Pond,

Table 2.2 Productivity and biomass in hypersaline lakes[a]

Lake and references	Productivity and standing crop
Marine origin	
SOLAR SALT PONDS Copeland and Jones, 1965 Rodriguez-Valera et al., 1985	**Photosynthesis** up to 5.25 g $O_2/m^3/d$; **Respiration** up to 10.7 g $O_2/m^3/d$ (see Table 13.2). *Dunaliella* most abundant (10^4 cells/ml) in 30% salinity; halobacteria most abundant (1.2×10^4 colony forming units/ml) in 45% TDS; highest microbial species diversity in 10.0% TDS.
SOLAR LAKE Cohen et al., 1977b, 1977c Jørgensen and Cohen, 1977 Jørgensen et al., 1983 Krumbein et al., 1977	**Photosynthesis** Mats: 1.2–17.6 mmol $O_2/m^2/h$ (= 14.4–211.2 mg $C/m^2/h$); Plankton: max = 8015 mg $C/m^2/d$, 4960 mg $C/m^3/d$; DARK CO_2 UPTAKE 1014 mg $C/m^3/d$. **Sulfate reduction rates:** 5.4 μmol $SO_4/cm^3/d$ at mud surface. Planktonic microorganisms $>10^6$ cells/ml during stratification; bacteria in surface sediments (colony-forming units/cc^3): 2×10^9 aerobes, 8×10^4 (or 6×10^6/g sediment) anaerobes, 10^3 phototrophic bacteria, 2×10^4 (or 2.5×10^6/g sediment) sulfate reducers, 2.3×10^3 sulfur oxidizers.
GAVISH SABKHA Gerdes et al.,1985a	**Photosynthesis** 150–520 mg $C/m^2/h$; 24.6–55.1 μg chl *a*/cm^3; 3.14–12.1 μg bchl *a*/cm^3.
DEEP LAKE Campbell, 1978	**Photosynthesis** 0–0.18 mg $C/m^3/h$. Rarely $>10^5$ cells/l.
LAKE SURIBATI Tominaga and Fukui, 1981	Very low primary productivity; photosynthesis detected below 0°C. 1 mg chl *a*/m^3.
Non-marine origin	
GREAT SALT LAKE Stephens and Gillespie, 1976 Post, 1977	**Photosynthesis** S. Arm: 145 g $C/m^2/yr$, 2.13 g $C/m^2/d$ maximum. N. Arm (summer): 5×10^6 bacteria/ml, 10^4 *Dunaliella salina* cells/ml, 2000 *D. viridis* cells/ml, 1 *Artemia*/m^3.
DEAD SEA Kaplan and Friedman, 1970 Oren and Shilo, 1981, 1982	Up to 4×10^5 *Dunaliella parva* cells/ml in 1964, up to 8800 cells/ml in 1980, currently low, ca. 10 cells/ml; Halobacteria up to 1.9×10^7 cell/ml in 1980.
PINK LAKE Hammer, 1981	**Photosynthesis** max = 48.1 mg $C/m^3/h$, avg = 184 mg $C/m^2/d$. Maximum phytoplankton = 1.08×10^5 cells/ml.
LITTLE MANITOU LAKE Haynes and Hammer, 1978	**Photosynthesis** 1188 mg $C/m^2/d$, 475 mg $C/m^3/d$. Chl *a* = 5.8–14.1 mg/m^3.
MONO LAKE Mason, 1967 Winkler, 1977	**Photosynthesis** 85–310 mg $C/m^2/h$ (summer), 0–25 mg$C/m^3/h$. 272–470 mg chl *a*/m^2, 0–10 mg chl *a*/m^3; Standing crop at 5 m depth (winter, cells/ml): 800 *Dunaliella*, 1500 *Nannochloris*, 500 large *Nitzschia*, 140 small *Nitzschia*, 8×10^4 bacteria.
BIG SODA LAKE Cloern et al., 1983a, 1983b	**CO_2 fixation** 500 g $C/m^2/yr$ (60% by algae, 30% by chemoautotrophs, 10% by phototrophic bacteria); winter productivity was highest = 2830 mg $C/m^2/d$. Chl *a* max >40 mg/m^3, bchl *a* max = ca. 200 mg/m^3.
WADI NATRUN Imhoff and Trüper, 1977	10^8–10^9 viable phototrophic bacteria/l.

[a]Chl *a* = chlorophyll *a*, bchl *a* = bacteriochlorophyll *a*.

Table 2.3 Upper salinity, temperature, and pH limits of selected halotolerant and halophilic taxa

Organism	Salinity (%)	Temperature (°C)	pH	References
Cyanobacteria				
Aphanothece halophytica	35	>43	–	Brock, 1976
Oscillatoria limnetica	25	48	–	Erlich and Dor, 1985
Microcoleus sp.	25	48	–	Erlich and Dor, 1985
Oscillatoria arenaria	33	–	–	Erlich and Dor, 1985
Dactylococcopsis salina	20	<46	–	Walsby et al., 1983
Nodularia spumigena	13	–	–	Felix and Rushforth, 1979
16 spp., Gavish Sabkha	18–20	–	–	Erlich and Dor, 1985
Phototrophic bacteria				
Ectothiorhodospira halochloris	40	>50	10	Imhoff and Trüper, 1977
E. halophila	30	47–50	>7.8	Raymond and Sistrom, 1969
E. vacuolata	10	>39	10	Imhoff et al., 1981
Chromatium violescens	>18	<60.5	–	Cohen et al., 1977a, 1977b
Prosthecochloris sp.	>18	<60.5	–	Cohen et al., 1977a, 1977b
Rhodospirillum salexigens	20	45	–	Drews, 1981
R. salinarum	24	45	>8.0	Nissen and Dundas, 1984
Other bacteria				
Halobacteria	35	50	–	Borowitzka, 1981
Alkaliphilic halobacteria	satd	>37	>10	Tindall et al., 1980
Alkaliphilic halophilic eubacteria	20	–	10	Weisser and Trüper, 1985
Denitrifiers	33	–	–	Elazari-Volcani, 1940

(continued next page)

Antarctica (Benoit and Hall, 1970; Cameron et al., 1972; Horowitz et al., 1972) and bitterns brines of solar salterns (Javor, 1983). The report of sterile saltern brines is not contradictory with the fact that halophilic bacteria are commonly recovered from solar salt, since the ion composition of the salt is markedly different from that of the remaining bitterns brines (Chapter 1).

Although many extremely hypersaline habitats lack macroscopic organisms and easily recognizable microorganisms (such as cyanobacteria and eukaryotic algae), bacterial floras may be large and complex. Numerous, diverse strains and species of halophilic, heterotrophic bacteria have been isolated from such environments (see Chapters 5 and 6). Phototrophic bacteria (Chapter 7) may also be important primary producers. At least in one case, dark CO_2 assimilation by chemoautotrophic, sulfur-oxidizing bacteria was found to be the dominant autotrophic process in the summer in a hypersaline, meromictic lake (Cloern et al., 1983a).

It appears to be a contradiction that brines can be good preservatives while many natural brines have rich and productive microbiotas. The remaining chapters in this book address both the biological and environmental

Table 2.3 Continued

Organism	Salinity (%)	Temperature (°C)	pH	References
Urea oxidizers	13	42–47	–	Rubentschik, 1926a, 1926b
Nitrifiers	15	–	–	Rubentschik, 1929
Sulfate reducers	satd	–	11	Hof, 1935; Imhoff et al.,1979; Klug et al., 1985
Sulfide oxidizers	30	–	–	Hof, 1935
Protozoa				
Ciliates	23.6	–	9.9	Jaschof and Schwartz, 1961
Ciliates, 9 spp.	>20	–	–	Post et al., 1983
Zoomastigophora, 5 spp.	>20	–	–	Post et al., 1983
Sarcodina, 5 spp.	>20	–	–	Post et al., 1983
Algae				
Dunaliella spp.	35	48	>9.2	Brock, 1975
Green flagellates	23.6	–	9.9	Jaschof and Schwartz, 1961
Ctenocladus circinnatus	11	28	>10.2	Cole et al., 1967; Blinn, 1971
Diatoms, 7 spp.	20.5	–	–	Erlich and Dor, 1985
Nitzschia spp.	>18		>9.7	Winkler, 1977
Invertebrates				
Rotifers				
Brachionus angularis	>16.1	–	–	Anderson, 1958
Keratella quadrata	>16.1	–	–	Anderson, 1958
Nematode sp.	12.5	–	–	Gerdes et al., 1985b
Turbellarian worm				
Macrostomum	13	–	–	Gerdes et al., 1985b
Gastropods				
Pirenella conica	10	–	–	Por, 1980
Batillariella estuarina	15.9	–	–	Bayly and Williams, 1966
Coxiella striata	11.3	–	–	Bayly and Williams, 1966
Anostracans				
Artemia salina	33	31	10	Cole and Brown, 1967; Mitchell and Geddes, 1977; Walsby et al., 1983
Parartemia zietziana	35.3	–	9.8	Geddes, 1976, 1981
Copepods				
Platycypris	17.6	35	9.8	Geddes, 1976
Diacypris	12.6	–	–	Geddes, 1976
Various spp.	13–14	–	–	Bayly, 1972; Por, 1980; Gerdes et al., 1985b
Ostracod spp.	13	–	–	Gerdes et al., 1985b
Isopods				
Haloniscus searleslei	16	–	–	Bayly, 1972
Insects				
Aedes detritus	10	–	–	Bayly, 1972
A. australis	12.5	–	–	Bayly, 1972
Ephydra cinerea	30	–	–	Bayly, 1972
E. hians	<18	–	>9.7	Winkler, 1977
5 spp. Gavish Sabkha	28–30	–	–	Gerdes et al., 1985b
Chironomids	28.5	–	–	Bayly and Williams, 1966

Table 2.4 Selected reviews and keynote papers on the ecology and biology of extremely hypersaline environments

Subject	References
Microbial ecology: general	Bauld (1981); Borowitzka (1981); Brock (1979); Larsen (1980); Rodriguez-Valera (1988)
Microbial ecology: specific environments	Nissenbaum (1975); Oren (1988); Post (1977); Wright and Burton (1981); Tindall (1988)
Bacteria: general reviews	Brown (1983); Kushner (1978); Morishita and Masui (1980)
Eubacteria	Imhoff (1986, 1988); Kushner and Kamekura (1988); Oren (1986); Rodriguez-Valera (1986)
Halobacteria	Dundas (1978); Hochstein (1988); Kushner (1985); Rodriguez-Valera (1988); Tindall and Trüper (1986)
Dunaliella and other eucaryotic microbes	Ben-Amotz and Avron (1983); Borowitzka and Borowitzka (1988); Brown (1976); Brown and Borowitzka (1979); Munns et al. (1983); Bayly (1972); Carpelan (1967); Geddes (1981)

factors that promote or hinder the success of microorganisms in hypersaline habitats. For simplified reference, Table 2.4 has been included to give a selected list of reviews and keynote papers concerning the microbial ecology of hypersaline lakes and the physiology and ecology of the organisms that inhabit them.

2.2 The upper salinity limits to life

The upper salinity limits to life in strong brines are often measured. Besides detailed studies of osmoregulation in a variety of salt-tolerant organisms, other chemical and physical factors that might restrict biotic activity are not as well documented. The low solubility of O_2 may limit the success of obligately aerobic organisms. The combination of low water activity, high ion content, and large concentrations of specific ions may preclude most organisms, including xerophilic yeasts and fungi that live in strong sugar solutions (Chapter 11). The lowest limit of water activity in brines for halobacteria as well as yeasts and fungi appears to be between a_w 0.6 and 0.7 (Chapter 1).

The ionic character of salty brines may impose electrochemical restrictions on cells with inefficient ion pumps. Many ions or ion complexes are highly hydrated and may compete with the cells for water. High Mg^{2+} concentrations found in potash-phase brines (bitterns) are probably largely responsible for excluding bacterial life (Javor, 1983). Mg^{2+} binds to the outside of bacteria and is required internally by the cells (Chapters 5 and 6), but regulation of its distribution within bacteria is poorly known.

The relative concentrations of monovalent and divalent ions were found to be important in halobacterial distribution (Edgerton and Brimblecombe, 1981). The relative and absolute concentrations of Ca^{2+} and Mg^{2+} affect the success of both *Dunaliella* and extremely halophilic bacteria (Baas-Becking,

1931; Tindall et al., 1980; Cohen et al., 1983). As divalent cations, the contribution of Mg^{2+} and Ca^{2+} to the ionic strength of a brine is four times that of the equivalent molal concentrations of monovalent ions. The inability of Don Juan Pond brines to support life may not only be due to the combination of low temperatures and low a_w, but also to the high ionic strength of $CaCl_2$ brines. The concept of ionic strength apart from major ion content or water activity has not been adequately addressed as a limiting factor to life in brines.

2.3 Fossil and living microbes in salt

Fossils have been recorded in many evaporites. Microbial mats have been noted in and between gypsum, anhydrite, and halite layers (Shearman, 1966; Friedman, 1982). Remains of vertebrates, invertebrates, and plants have been recorded in ancient gypsum and salt as well as in interbedded shales (Tasch, 1960, 1963). Similar organic remains were also found in modern salt (Tasch, 1969). Microscopic bodies resembling *Gallionella* and *Leptothrix* (iron bacteria) and filaments of fungi have been observed in cores of Zechstein ferruginous salt of Permian age (Great Britain and Germany) (Tasch, 1963).

Salt often crystallizes imperfectly, trapping brines and bacteria in pockets ("negative crystals") within the growing crystals. NaCl crystals formed in solutions with high concentrations of halobacterial cells contained more and larger brine inclusions than crystals grown in sterile solutions (Norton and Grant, 1988). Inclusion fluids may be a valuable source of geochemical information since analyses of their ionic content, trace elements, dissolved gases, and stable isotopic signatures can be used to reconstruct the nature of the ancient atmosphere and oceans (Roedder, 1984; Knauth and Beeunas, 1986). Analyses of ionic content and trace elements have been used to interpret whether brine inclusions are primary (formed when the salt initially crystallized) or secondary (formed after dissolution and recrystallization of the original salt) (Holser, 1963; Chapter 1).

Many authors have noted the remains of bacterial-like particles, insoluble organic matter, and biogenic gases (e.g., CH_4 and H_2S) in brine inclusions (literature summarized by Sonnenfeld, 1984). Norton and Grant (1988) found that brine inclusions trapped both viable and non-viable cells in crystals formed under laboratory conditions. Viability was tested after six months. The viability of bacteria trapped in ancient salt remains a debatable question. Viable fossil bacteria from primary brine inclusions of Paleozoic salts were reported by Reiser and Tasch (1960) and Dombrowski (1961, 1966). It remains to be demonstrated whether such bacteria could be revived under sterile conditions acceptable in a modern microbiological laboratory. Bien and Schwartz (1965) noted fossil bacteria in about 30 ancient rock salt samples but they were unsuccessful in any enrichments for heterotrophic, aerobic bacteria, sulfate-reducing bacteria, or thiobacilli.

Do the preservative qualities of salt and brines permit non-growing bacterial cells to retain their essential structures intact over the course of 200 million years or more? Such claims have far-reaching implications, ranging from theories of panspermia and evolution to the ideas of suspended animation and eternal life. In the view of this author, all assertions of viable bacteria from Paleozoic evaporites should be taken *cum grano salis*.

References

Anderson, G.C. 1958. Some limnological features of a shallow saline meromictic lake. *Limnology and Oceanography* 3: 259–270.

Baas-Becking, L.G.M. 1931. Salt effects on swarmers of *Dunaliella viridis* Teod. *Journal of General Physiology* 14: 765–779.

Baas-Becking, L.G.M. and **Kaplan, I.R.**. 1956. The microbiological origin of the sulphur nodules of Lake Eyre. *Transactions of the Royal Society of South Australia* 79: 52–65.

Bauld, J. 1981. Occurrence of microbial mats in saline lakes. *Hydrobiologia* 81: 87–111.

Bayly, I.A.E. 1972. Salinity tolerance and osmotic behavior of animals in athalassic saline and marine hypersaline waters. *Annual Review of Ecology and Systematics* 3: 233–268.

Bayly, I.A.E. and **Williams, W.D.** 1966. Chemical and biological studies of some saline lakes of south-east Australia. *Australian Journal of Marine and Freshwater Research* 17: 177–228.

Ben-Amotz, A. and **Avron, M.** 1983. Accumulation of metabolites by halotolerant algae and its industrial potential. *Annual Review of Microbiology* 37: 95–119.

Benoit, R.E. and **Hall, Jr., C.L.** 1970. The microbiology of some Dry Valley soils of Victoria Land, pp. 697–701 in Holdgate, M.W. (editor), *Antarctic Ecology*, vol. 2, Academic Press, New York.

Bien, E. and **Schwartz, W.** 1965. Geomikrobiologische Untersuchungen. VI. Über das Vorkommen konservierter toter und lebender Bakterienzellen in Salzgesteinen. *Zeitschrift für Allgemeine Mikrobiologie* 5: 185–205.

Blinn, D.W. 1971. Autecology of a filamentous alga, *Ctenocladus circinnatus* (Chlorophyceae), in saline environments. *Canadian Journal of Botany* 49: 735–743.

Borowitzka, L. 1981. The microflora: adaptations to life in extremely saline lakes. *Hydrobiologia* 81: 33–46.

Borowitzka, M.A. and **Borowitzka, L.J.** 1988. *Dunaliella*, p. 27–58 in Borowitzka, M.A., and Borowitzka, L.J. (editors), *Micro-algal Biotechnology*, Cambridge University Press, Cambridge.

Brock, T.D. 1975. Salinity and the ecology of *Dunaliella* from Great Salt Lake. *Journal of General Microbiology* 89: 285–292.

Brock, T.D. 1976. Halophilic blue-green algae. *Archives of Microbiology* 107: 109–111.

Brock, T.D. 1979. Ecology of saline lakes, pp. 29–47 in Shilo, M. (editor), *Strategies of Microbial Life in Extreme Environments*, Dahlem Konferenzen, Berlin.

Brown, A.D. 1976. Compatible solutes and extreme water stress in eukaryotic microorganisms. *Advances in Microbial Physiology* 17: 181–242.

Brown, A.D. 1983. Halophilic prokaryotes, pp. 137–162 in Lange, O.L., Nobel, P.S., Osmond, C.B., and Ziegler, H. (editors), *Physiological Plant Ecology* III, Springer-Verlag, New York.

Brown, A.D. and **Borowitzka, L.J.** 1979. Halotolerance of *Dunaliella*, pp. 139–190 in Levandowsky, M. and Hutner, S.H. (editor), *Biochemistry and Physiology of Protozoa*, vol. 1, Academic Press, New York.

Cameron, R.E., Morelli, F.A., and **Randall, L.P.** 1972. Aerial, aquatic and soil microbiology of Don Juan Pond, Antarctica. *Antarctic Journal of the United States* 7: 254–258.
Campbell, P.J. 1978. Primary productivity of a hypersaline Antarctic lake. *Australian Journal of Marine and Freshwater Research* 29: 717–724.
Carpelan, L.H. 1967. Invertebrates in relation to hypersaline habitats. *Texas University Contributions to Marine Science* 12: 219–229.
Cloern, J.E., Cole, B.E., and **Oremland, R.S.** 1983a. Autotrophic processes in meromictic Big Soda Lake, Nevada. *Limnology and Oceanography* 28: 1049–1061.
Cloern, J.E., Cole, B.E., and **Oremland, R.S.** 1983b. Seasonal changes in the chemistry and biology of a meromictic lake (Big Soda Lake, Nevada, U.S.A.). *Hydrobiologia* 105: 195–206.
Cohen, S., Oren, A., and **Shilo, M.** 1983. The divalent cation requirement of Dead Sea halobacteria. *Archives of Microbiology* 136: 184–190.
Cohen, Y., Krumbein, W.E., Goldberg, M., and **Shilo, M.** 1977a. Solar Lake (Sinai). 1. Physical and chemical limnology. *Limnology and Oceanography* 22: 597–608.
Cohen, Y., Krumbein, W.E., and **Shilo, M.** 1977b. Solar Lake (Sinai). 2. Distribution of photosynthetic microorganisms and primary production. *Limnology and Oceanography* 22: 609–620.
Cohen, Y., Krumbein, W.E., and **Shilo, M.** 1977c. Solar Lake (Sinai). 3. Bacterial distribution and production. *Limnology and Oceanography* 22: 621–634.
Cole, G.A. and **Brown, R.J.** 1967. The chemistry of *Artemia* habitats. *Ecology* 48: 858–861.
Cole, G.A., Whiteside, M.C., and **Brown, R.J.** 1967. Unusual monomixis in two saline Arizona ponds. *Limnology and Oceanography* 12: 584–591.
Copeland, B.J. 1967. Environmental characteristics of hypersaline lagoons. *Texas University Contributions to Marine Science* 12: 207–218.
Cronin, E.A. and **Post, F.J.** 1977. Report of a dematiaceous hyphomycete from the Great Salt Lake, Utah. *Mycologia* 69: 846–847.
Davis, J.S. 1978. Biological communities of a nutrient enriched salina. *Aquatic Botany* 4: 23–42.
De Rosa, M., Gambacorta, A., Nicolaus, B., and **Grant, W.D.** 1983. A C_{25},C_{25} diether core lipid from archaebacterial haloalkalophiles. *Journal of General Microbiology* 129: 2333–2337.
Dombrowski, H.F. 1961. *Bacillus circulans* aus Zechsteinsalzen. *Zentralblatt für Bakteriologie, Mikrobiologie, und Hygiene* Abt. 1 Originale 183: 173– 179.
Dombrowski, J.G. 1966. Geological problems in the question of living bacteria in Paleozoic salt deposits, pp. 215–220 in Rau, J. (editor), *Second Symposium on Salt*, vol. 1, Northern Ohio Geological Society, Cleveland.
Drews, G. 1981. *Rhodospirillum salexigens*, spec. nov., an obligately halophilic phototrophic bacterium. *Archives of Microbiology* 130: 325–327.
Dundas, I. 1978. The normal Halobacteriaceae, the general physiology and taxonomy of a well adapted and highly successful group of microorganisms, pp. 641–651 in Caplan, S.R. and Ginzburg, M. (editors), *Energetics and Structure of Halophilic Microorganisms*, Elsevier-North-Holland Biomedical Press, New York.
Edgerton, M.E. and **Brimblecombe, P.** 1981. Thermodynamics of halobacterial environments. *Canadian Journal of Microbiology* 27: 899–909.
Elazari-Volcani, B. 1940. *Studies on the microflora of the Dead Sea*. Ph.D. Thesis, Hebrew University of Jerusalem, 119 pp.
Elazari-Volcani, B. 1943a. A dimastigamoeba in the bed of the Dead Sea. *Nature (London)* 152: 275–277.
Elazari-Volcani, B. 1943b. Bacteria in the bottom of the Dead Sea. *Nature (London)* 152: 274–275.

Ehrlich, A. and **Dor, I.** 1985. Photosynthetic microorganisms of the Gavish Sabkha, pp. 296–321, in Friedman, G.M. and Krumbein, W.E. (editors), *Hypersaline Ecosystems. The Gavish Sabkha,* Ecological Studies 53, Springer-Verlag, New York.

Eugster, H.P. 1980. Lake Magadi, Kenya, and its precursors, pp. 195–232 in Nissenbaum, A. (editor), *Hypersaline Brines and Evaporitic Environments,* Developments in Sedimentology 28, Elsevier Scientific, New York.

Felix, E.A. and **Rushforth., S.R.** 1979. The algal flora of the Great Salt Lake, Utah, U.S.A. *Nova Hedwigia* 31: 163–194.

Friedman, G.M. 1982. Evaporites as source rock for petroleum, pp. 385–395 in Handford, C.R., Loucks, R.G. and Davies, G.R. (editors), *Depositional and Diagenetic Spectra of Evaporites–a Core Workshop,* Society of Economic Paleontologists and Mineralogists, no. 3, Calgary.

Geddes, M.C. 1976. Seasonal fauna of some ephemeral saline waters in western Victoria with particular reference to *Parartemia zietziana* Sayce (Crustacea: Anostraca). *Australian Journal of Marine and Freshwater Research* 27: 1–22.

Geddes, M.C. 1981. The brine shrimps *Artemia* and *Parartemia. Hydrobiologia* 81: 169–179.

Gerdes, G., Krumbein, W.E., and **Holtkamp, E.** 1985a. Salinity and water activity related zonation of microbial communities and potential stromatolites of the Gavish Sabkha, pp. 238–266 in Friedman, G.M. and Krumbein, W.E. (editors), *Hypersaline Ecosystems. The Gavish Sabkha,* Ecological Studies 53, Springer-Verlag, New York.

Gerdes, G., Spira, J., and **Dimentan, C.** 1985b. The fauna of the Gavish Sabkha and the Solar Lake--a comparative study, pp. 322–345 in Friedman, G.M. and Krumbein, W.E. (editor), *Hypersaline Ecosystems. The Gavish Sabkha,* Ecological Studies 53, Springer-Verlag, New York.

Giani, E., Giani, L., Cohen, Y., and **Krumbein, W.E.** 1984. Methanogenesis in the hypersaline Solar Lake (Sinai). *FEMS Microbiology Letters* 25: 219–224.

Golubic, S. 1980. Halophily and halotolerance in cyanophytes. *Origins of Life* 10: 169–183.

Hammer, U.T. 1981. A comparative study of primary production and related factors in four saline lakes in Victoria, Australia. *Internationale Revue der gesamten Hydrobiologie* 66: 701–743.

Hammer, U.T., Shamess, J., and **Haynes. R.C.** 1983. The distribution and abundance of algae in saline lakes of Saskatchewan, Canada. *Hydrobiologia* 105: 1–26.

Hand, R.M. 1980. Bacterial populations in two saline Antarctic lakes, pp. 123–129 in Trudinger, P.A. and Walter, M.R. (editors), *Biogeochemistry of Ancient and Modern Environments,* Springer-Verlag, New York.

Haynes, R.C. and **Hammer, U.T.** 1978. The saline lakes of Saskatchewan IV. Primary production by phytoplankton in selected saline ecosystems. *Internationale Revue der gesamten Hydrobiologie* 63: 337–351.

Hill, J.H. and **White, E.C.** 1929. Sodium chloride media for the separation of certain Gram-positive cocci from Gram-negative bacilli. *Journal of Bacteriology* 18: 43–57.

Hirsch, P. 1980. Distribution and pure culture of morphologically distinct Solar Lake microorganisms, pp. 41–60 in Nissenbaum, A. (editor), *Hypersaline Brines and Evaporitic Environments,* Developments in Sedimentology 28, Elsevier Scientific, New York.

Hochstein, L.I. 1988. The physiology and metabolism of the extremely halophilic bacteria, p. 67–83 in Rodriguez-Valera, F. (editor), *Halophilic Bacteria,* Vol. II, CRC Press, Boca Raton.

Hof, T. 1935. Investigations concerning bacterial life in strong brines. *Extrait du Recueil des Travaux Botaniques Nérlandais* 32: 92–173.

Holser, W.T. 1963. Chemistry of brine inclusions in Permian salt from Hutchinson,

Kansas, pp. 86–95 in Bersticker, A.C., Hoekstra, K.E. and Hall, J.F. (editors), *Symposium on Salt,* Northern Ohio Geological Society, Cleveland.

Horowitz, N.H., Cameron, R.E., and **Hubbard, J.S.** 1972. Microbiology of the Dry Valleys of Antarctica. *Science* 176: 242–245.

Imhoff, J.F. 1986. Osmoregulation and compatible solutes in eubacteria. *FEMS Microbiology Reviews* 39: 57–66.

Imhoff, J.F. 1988. Halophilic phototrophic bacteria, p. 85–108 in Rodriguez-Valera, F. (editor), *Halophilic Bacteria,* Vol. II, CRC Press, Boca Raton.

Imhoff, J.F. and **Trüper, H.G.** 1977. *Ectothiorhodospira halochloris* sp. nov., a new extremely halophilic phototrophic bacterium containing bacteriochlorophyll *b. Archives of Microbiology* 114: 115–121.

Imhoff, J.F., Hashwa, F., and **Trüper, H.G.** 1978. Isolation of extremely halophilic phototrophic bacteria from the alkaline Wadi Natrun, Egypt. *Archiv für Hydrobiologie* 84: 381–388.

Imhoff, J.F., Sahl, H.G., Soliman, G.S.H., and **Trüper, H.G.** 1979. The Wadi Natrun: chemical composition and microbial mass developments in alkaline brines and eutrophic desert lakes. *Geomicrobiology Journal* 1: 219–234.

Imhoff, J.F., Tindall, B.J., Grant, W.D., and **Trüper, H.G.** 1981. *Ectothiorhodospira vacuolata* sp. nov., a new phototrophic bacterium from soda lakes. *Archives of Microbiology* 130: 238–242.

Jaschof, H. and **W. Schwartz.** 1961. Untersuchungen über Lebensgemeinschaften halophiler Mikroorganismen II. Über die Mikroben assoziationen einer alkalischen Sole aus dem Hochtal von Mexiko. *Zeitschrift für Allgemeine Mikrobiologie* 1: 258–273.

Javor, B.J. 1983. Planktonic standing crop and nutrients in a saltern ecosystem. *Limnology and Oceanography* 28: 153–159.

Jones, B.F., Eugster, H.P., and **Rettig, S.L.** 1977. Hydrochemistry of the Lake Magadi basin, Kenya. *Geochimica et Cosmochimica Acta* 41: 53–72.

Jørgensen, B.B. and **Cohen, Y.** 1977. Solar Lake (Sinai). 5. The sulfur cycle of the benthic cyanobacterial mats. *Limnology and Oceanography* 22: 657–666.

Jørgensen, B.B., Revsbech, N.P., and **Cohen, Y.** 1983. Photosynthesis and structure of benthic microbial mats: microelectrode and SEM studies of four cyanobacterial communities. *Limnology and Oceanography* 28: 1075–1093.

Kaplan, I.R. and **Friedman, A.** 1970. Biological productivity in the Dead Sea. Part 1: Microorganisms in the water column. *Israel Journal of Chemistry* 8: 513–528.

Kerry, K.R., Grace, D.R., Williams, R., and **Burton, H.R.** 1977. Studies on some saline lakes of the Vestfold Hills, Antarctica, pp. 839-858 in Llano, G.A. (editor), *Adaptations Within Antarctic Ecosystems,* Smithsonian Institution, Washington, D.C.

Klug, M., Boston, P., Francois, R., Gyure, R., Javor, B., Tribble, G., and **Vairavamurthy, A.** 1985. Sulfur reduction in sediments of marine and evaporite environments, pp. 128-157 in Sagen, D. (editor), *The Global Sulfur Cycle,* National Aeronautics and Space Administration Technical Memorandum 87570, Washington, D.C.

Knauth, L.P. and **Beeunas, M.A.** 1986. Isotope geochemistry of fluid inclusions in Permian halite with implications for the isotopic history of ocean water and the origin of saline formation waters. *Geochimica et Cosmochimica Acta* 50: 419–433.

Krumbein, W.E., Cohen, Y., and **Shilo, M.** 1977. Solar Lake (Sinai). 4. Stromatolitic cyanobacterial mats. *Limnology and Oceanography* 22: 635–656.

Kushner, D.J. 1978. Life in high salt and solute concentrations, pp. 317– 368 in Kushner, D.J. (editor), *Microbial Life in Extreme Environments.* Academic Press, London.

Kushner, D.J. 1985. The Halobacteriaceae, pp. 171–214 in Woese, C.R. and Wolfe, R.S. (editors), *The Bacteria. A Treatise on Structure and Function,* Vol. VIII, *Archaebacteria,* Academic Press, New York.

Kushner, D.J. and **Kamekura, M.** 1988. Physiology of halophilic eubacteria, p. 109–

138 in Rodriguez-Valera, F. (editor), *Halophilic Bacteria*, Vol. I, CRC Press, Boca Raton.

Larsen, H. 1980. Ecology of hypersaline environments, pp. 23–39 in Nissenbaum, A. (editor), *Hypersaline Brines and Evaporitic Environments*, Developments in Sedimentology 28, Elsevier Scientific, New York.

Larsen, H. 1981. The family Halobacteriaceae, pp. 985–994 in Starr, M.P., Stolp, H., Trüper, H.G., Balows, A. and Schlegel, H.G. (editors), *The Prokaryotes. A Handbook on Habitats, Isolation, and Identification of Bacteria*, Springer-Verlag, New York.

Mason, D.T. 1967. Limnology of Mono Lake. *University of California Publications in Zoology* 83: 1–102.

Mathrani, I.M. and **Boone, D.R.** 1985. Isolation and characterization of a moderately halophilic methanogen from a solar saltern. *Applied and Environmental Microbiology* 50: 140–143.

Melack, J.M. 1983. Large, deep salt lakes: a comparative limnological analysis. *Hydrobiologia* 105: 223–230.

Mitchell, B.D. and **Geddes, M.C.** 1977. Distribution of the brine shrimps *Parartemia zietziana* Sayce and *Artemia salina* (L.) along a salinity and oxygen gradient in a South Australian salt field. *Freshwater Biology* 7: 461–467.

Morishita, H. and **Masui, M. (editors).** 1980. *Saline Environment. Physiological and Biochemical Adaptation in Halophilic Microorganisms.* Japanese Conference on Halophilic Microbiology. Osaka. 200 pp.

Munns, R., Greenway, H., and **Kirst, G.O.** 1983. Halotolerant eukaryotes, pp. 59–135 in Lange, O.L., Nobel, P.S., Osmond, C.S. and Ziegler, H. (editors), *Physiological Plant Ecology III*, Springer-Verlag, New York.

Nissen, H. and **Dundas, I.D.** 1984. *Rhodospirillum salinarium* sp. nov., a halophilic photosynthetic bacterium isolated from a Portuguese saltern. *Archives of Microbiology* 138: 251–256.

Nissenbaum, A. 1975. The microbiology and biogeochemistry of the Dead Sea. *Microbial Ecology* 2: 139–161.

Norton, C.F. and **Grant, W.D.** 1988. Survival of halobacteria within fluid inclusions in salt crystals. *Journal of General Microbiology* 134: 1365–1373.

Oremland, R.S., Marsh, L., and **Des Marais, D.J.** 1982. Methanogenesis in Big Soda Lake, Nevada: an alkaline, moderately hypersaline desert lake. *Applied and Environmental Microbiology* 43: 462–468.

Oren, A. 1981. Approaches to the microbial ecology of the Dead Sea. *Kieler Meeresforschungen* 5: 416–424.

Oren, A. 1983. *Clostridium lortetii* sp. nov, a halophilic obligatory anaerobic bacterium producing endospores with attached vacuoles. *Archives of Microbiology* 136: 42–48.

Oren, A. 1986. The ecology and taxonomy of anaerobic halophilic eubacteria. *FEMS Microbiology Reviews* 39: 23–29.

Oren, A. 1988. The microbial ecology of the Dead Sea. *Advances in Microbial Ecology* 10: 193–229.

Oren, A. and **Shilo, M.** 1981. Bacteriorhodopsin in a bloom of halobacteria in the Dead Sea. *Archives of Microbiology* 130: 185–187.

Oren, A. and **Shilo, M.** 1982. Population dynamics of *Dunaliella parva* in the Dead Sea. *Limnology and Oceanography* 27: 201–211.

Pack, D.A. 1919. Two ciliata of Great Salt Lake. *Biological Bulletin* 36: 273–282.

Paterek, J.R. and **Smith, P.H.** 1985. Isolation and characterization of a halophilic methanogen from Great Salt Lake. *Applied and Environmental Microbiology* 50: 877–881.

Por, F.D. 1972. Hydrobiological notes on the high-salinity waters of the Sinai Peninsula. *Marine Biology* 14: 111–119.

Por, F.D. 1980. A classification of hypersaline waters, based on trophic criteria. *Marine Ecology* 1: 121–131.

Post, F.J. 1977. The microbial ecology of the Great Salt Lake. *Microbial Ecology* 3: 143–165.

Post, F.J. 1981. Microbiology of the Great Salt Lake north arm. *Hydrobiologia* 81: 59–69.

Post, F.J., Borowitzka, L.J., Borowitzka, M.A., Mackay, B. and **Moulton, T.** 1983. The protozoa of a Western Australian hypersaline lagoon. *Hydrobiologia* 105: 95–113.

Potts, M. 1980. Blue-green algae (Cyanophyta) in marine coastal environments of the Sinai Peninsula; distribution, zonation, stratification and taxonomic diversity. *Phycologia* 19: 60–73.

Priscu, J.C., Axler, R.P., Carlton, R.G., Reuter, J.E., Arneson, P.A., and **Goldman, C.R.** 1982. Vertical profiles of primary productivity, biomass and physico-chemical properties in meromictic Big Soda Lake, Nevada, U.S.A. *Hydrobiologia* 96: 113–120.

Quesada, E., Ventosa, A., Ruiz-Berraquero, F., and **Ramos-Cormenzana, A.** 1984. *Deleya halophila*, a new species of moderately halophilic bacteria. *International Journal of Systematic Bacteriology* 34: 287–292.

Raymond, J.C. and **Sistrom, W.R.** 1969. *Ectothiorhodospira halophila*: a new species of the genus *Ectothiorhodospira*. *Archiv für Mikrobiologie* 69: 121–126.

Reiser, R. and **Tasch, P.** 1960. Investigation of the viability of osmophile bacteria of great geological age. *Transactions of the Kansas Academy of Science* 63: 31–34.

Rodriguez-Valera, F. 1986. The ecology and taxonomy of aerobic chemoorganotrophic halophilic eubacteria. *FEMS Microbiology Reviews* 39: 17–22.

Rodriguez-Valera, F. 1988. Characteristics and microbial ecology of hypersaline environments, p. 3–30 in Rodriguez-Valera, F. (editor), *Halophilic Bacteria*, Vol. I, CRC Press, Boca Raton.

Rodriguez-Valera, F., Ruiz-Berraquero, F., and **Ramos-Cormenzana, A.** 1981. Characteristics of the heterotrophic bacterial populations in hypersaline environments of different salt concentrations. *Microbial Ecology* 7: 235–243.

Rodriguez-Valera, F., Ventosa, A., Juez, G., and **Imhoff, J.F.** 1985. Variation of environmental features and microbial populations with salt concentration in a multipond saltern. *Microbial Ecology* 11: 107–115.

Roedder, E. 1984. The fluids of salts. *American Mineralogist* 69: 413–439.

Rubentschik, L. 1926a. Über einige neue Urobakterienarten. *Zentralblatt für Bakteriologie II* 66: 161–180.

Rubentschik, L. 1926b. Über die Einwirkung von Salzen auf die Lebenstatigkeit der Urobakterien. *Zentralblatt für Bakteriologie II* 67: 167–194.

Rubentschik, L. 1929. Zur Nitrifikation bei hohen Salzkonzentrationen. *Zentralblatt für Bakteriologie II* 77: 1–18.

Shearman, D.J. 1966. Origin of marine evaporites by diagenesis. *Transactions of the Institute of Mining and Metallurgy* 75: 208–215.

Sonnenfeld, P. 1984. *Brines and Evaporites*. Academic Press. New York. 613 pp.

Stephens, D.W., and **Gillespie, D.M.** 1976. Phytoplankton production in the Great Salt Lake, Utah, and a laboratory study of algal response to enrichment. *Limnology and Oceanography* 21: 74–87.

Tasch, P. 1960. Paleoecological observations of the Wellington Salt (Hutchinson Member). *Transactions of the Kansas Academy of Science* 63: 24–30.

Tasch, P. 1963. Fossil content of salt and association evaporites, pp. 96–102 in Bersticker, A.C., Hoekstra, K.E. and Hall, J.F. (editors), *Symposium on Salt*, Northern Ohio Geological Society, Cleveland.

Tasch, P. 1969. Biochemical and geochemical aspects of the White Salt Pan—Bonaire,

Netherlands Antilles, pp. 204–210 in Rau, J.L. and Dellwig, L.F. (editors), *Third Symposium on Salt,* vol. 1, Northern Ohio Geological Society, Cleveland.

Timms, B.V. 1983. A study of benthic communities in some shallow saline lakes of western Victoria, Australia. *Hydrobiologia* 105: 165–177.

Tindall, B.J. 1988. Prokaryotic life in the alkaline, saline, athalassic environment, p. 31–67 in Rodriguez-Valera, F. (editor), *Halophilic Bacteria,* Vol. I, CRC Press, Boca Raton.

Tindall, B.J. and **Trüper, H.G.** 1986. Ecophysiology of the aerobic halophilic archaebacteria. *Systematic and Applied Microbiology* 7: 202–212.

Tindall, B.J., Mills, A.A., and **Grant, W.D.** 1980. An alkalophilic red halophilic bacterium with a low magnesium requirement from a Kenyan soda lake. *Journal of General Microbiology* 116: 257–260.

Tindall, B.J., Ross, H.N.M., and **Grant, W.D.** 1984. *Natronobacterium,* gen. nov. and *Natronococcus,* gen. nov., two new genera of haloalkalophilic archaebacteria. *Systematic and Applied Microbiology* 5: 41–57.

Tominaga, H., and **Fukui, F.** 1981. Saline lakes at Syowa Oasis, Antarctica. *Hydrobiologia* 82: 375–389.

Walsby, A.E., van Rijn, J., and **Cohen, Y.** 1983. The biology of a new gas-vacuolate cyanobacterium, *Dactylococcopsis salina* sp. nov, in Solar Lake. *Proceedings of the Royal Society of London B* 217: 417–447.

Weisser, J. and **Trüper, H.** 1985. Osmoregulation in a new haloalkaliphilic *Bacillus* from the Wadi Natrun (Egypt). *Systematic and Applied Microbiology* 6: 7–11.

Wilbert, N. and **Kahan, D.** 1981. Ciliates of Solar Lake on the Red Sea shore. *Archiv für Protistenkunde* 124: 70–95.

Winkler, D.W. (editor). 1977. An ecological study of Mono Lake, California. *Institute for Ecology Publications* 12, University of California, Davis, 190 pp.

Wright, S.W. and **Burton, H.R.** 1981. The biology of Antarctic saline lakes. *Hydrobiologia* 82: 319–338.

Zeikus, J.G., Hegge, P.W., Thompson, T.E., Phelps, T.J., and **Langworthy, T.A.** 1983. Isolation and description of *Haloanaerobium praevalens* gen. nov. and sp. nov., an obligately anaerobic halophile common to Great Salt Lake sediments. *Current Microbiology* 9: 225–234.

ZoBell, C.E., Anderson, D.Q., and **Smith, W.W.** 1937. The bacteriostatic and bactericidal action of Great Salt Lake water. *Journal of Bacteriology* 33: 253–262.

3

The Effects of Dissolved Organic Carbon on Evaporite Minerals

Because evaporites are salts precipitated from electrolyte solutions, it logically follows that the physical chemistry of natural brines should be similar to that of pure solutions of mixed electrolytes prepared in the laboratory. However, such laboratory models do not acknowledge the contribution of organisms to brine chemistry, and thus they ignore the potential role of dissolved organic matter (DOC) as natural chelators or as competitive, inhibitory, or catalytic substances. The role DOC plays in calcium carbonate and gypsum precipitation is much better known than its role in the formation of more economically important halite and potash minerals. This gap in knowledge will remain until more is known about the types and concentrations of DOC in natural brines saturated with respect to these minerals. The following discussion outlines the known or presumed role of DOC in modifying the nature of several evaporite minerals.

3.1 Calcium carbonate

Calcium carbonate (calcite and aragonite) solubility in brines is affected by a variety of chemical factors including salinity, pH, ion complexing, ion activity coefficients, and apparent dissociation constants of the carbonate system (Chapter 1). $CaCO_3$ is much more soluble in brines such as evaporated seawater than in normal seawater. In normal seawater (3.5% salinity), Mg^{2+} hinders the growth of calcite crystals and it may affect the nucleation of aragonite crystals (Berner, 1975; Pytkowicz, 1975). The same effect is probably true in evaporating brines.

Another component of seawater that strongly affects the precipitation and dissolution of $CaCO_3$ is DOC. Studies in non-hypersaline marine environments have shown that organic compounds inhibit reactions between carbonate minerals and seawater by adsorption and inactivation of $CaCO_3$ nuclei that are constantly forming (Chave, 1965; Chave and Suess, 1967, 1970). Suess (1970) determined experimentally that carbonate mineral surfaces are saturated with adsorbed organic carbon when they are coated with a monomolecular layer. It was concluded that organo-carbonate associations in seawater appear to reduce reaction rates of the inorganic carbonate equilibrium, and that they may inhibit the reaction entirely if the isolation of the mineral by this association is complete.

Particular classes of organic substances are associated with $CaCO_3$ grains. Carbonate sediments are particularly enriched in the acidic amino acids such as aspartic and glutamic acid, unlike non-carbonate sediments (Mitterer, 1968, 1972). Fatty acids are a major component of lipid coating on calcite in the sea (Meyers and Quinn, 1971). Humic acids also influence $CaCO_3$ precipitation (Suess and Fütterer, 1972; Otsuki and Wetzel, 1973).

The type of organic matter influences the solubilities and crystal morphologies of carbonates (Kitano et al., 1969). Certain Ca^{2+}-complexing compounds such as citrate, malate, pyruvate, and glycogen increased the proportion of calcite precipitating by increasing the solubility of $CaCO_3$ and reducing the rate of precipitation. However, the addition of Mg^{2+} decreased the proportion of calcite.

The combined inhibitory effects of Mg^{2+} and DOC are believed to be responsible for the observed supersaturation of surface seawater with respect to aragonite (200% to over 300%) (Ben-Yaakov and Kaplan, 1969; Pytkowicz, 1971). Pytkowicz used a method of saturometry based on the measurement of the hydrogen ion activity (pH) of seawater before and after introduction of an excess of clean aragonite powder. By providing essentially an infinite number of nucleation sites, supersaturated $CaCO_3$ precipitates on the aragonite.

Precipitation and dissolution of $CaCO_3$ in seawater theoretically accompany photosynthesis and respiration, respectively. In organic-rich brine (9.6% salinity) overlying microbial mats in a salt marsh pool, total alkalinity showed the expected decrease during the course of daylight, but supersaturation with respect to aragonite showed unexpected, widely fluctuating behavior (Javor, 1979) (Figure 3.1). After slowly decreasing to 650% saturation between 09:00 and 11:00, the overlying brines suddenly became nearly 1100% saturated with respect to aragonite for 2 hours at mid-day. In the early afternoon, the per cent saturation precipitously dropped to morning values and lower. These observations suggest that DOC and perhaps inorganic factors promoted a mid-day reaction that prevented $CaCO_3$ nuclei from growing.

On the day following the measurements made in Figure 3.1, tidal flooding of the pool with the microbial mats caused temporary salinity stratification (4.5% at the surface, 8.5% at the bottom). The more hypersaline brine showed

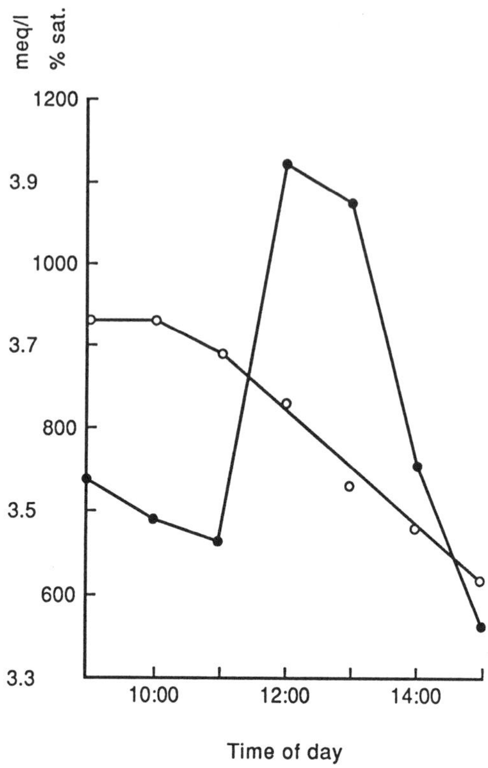

Figure 3.1 Total alkalinity as milliequivalents per liter (open circles) and aragonite saturation as percent saturation (closed circles, % sat) of seawater (9.6% salinity) covering microbial mats in Laguna Guerrero Negro, Mexico, on 12 August, 1977 (from Javor, 1979).

fluctuating aragonite saturation behavior (although not as pronounced as the previous day) while the fresh seawater had low saturation values (110–240%) that behaved as would be predicted by daily total alkalinity changes (Figure 3.2). Five days after tidal flooding, the seawater covering the microbial mats was completely mixed (4.9% salinity). Although the salinity was well below that earlier in the week, the daily aragonite saturation behavior of the 4.9% salinity brine resembled that measured in 8.5% salinity brine.

The results of these saturometry experiments suggest that in the organic-rich brines overlying intensely active microbial mats, $CaCO_3$ may become even more supersaturated than that predicted by models based on the composition of normal seawater. The influence of DOC on $CaCO_3$ solubility may also explain the difference between total alkalinity measurements in saltern brines (Figure 1.4) and seawater evaporated in the laboratory (Lazar et al., 1983). Hypersalinity and DOC-$CaCO_3$ interactions may also explain why

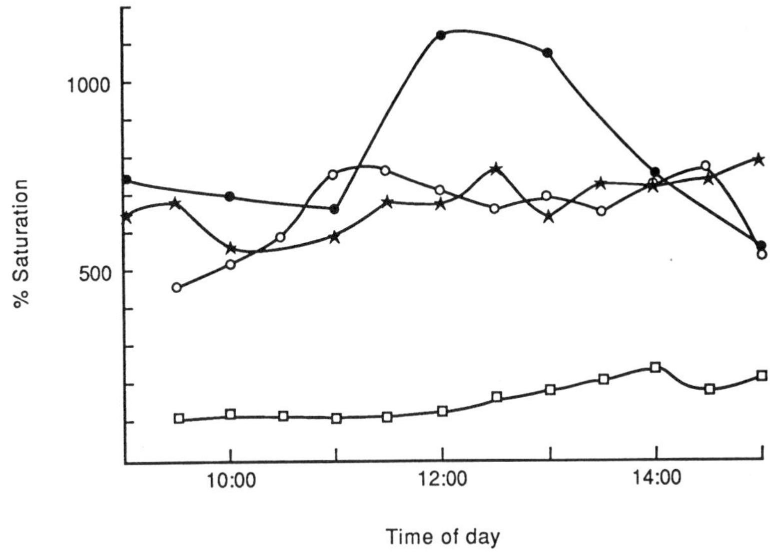

Figure 3.2 Aragonite saturation at different saturation levels in seawater covering the microbial mats described in Figure 3.1. Closed circles: 9.6% salinity, 12 August. Open circles: bottom brines, 8.5% salinity, 13 August. Squares: top brines, 4.5% salinity, 13 August. Stars: 4.9% salinity, 18 August (from Javor, 1979).

saturometry measurements are time-dependent after the initial fast reaction (Figure 3.3). In those measurements, seawater allowed to evaporate in the laboratory showed the predicted supersaturation with respect to aragonite at the beginning of the experiments, but after one day the newly-precipitated aragonite began to return to solution. The dissolution was complete in the control sample (3.2% salinity) whereas in the 9.2% and the 14.5% salinity samples, a second oscillation in aragonite precipitation was observed after one or two days. These oscillations in aragonite precipitation and dissolution in hypersaline marine brines apparently occurred in the absence of any significant microbial activity. It is suspected that DOC present in the samples caused the redissolution of aragonite. The oscillatory behavior in the stronger brines remains enigmatic.

3.2 Dolomite

Dolomite (magnesium calcium carbonate) is a common carbonate mineral in both ancient evaporite and non-evaporite sedimentary sequences. In modern carbonate deposits, dolomite typically constitutes a small fraction of the total carbonates. Dolomitic layers in ancient evaporites are commonly bituminous,

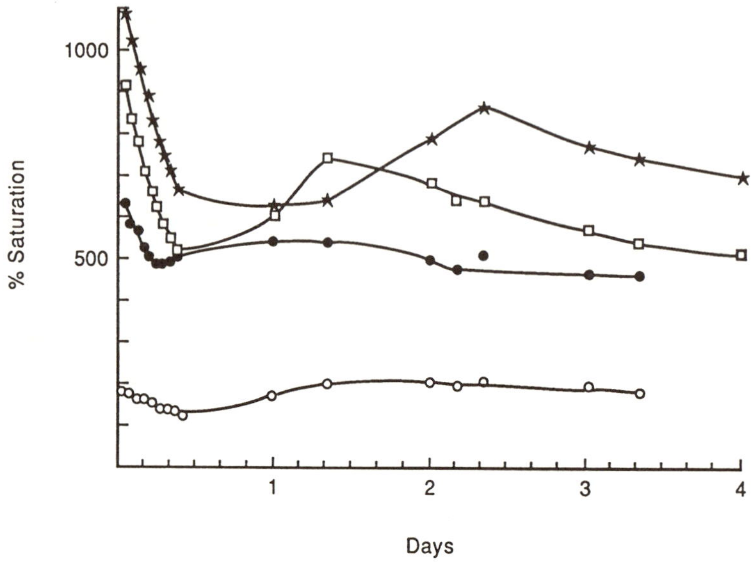

Figure 3.3 Saturation with respect to aragonite in seawater evaporated in the laboratory. Samples were constantly agitated with excess aragonite in screw-cap bottles during the course of 4 days. Open circles: 3.2% salinity. Closed circles: 5.4% salinity. Squares: 9.2% salinity. Stars: 14.5% salinity.

but there is no proof as that original carbonate was $CaCO_3$ or dolomite. Like that of $CaCO_3$ (calcite and aragonite), the solubility of dolomite is exceeded in modern seawater. However, dolomite does not precipitate except under a narrow range of ill-defined conditions. Modern dolomite and protodolomite (unordered dolomite) are often associated with organic-rich niches and/or hypersaline marine environments.

Whether dolomite can occur as a primary precipitate in sabkhas or only as a replacement mineral has been critically reviewed by Hardie (1987). The model of Kastner (Baker and Kastner, 1981; Kastner, 1984) states that dolomite formation is favored by low sulfate concentrations and that high Mg/Ca ratios are not essential. The most effective means of reducing the sulfate content of brines would be by bacterial sulfate reduction. However, Hardie points out that some modern environments where dolomite is found are characterized by elevated sulfate concentrations and that the actual role of sulfate-reducing bacteria in dolomite formation is the increase in dissolved HCO_3^- (see Chapter 4).

Gebelein and Hoffman (1973) pointed out that stromatolites from the lower Paleozoic and Proterozoic calcite-dolomite laminations (on a cm- or mm-scale) were probably formed similarly to modern marine stromatolites composed of algal-rich and sediment-rich laminae. In ancient stromatolites, the dolomite layers contain or are capped by bituminous material. The authors showed that the Mg/Ca ratio of sheath material from a stromatolite-building cyanobacterium, *Schizothrix calcicola*, was enriched 3- to 4-fold over that of seawater when the cells were grown in seawater. The addition of dissolved carbonate caused the precipitation of Mg-calcite (but not dolomite) in the sheath material.

Davies et al. (1975) reported dolomite precipitation after 9 months in an experimental tank with calcite, magnesium carbonate, and decaying algae. Dolomite was associated with the decaying algae. A similar association was noted by these authors in a zone of decomposition below a microbial mat near the Great Barrier reef. The authors stressed that the increase in alkalinity caused by organic decomposition drives the precipitation of dolomite. Although they recognized the potential role of organic matter in retarding aragonite precipitation, they did not suggest that there was any direct DOC-dolomite interaction involved in dolomite precipitation or supersaturation.

Recent dolomite precipitation may be both diagenetic and primary (Guntalilaka et al., 1984, 1987). In a slightly hypersaline lagoon in Kuwait characterized by organic-rich, highly reducing aragonitic sediments, dolomite (described as diagenetic) occurs as clusters within minute cavities on the surface of pellets (presumably of animal origin). Dolomite (reported as primary) precipitates as spherules in the walls of crustacean burrows and as microdolomite in the core of the burrows. The authors concluded that both the occurrence of dolomite and its distinct mineral textures most likely result from a combination of porewater composition and flow, organic matter, and bacterial activity.

It is most likely that the constraints on $CaCO_3$ precipitation imposed by DOC are equally important in dolomite precipitation, but they have never been shown experimentally. Like Ca^{2+}, Mg^{2+} can potentially bind to acidic organic residues. Mg^{2+} is the dominant cation involved in bicarbonate and carbonate complexes in seawater (Garrels and Thompson, 1962). About equal proportions of Mg^{2+} (13%) and Ca^{2+} (9%) are bound to anions. There is little reason to doubt that both Ca^{2+}- and Mg^{2+}-complexes are even more abundant in anaerobic porewaters characterized by elevated total alkalinity and increased DOC. Perhaps more important that the molar ratio of Mg/Ca in assessing the conditions of dolomite precipitation are the concentrations of uncomplexed Mg^{2+} and Ca^{2+}. If the affinity of specific classes of DOC is about the same for both cations, the relatively low total concentration of free Ca^{2+} in organic-rich porewater may permit ordered or disordered dolomite to form.

Thus, anaerobic decomposition and DOC may promote local, primary dolomite precipitation by a combination of factors: 1) Local elevated concentrations of Mg^{2+} by algae or microbial mats during active growth. 2) Formation of potential divalent cation-binding organic acids during degradation by fermentation (Chapter 4), resulting in decreased absolute concentrations of uncomplexed Ca^{2+} or elevated ratios of uncomplexed Mg^{2+}/Ca^{2+}. 3) Increase in total alkalinity and decrease in total sulfate during degradation by sulfate reduction (Chapter 4). 4) DOC binding to growing dolomite crystals, resulting in distinct crystal morphologies.

3.3 Gypsum and celestite

Both the solubility and crystal morphology of gypsum (calcium sulfate) are strongly affected by DOC. Barcelona and Atwood (1978) demonstrated that lipid material found in saltern brines inhibited nucleation of $CaSO_4$ crystals. Long-chain fatty acids became selectively bound to Ca^{2+}, as did succinate and mannitol. Glycerol and pyruvate (3-carbon compounds) did not affect nucleation time. It is interesting that glycerol, which is the major compatible solute in the brine alga *Dunaliella* (Chapter 9), does not affect the rate of precipitation of gypsum. *Dunaliella* typically thrives best in saltern ponds at the gypsum stage of brine concentration.

Free fatty acids constitute 50–84% of the lipids recovered from saltern ponds (Barcelona and Atwood, 1979). Other lipids include long-chain hydrocarbons. The gypsum selectively removes the longer-chain fatty acids (C_{16}–C_{22}) and hydrocarbons (C_{32} preferred over C_{28} and shorter alkanes) from the brines. Total lipid material extracted from saltern gypsum was 9–38 $\mu g \cdot kg^{-1}$.

Organic matter also affects crystal morphology of gypsum. Adsorbed lipids caused the growth of more tabular, equidimensional crystals (Barcelona and Atwood, 1978). Cody (1979) showed that lens-shaped crystals grew only in the presence of organic matter. Lenticular crystals are common in both modern and ancient evaporites.

A summary of the morphologies of gypsum crystals grown under a variety of conditions in silica gel was given by Van Rosmalen et al. (1976). The presence of inorganic compounds or low pH led to needle-shaped crystals while the presence of various organic compounds caused stubbier crystals to form. Some of the organic substances tested were lactic acid (which brought about the formation of short, broad, rod-shaped crystals), tartrate (broad rods), and acetate and citrate (platy crystals). Acetate is the dominant free fatty acid in normal marine and hypersaline sediments (Chapter 4), which could help explain why flat, lenticular gypsum crystals are formed in many evaporite environments.

Celestite ($SrSO_4$) is sometimes found as a co-precipitate with aragonite and gypsum, and even in halite in marine evaporites (Braitsch, 1971). In solar salterns, a fraction of the dissolved Sr^{2+} precipitates with gypsum while the bulk precipitates as distinct celestite crystals in NaCl-saturated brines (Figure 13.2 and Javor, 1983). The crystal morphology of celestite is greatly affected by DOC. $SrSO_4$ (and $SrCO_3$, strontianite) precipitate as dumbbell-shaped bundles of acicular crystals in artificial solutions with DOC and in saltern brines (which contain DOC), but as individual acicular crystals in pure saline solutions. It is unknown whether DOC can affect the kinetics of celestite precipitation, and whether Sr^{2+}-organic complexes similar to Ca^{2+}-organic complexes play a role in determining the concentration of brine that permits celestite precipitation.

3.4 Halite

Ancient sodium chloride (halite) deposits are often interlaminated with clay or less soluble evaporites (gypsum and $CaCO_3$) as a result of periodic brine freshening. Laminae of bituminous matter in halite deposits may result from washed-in organic matter, the accumulation of organic matter produced at lower salinities in the evaporite system, or the catastrophic deposit of organic material caused by death and lysis during a brine freshening event. Droplets of hydrocarbons ("mountain tar," "paraffin earth") can accumulate in halite deposits in primary brine inclusions and small veins. A review of these occurrences in both halite and potash minerals, cited from German and Russian literature, was presented by Sonnenfeld (1984). It is noteworthy that hydrocarbons accumulate, as this class of organic matter has been shown to be recalcitrant to oxidation by natural populations of halophilic microorganisms (Chapter 18). Gases documented in microinclusions of halite include nitrogen, hydrogen, methane, higher gaseous alkanes, and H_2S.

DOC alters the surface tension of the brine and can alter the rate of crystallization and the crystal morphology of halite. Gelatinous sheath material from the cyanobacterium *Aphanothece halophytica* (Chapter 8) or a related organism that accumulated in a solar saltern during a rainy period (brine freshening) was shown to cause the growth of small, fragile (brine-filled) halite crystals (Baha Al-Deen and Baha Al-Deen, 1972). Shuman (1965) also noted that viscous substances such as methyl cellulose, gelatin, and gum arabic did not reduce the volume of brine inclusions in salt. However, polyphosphate (which traps Ca^{2+}), heavy metals, and an undefined extract from seaweed decreased the volume of brine inclusions in halite.

The crystal habit of halite can be modified from cubic to other forms by a variety of both organic and inorganic substances (Milone and Ferraro, 1947; Ploss, 1964; Shuman, 1965). Octahedral, dodecahedral, and dendritic crystals were reported to arise when NaCl solutions were mixed with various concentrations of substances, including some, such as ferrocyanide and molasses, that are not found in natural evaporite environments. No systematic study of the effects of bacterial decomposition products on halite precipitation, brine inclusions, and crystal morphology, particularly in the range of concentrations such substances naturally occur, have been published.

3.5 Potash minerals

Similar kinds of organic matter and gases have been found in both halite and potash deposits. Trapped bituminous matter in evaporite salts can render them grey and malodorous (Borchert and Muir, 1964). Like sylvite, carnallite may contain large amounts of trapped gases. Dissolution or heating of such "crackle salts" results in gas release with a crackling sound (decrepitation). Besides CH_4, higher alkanes are sometimes found in the inclusions of both potash and halite (literature cited by Sonnenfeld, 1984). However, the gases may enter the salt deposits long after deposition, so the interpretation of the origin of evaporite-trapped gases requires careful basin analysis. For example, the high pressure of CO_2 found in some Permian Zechstein potash deposits in Germany is believed to be due to gases that entered the evaporite from Tertiary magmatic rocks (Giesel, 1972).

Many of the available data documenting the effects of DOC on potash mineral precipitation have been collected for the potash mining industry in the development of methods of separating minerals. Minerals are commonly separated by flotation in mixed-salt brines using various organic agents (Noyes, 1966). "Collectors" selectively attach to the surface of small crystals and render them hydrophobic. Air bubbles generated in the flotation tanks selectively attach to the non-wetted surfaces and cause the precipitate to accumulate in the froth. Collectors, which are usually used in low ($mg \cdot l^{-1}$) concentrations, include Na^+ and K^+ salts of fatty acids and amine acetates. For example, floating halite can be collected from a NaCl-KCl mixture with oleic acid while sylvite is typically collected with cationic primary aliphatic amines (C_6–C_{24}). Natural potassium salts contain slimy, clay-like and other water-soluble materials that can be removed from flotation brines with various polysaccharides, cellulose, and lignin derivatives. Organic frothing agents that are also used to lower the surface tension of flotation brines include alcohols and oils. Because many of these organic compounds are concentrated in evaporite basin brines, it is most likely that such DOC compounds have played a role in separating or moderating potash deposition in the natural evaporite deposits.

References

Baha Al-Deen, B. and **Baha Al-Deen, A.H.** 1972. Posible efecto de microalgas en la forma de cristalización de cloruro de sodio en la Salina de Araya. *Boletiń del Instituto Oceanográfico Universidad de Oriente* 11: 35–38.

Baker, P.A. and **Kastner, M.** 1981. Constraints on the formation of sedimentary dolomite. *Science* 213: 214–216.

Barcelona, M.J. and **Atwood, D.K.** 1978. Gypsum-organic interactions in natural seawater: effect of organics on precipitation kinetics and crystal morphology. *Marine Chemistry* 6: 99-115.

Barcelona, M.J. and **Atwood, D.K.** 1979. Gypsum-organic interactions in the marine environment: sorption of fatty acids and hydrocarbons. *Geochimica et Cosmochimica Acta* 43: 47–53.

Ben-Yaakov, S. and **Kaplan, I.R.** 1969. Determination of carbonate saturation of seawater with a carbonate saturometer. *Limnology and Oceanography* 14: 874–882.

Berner, R.A. 1975. The role of magnesium in the crystal growth of calcite and aragonite from sea water. *Geochimica et Cosmochimica Acta* 39: 489–504.

Borchert, H. and **Muir, R.O.** 1964. *Salt Deposits. The Origin, Metamorphism and Deformation of Evaporites.* D. Van Nostrand Co., New York, 338 p.

Braitsch, O. 1971. *Salt Deposits. Their Origin and Composition.* Springer-Verlag, Berlin, 297 p.

Chave, K. 1965. Carbonates: association with organic matter in surface seawater. *Science* 148: 1723–1724.

Chave, K. and **Suess, E.** 1967. Suspended minerals in seawater. *Transactions of the New York Academy of Science,* ser. 2, vol. 29: 991–1000.

Chave, K. and **Suess, E.** 1970. Calcium carbonate saturation in seawater: effects of dissolved organic matter. *Limnology and Oceanography* 15: 633–637.

Cody, R.D. 1979. Lenticular gypsum: occurrences in nature, and experimental determinations of effects of soluble green plant material on its formation. *Journal of Sedimentary Petrology* 49: 1015–1028.

Davies, P.J., Ferguson, J., and **Bubela, B.** 1975. Dolomite and organic material. *Nature (London)* 255: 472–474.

Garrels, R.M. and **Thompson, M.E.** 1962. A chemical model for sea water at 25°C and one atmosphere total pressure. *American Journal of Science* 260: 57–66.

Gebelein, C.D. and **Hoffman, P.** 1973. Algal origin of dolomitic laminations in stromatolitic limestone. *Journal of Sedimentary Petrology* 43: 603–613.

Giesel, W. 1972. Outbursts of carbon dioxide in potash mines—fundamentals and possibilities of forecast, p. 235–239 in Richter-Bernburg, G. (editor), *Geology of Saline Deposits.* Unesco, Paris.

Gunatilaka, A., Saleh, A., Al-Temeemi, A., and **Nassar, N.** 1984. Occurrence of subtidal dolomite in a hypersaline lagoon, Kuwait. *Nature (London)* 311: 450–452.

Gunatilaka, A., Al-Zamel, A., Shearman, D., and **Reda, A.** 1987. A spherulitic fabric in selectively dolomitized siliciclastic crustacean burrows, northern Kuwait. *Journal of Sedimentary Petrology* 57: 922–927.

Hardie, L.A. 1987. Perspectives. Dolomitization: a critical view of some current views. *Journal of Sedimentary Petrology* 57: 166–183.

Javor, B.J. 1979. *Ecology, Physiology, and Carbonate Chemistry of Blue-Green Algal Mats, Laguna Guerrero Negro, Mexico.* Ph.D. Thesis. University of Oregon, Eugene. 260 pp.

Javor, B.J. 1983. Nutrients and ecology of the Western Salt and Exportadora de Sal saltern brines, pp. 195–205 in Schreiber, B.C. and Harner, H.L. (editors), *Sixth International Symposium on Salt,* vol. 1, The Salt Institute, Toronto.

Kastner, M. 1984. Control of dolomite formation. *Nature (London)* 311: 410–411.

Kitano, Y. et al. 1969. Effects of organic matter on solubilities and crystal form of carbonates. *American Zoologist* 9: 681–688.

Lazar, B., Starinsky, A., Katz, A., Sass, E., and **Ben-Yaakov, S.** 1983. The carbonate system in hypersaline solutions: alkalinity and $CaCO_3$ solubility of evaporated seawater. *Limnology and Oceanography* 28: 978–986.

Meyers, P.A. and **Quinn, J.G.** 1971. Interaction between fatty acids and calcite in seawater. *Limnology and Oceanography* 16: 992–997.

Milone, M. and **Ferrero, F.** 1947. The relations between surface tension and crystalline habit. II. *Gazzetta Chimica Italiana* 77: 348–352.

Mitterer, R.M. 1968. Amino acid composition of organic matrix in calcareous oolites. *Science* 162: 1498–1499.
Mitterer, R.M. 1972. Biogeochemistry of aragonite mud and oolites. *Geochimica et Cosmochimica Acta* 36: 1407–1422.
Noyes, R. 1966. *Potash and Potassium Fertilizers.* Noyes Development Corporation. Park Ridge. 210 pp.
Otsuki, A. and **Wetzel, R.G.** 1973. Interaction of yellow organic acids with $CaCO_3$ in freshwater. *Limnology and Oceanography* 18: 490–493.
Ploss, R.S. 1964. Sodium chloride: modification of crystal habit by chemical agents. *Science* 144: 169–170.
Pytkowicz, R.M. 1971. Sand-seawater interactions in Bermuda beaches. *Geochimica et Cosmochimica Acta* 35: 509–515.
Pytkowicz, R.M. 1975. Some trends in marine chemistry and geochemistry. *Earth-Science Reviews* 11: 1–46.
Shuman, A.C. 1965. Gross imperfections and habit modification in salt crystals, pp. 246–253 in Rau, J.L. (editor), *Second Symposium on Salt,* vol. 2, Northern Ohio Geological Society, Cleveland.
Sonnenfeld, P. 1984. *Brines and Evaporites.* Academic Press. New York. 613 pp.
Suess, E. 1970. Interaction of organic compounds with $CaCO_3$. I. Association phenomena and geochemical implications. *Geochimica et Cosmochimica Acta* 34: 157–168.
Suess, E. and **Fütterer, D.** 1972. Aragonitic ooids: experimental precipitation from seawater in the presence of humic acid. *Sedimentology* 19: 129–139.
Van Rosmalen, G.M., Marchée, W.G.J., and **Bennema, P.** 1976. A comparison of gypsum crystals grown in silica gel and agar in the presence of additives. *Journal of Crystal Growth* 35: 169–176.

4

The Fate of Carbon and Sulfur in Hypersaline Environments

Although hypersaline environments are often thought of as terminal habitats, they play an active role in the interconversion of organic and inorganic carbon, a process that is interfaced microbiologically with the reduction of sulfate to sulfide. In both normal marine and hypersaline marine environments, carbon cycle transformations result in changing pools of dissolved $\Sigma\ CO_2$ (CO_2 + HCO_3^- + CO_3^{2-}), solid carbonates, and organic carbon. In the marine environment, sulfur cycle transformations result in changing pools of dissolved sulfate and sulfide, solid sulfides and sulfur, and organic sulfur, while in marine-derived hypersaline environments, solid sulfate (gypsum) may provide an additional large pool of potential oxidant of organic matter. Given the simplicity of food chains in hypersaline environments relative to those in normal marine habitats, the high concentrations of sulfate, and the possibly limited activity of anaerobic bacteria in extremely saline sediments, it might be expected that rates and kinds of interconversions between reduced and oxidized pools of carbon and sulfur in hypersaline environments may differ from those in normal marine sediments. The following discussion outlines aspects of carbon and sulfur cycles from sediment and porewaters of hypersaline environments, using evidence from rates and types of transformations, organic and inorganic geochemistry, and stable isotope fractionations.

4.1 Carbon cycle

Degradation processes and hydrocarbon accumulation Evaporite deposits are often the reservoirs of vast petroleum deposits. The interrelationships among evaporites, the biological carbon cycle, and petroleum genesis have been well-established. Discussions on these topics have been compiled in a volume by B. C. Schreiber (1988). Degens and Paluska (1979) interpreted the paleoenvironment of hydrocarbon-bearing shales of the Caspian Sea and the North Sea to have been characterized by deep anoxic basins with salinity-stratified seawater. High organic productivity in the surface waters led to the accumulation of organic matter in the hypersaline depths. Moderately hypersaline environments (ca. 4–12% salinity) characterized by carbonate deposition and fluctuating salinities apparently were the source of much of the world's petroleum reserves (Kirkland and Evans, 1981; Eugster, 1985; Evans and Kirkland, 1988). Such hypersaline environments evidently produced the organic matter associated with the petroleum reserves of the Persian Gulf as well as those of the Michigan and Paradox basins of North America. The importance of carbonate-precipitating sedimentary environments over geologic time cannot be overemphasized. About 5200×10^{18} moles carbon are locked in carbonates while about 1300×10^{18} moles organic carbon are calculated to be in sedimentary reservoirs (Garrels and Lerman, 1984).

Much of the organic matter associated with evaporite sedimentary cycles is found in shaly carbonate beds rather than in anhydrite or halite layers, either because the organic matter was originally produced under relatively less saline conditions, or the DOC of the stronger brines migrated to the more porous zones in the sediments. Alternatively, high rates of precipitation and sedimentation of gypsum and halite may effectively dilute the organic matter in those sediments. Normal progressive evaporite development in marine environments is characterized by the formation of a basin of restricted circulation (where organic productivity is high) followed by a succession of evaporite facies* (Borchert and Muir, 1964). In the case of the Paradox basin evaporites, however, much of the organic matter may have entered from terrestrial sources and then undergone degradation in the anoxic evaporite environment of the basin (Hite et al., 1984).

One of the themes of this book is that only some microbial activities are interrupted at high salinities (see Table 5.1). In order to accumulate hydrocarbons in anaerobic evaporitic sediments, the organic matter must first be produced (*in situ* or imported primary productivity) and partially degraded. The most important anaerobic degradation processes in marine sediments are fermentation, sulfate respiration (sulfate reduction), and methanogenesis (although some marine methanogens and sulfate reducers may actually generate organic carbon from CO_2 or CO plus H_2). Some primary producers can fix CO_2 in salinities up to halite saturation, although optimal rates occur in lower salinities (see Chapters 7, 8, and 9). Fermentation includes a broad variety of anaerobic, degradative reactions in which organic matter is partially broken down, releasing CO_2, H_2, H_2S, NH_3, low molecular weight alcohols, acetate

*A *facies* is a suite of environments in a single geographic setting (for example, an intertidal region plus a beach plus a salt flat).

and other volatile fatty acids, a variety of organic acids, and the residual organic matter that is not cleaved or metabolized. Low molecular weight fermentation products are the main source of substrates for both methanogens (Chapter 6) and sulfate-reducers (Section 4.2). They also apparently play a role in determining the crystal morphology and extent of saturation of evaporite minerals (Chapter 3). Therefore, fermentation is a critical metabolic link between the original organic matter and its final oxidation by sulfate-reducers, and between the inorganic composition of the brine and the mineral composition of the primary (and perhaps secondary) evaporites.

Fermentation occurs in extremely hypersaline environments, even when other metabolic processes are not functioning. While Klug et al. (1985) demonstrated that sulfate-reduction rates decreased with increasing salinity in shallow solar saltern sediments (see Section 4.2), volatile fatty acids (particularly acetate) accumulated to a much greater extent in the extremely hypersaline sediments (Figure 4.1). Baas-Becking and Kaplan (1956) showed that H_2 accumulated to the greatest extent in Lake Eyre sediments incubated in very high salt concentrations (see Chapter 17). These results suggest that the fermentation products accumulate because they cannot be used by sulfate reducers and methanogens at such high salinities, or that fermentation rates greatly exceed sulfate reduction rates. Likewise, the often-reported occurrence of sulfide in some extremely hypersaline habitats may be more a product of the breakdown of sulfur-containing amino acids by putrefactive bacteria than of the activities of sulfate-reducing bacteria.

Although organic chemists have identified and quantified the rich variety of organic compounds in a number of sediment and brine profiles from modern hypersaline environments, little attempt has been made to study accumulations across salinity gradients. Figures 4.2 and 4.3 show the concentrations of total carbohydrates and primary amines (largely amino acids) in surface brines of an oligotrophic saltern (Exportadora de Sal, Mexico) and a eutrophic saltern (Western Salt, California, U.S.A.) (Javor, 1983). Variations in the microbial flora apparently caused the carbohydrate and amino acid concentrations to differ in the two salterns. Bitterns brines (>1 M Mg^{2+}) were sterile and characterized by extremely high accumulations of dissolved organic substances. Bitterns brines of the eutrophic saltern were yellow, presumably due to the presence of humic-like substances. Both carbohydrates and DOC show a high degree of preservation in Solar Lake sediments (see Chapter 14). Comparable concentrations of dissolved amino acids were detected in brines (7.7–26.9% salinity) of the Eocene Green River Formation (Degens et al., 1964). Paleozoic brines were analyzed for sugars with negative results.

While rates of sulfate reduction across a salinity gradient were measured by Klug et al. (1985), fermentation rates were not. Rates of glutamate and hydrocarbon oxidation in Great Salt Lake samples across a salinity gradient were measured by Ward and Brock (1978) (see Chapter 18). Glutamate degradation was greatly attenuated above 12% salinity but was still detectable

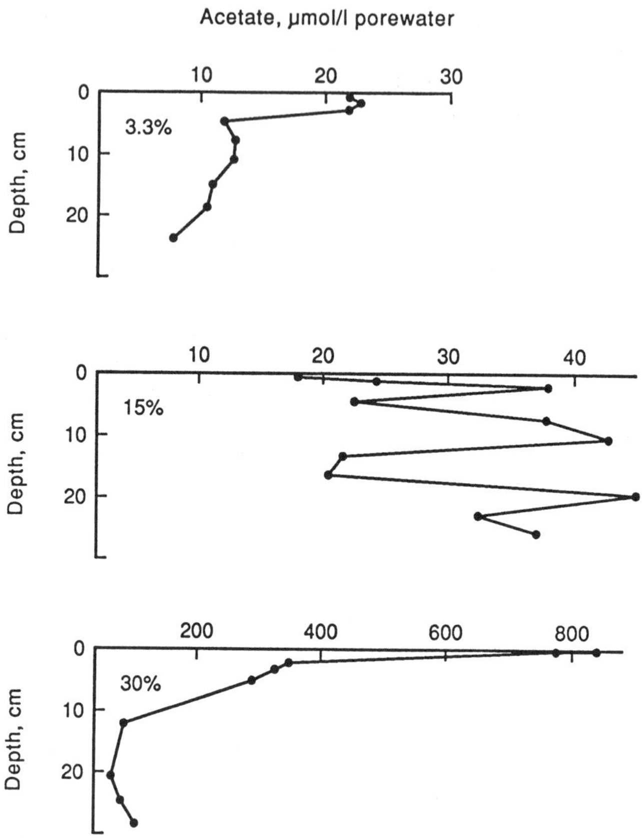

Figure 4.1 Dissolved acetate measured in porewaters of the San Francisco saltern (California, U.S.A.) during August, 1984, in samples from water with 3.3%, 15.0%, and 30.0% salinity (from Klug et al., 1985).

in 28% salinity. Hexadecane degradation was even more strongly affected by salinity, and measurable rates could not be detected at the highest salinities. Enrichments for mineral oil-oxidizers were only successful in samples with less than 20% salinity. Other measurements of anaerobic degradation processes across salinity gradients are summarized in Table 5.1.

Lipids in recent and ancient hypersaline environments Because hydrocarbons eventually accumulate in extremely hypersaline environments, the identification of the organisms responsible for the synthesis of unique lipids are of interest to organic geochemists and petroleum geologists. Such organic "fingerprints" may be the only paleontological indicators when morphological fossils are not preserved. Table 4.1 lists some potential unique

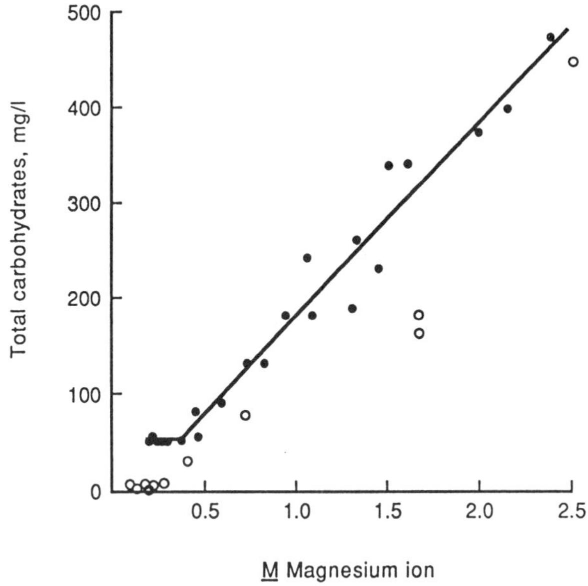

Figure 4.2 Dissolved total carbohydrates (determined with the phenolhydrazine sulfate technique) of surface brines of the Exportadora de Sal saltern (closed circles) and the Western Salt saltern (open circles). The degree of concentration of the brine is expressed as the concentration in molarity of soluble magnesium ion. The line drawn through the closed circles between 0.4 and 2.5 M Mg has an r^2 value of 0.971.

biomarkers (both lipid and non-lipid) synthesized by cultures of halophiles. For comparison, an extensive list of preserved lipid compounds from the Solar Lake microbial mat environment is given in Table 14.5.

Lipid analyses for modern hypersaline environments such as Solar Lake and Gavish Sabkha show that the spectrum of compounds is much more complicated than that which can be determined for single bacterial isolates. For example, de Leeuw et al. (1985) recognized at least 53 different hydrocarbons, 42 different fatty acids, 23 different alcohols, and 18 different sterols from the top section of microbial mats of the Gavish Sabkha. In both cultured cells and natural materials, unidentified lipids or other compounds may be important components. Conversely, major identified lipids may be produced by unidentified microorganisms. In microbial mats from Abu Dhabi, an unidentified alkane was even more abundant than phytane and pristane (Cardoso et al., 1978). A branched hydrocarbon identified as 2,6-dimethylhexadecane constituted 75% of the hydrocarbons extracted from the bottom of Lake Vanda (Antarctica), but it has never been reported in any living organism (Matsumoto et al., 1984). Hydrocarbons were found only at the bottom of the lake and never in the water column (see Chapter 16). Material that could not be identified as either protein, carbohydrate, or lipid constituted about

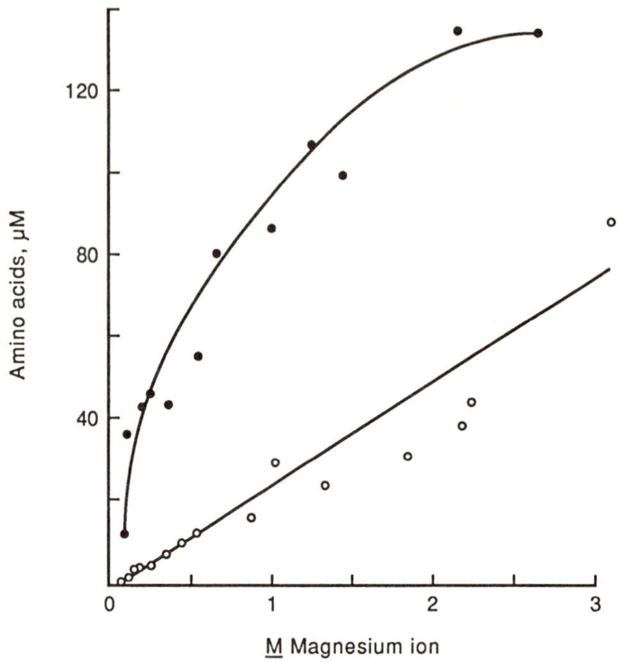

Figure 4.3 Dissolved total amino acids in the surface brines of the Exportadora de Sal saltern (closed circles) and the Western Salt saltern (open circles). Total amino acids were determined in acid-hydrolyzed samples with fluorescamine against glycine standards.

20% of the dry weight of *Dunaliella salina* and *D. bardawil* grown under low salt or conditions of nitrogen deficiency, but only 5% of the dry weight when cells were cultured in 2 M NaCl (Ben-Amotz et al., 1985). Half the dry weight of the diatom *Nitzschia* from Mono Lake grown in 1.4 M NaCl was of unknown composition, although Ben-Amotz et al. did not state whether this measurement included the silica walls.

Yeasts and fungi are not commonly recognized in extremely hypersaline environments but their presence may sometimes be detected in lipid analyses. The carotenoid torulene, which is unique to fungi, was recognized in the Abu Dhabi microbial mats, although fungal filaments were not observed (Cardoso et al., 1978). Caution must be exercised when ergosterol is detected, however. This sterol, which is usually a telltale marker for fungi, has been detected in small quantities in *Dunaliella tertiolecta* (Wright, 1979).

With the discovery of unique lipids produced by archaebacteria came the possibility of using geochemical fossils of halobacteria as important indicators of extreme salinity in geological deposits. Glycerol phytanyl diether isolated from reducing sediments of the Dead Sea was believed to be derived from halobacteria (Kaplan and Baedecker, 1970). The abundance of phytanic acid, dihydrogen phytol, and phytane in those sediments led to the suggestion that phytane was derived from the halobacterial lipid rather than from chlorophyll. Anderson et al. (1977) demonstrated from stereoisometric measurements that the phytanic acid was most likely derived from halobacteria rather than from algal chlorophyll. However, no bacterioruberin (the dominant carotenoid of halobacteria) was detected, presumably because it is rapidly degraded in the extremely hypersaline sediments (Nissenbaum et al., 1972).

In hypersaline sediments, archaebacteria (halobacteria and methanogens) are the only known potential organisms to produce diphytanylglycerol diethers (Hahn and Haug, 1985). The presence of symmetric C_{20},C_{20} diether isoprenoids in sediments would not indicate which archaebacteria produced them. The presence of asymmetric C_{20},C_{25} diether isoprenoids would indicate the presence of natronobacteria or methanogens, while symmetric C_{25},C_{25} diether isoprenoids would be indicative of natronobacteria alone (Table 4.1). Some methanogens synthesize C_{40} tetraethers consisting of two C_{20} isoprenoids linked head-to-head, although these lipids have not been described from hypersaline sediments or in halophilic methanogens. The presence of certain neutral isoprenoids (C_{15} to C_{30} regular isoprenoids and tail-to-tail linked C_{25} and C_{30} [squalene] isoprenoids) may indicate there is a contribution by archaebacteria but not necessarily halobacteria. Halobacteria contain C_{30} squalenes while methanogens may produce C_{18} to C_{30} isoprenoids (Hahn, 1982). Certain menaquinones may be microbial indicators if they are not subject to degradation (Table 4.1).

The interpretation of the hydrocarbon composition of ancient evaporites is extremely complex due to the variety of alterations that can occur in the hydrocarbon fractions over geologic time, and the overlap of lipids produced by different microorganisms. In the hypersaline Ghareb Formation near the Dead Sea (of Cretaceous age), black calcareous shales with gypsum-filled foraminifera are believed to have been deposited in the anaerobic bottom of a hypersaline, stratified lagoon (Spiro, 1977). The predominance of C_{26} to C_{32} n-alkanes, the preference of even- over odd-number alkanes, the high concentrations of C_{19} alkane, and the high concentration of C_{17} n-alkane in the higher strata of the formation were interpreted to be indicative of algae, phototrophic bacteria, and non-phototrophic, anaerobic bacteria. The oils and asphalts of the formation showed a similar hydrocarbon distribution. The dominance of even-number, long-chain alkanes and the dominance of phytane over pristane indicated an algal source deposited under hypersaline conditions (Spiro et al., 1983). Bituminous rocks with a dominance of odd-number alkanes would indicate a contribution by land plants. Other aspects of the Dead Sea oils and asphalts were discussed by Amit and Bein (1979) and Nissenbaum and Goldberg (1980).

The predominance of even-number, C_{20} to C_{30} alkanes has also been interpreted to represent aerobic and anaerobic hypersaline environments where

Table 4.1 Some potential biomarker compounds produced by halophilic microorganisms

Compound	Organisms and environments	Reference
FATTY ACIDS		
C_{17} and C_{19} cyclopropanoic fatty acids	Moderately halophilic bacterium B_{a1} from the Dead Sea; *Pseudomonas halosaccharolytica*; phototrophic bacteria *Ectothiorhodospira* spp. of hypersaline, alkaline environments	Peleg and Tietz, 1976 Ohno et al., 1979 Asselineau and Trüper, 1982
Non-cyclic $C_{17:0}$ and $C_{17:1}$ fatty acids	Moderately halophilic bacterium *Vibrio costicola*; eucaryotic alga *Dunaliella salina*	Ben-Amotz et al., 1985 Hanna et al., 1984
Non-cyclic $C_{17:2}$ or $C_{17:3}$ fatty acids	*Ectothiorhodospira* spp. from hypersaline, alkaline environments	Asselineau and Trüper, 1982
Polyunsaturated fatty acids: $C_{16:4}$, $C_{20:5}$, $C_{20:6}$, $C_{22:2}$, $C_{22:5}$	*Dunaliella salina*; aerobic environment	Tornabene et al., 1980 Ben-Amotz et al., 1985
No C_{17}, C_{20}, or C_{22} fatty acids	*Dunaliella bardawil, D. tertiolecta, D. parva*; aerobic environment	Evans et al., 1982 Fried et al., 1982
$C_{20:6}$ and $C_{21:0}$ fatty acids	Diatom *Nitzschia* sp. from Mono Lake; aerobic, alkaline environment	Ben-Amotz et al., 1985
$C_{16:2}$ fatty acids	Halophilic cyanobacteria; some eucaryotic algae	Colclasure et al., 1974 Oren et al., 1985 Ben-Amotz et al., 1985
ISOPRENOIDS		
Phytoene	Halobacteria; moderately halophilic bacterium *Planococcus halophilus*	Kushwaha et al., 1974
Symmetric C_{20},C_{20} diether isoprenoids	All archaebacteria	Kushner, 1985 Tindall, 1985
Symmetric C_{25},C_{25} diether isoprenoids	Alkaliphilic halobacteria and methanogens	De Rosa et al., 1982 Tindall, 1985
Asymmetric C_{20},C_{25} diether isoprenoids	Alkaliphilic halobacteria and methanogens	De Rosa et al., 1982 Tindall, 1985 Grant et al., 1985
Tetrahydrogenated menaquinone MK-9	*Actinopolyspora halophila*; aerobic environment	Collins et al., 1981
Dihydrogenated menaquinone MK-8 methylated at 8th isoprenyl unit	*Natronobacterium gregoryi*; alkaline environment	Collins and Tindall, 1987
NON-LIPIDS		
Ectoine (cyclic amino acid)	*Ectothiorhodospira* spp.; *Micrococcus* sp.	Galinski et al., 1985 Imhoff, 1986
Bacteriorhodopsin and related retinal pigments	Halobacteria	Kushner, 1985 Bivin and Stoeckenius, 1986

bacteria degrade the remains of cyanobacteria. Dembicki et al. (1976) used this interpretation to describe the environment of deposition of some hypersaline, organic-rich carbonates of Mississippian age. However, the Green River Formation, an athalassic, lacustrine deposit believed to have been deposited in a meromictic, hypersaline, alkaline environment, shows a preference of odd-number n-alkanes (Robinson, 1979). It contains both oil shales and the world's largest deposit of Na_2CO_3. It is believed to have been a playa-lake environment with exceptionally high productivity in the form of microbial mats. Parts of the basin were subject to alternating hypersaline (trona-halite deposits) and more dilute (microbial mat-$CaCO_3$) conditions (Eugster and Hardie, 1975; Surdam and Wolfbauer, 1975). Possibly due to the low concentration of sulfate (and thus relatively low activity by sulfate reducers) in the waters associated with the Green River Formation, the oil shales became a repository for the recalcitrant organic matter produced in the microbial mats.

High phytane:pristane ratios may or may not be indicators of the influence of halobacteria. Phytane formation is favored in a reducing environment (Robinson, 1979). In the Green River shale the ratio of phytane:pristane ranged from one to ten. A high phytane:pristane ratio (7–20) and the abundance of a regular C_{25} isoprenoid in hypersaline Tertiary sediments in Germany most likely indicate the presence of methanogens and thus reducing conditions, but they do not point to the presence of halobacteria (Waples et al., 1974). Analysis of the stereochemistry of phytane may indicate whether this compound was derived from an algal or halobacterial source. Likewise, the presence of regular C_{13} to C_{25} isoprenoids and squalane in oils and carbonates of a lacustrine evaporite sequence of Cambrian age may indicate either methanogenic bacteria, halobacteria, or chlorophyll-containing organisms (McKirdy and Kantsler, 1980).

Other potential biomarkers include C_{17} and C_{19} cyclopropanoic fatty acids produced by eubacterial halophiles and polyunsaturated fatty acids (with up to six unsaturated carbon bonds) in the green alga *Dunaliella* (Table 4.1). Other compounds unique to halophiles, such as bacteriorhodopsin and ectoine, could be potential biomarkers, although these nitrogenous compounds may be quickly scavenged by bacteria in the sediments.

ten Haven et al. (1985) conducted a study comparing the hydrocarbon chemistry of marls (fine-grained carbonates) and a gypsum layer from a late Miocene evaporite unit. The marl samples had a moderate preference of odd-number n-alkanes while the gypsum showed a slight preference of even-number n-alkanes. Both sediments contained pristane and phytane. In gypsum, the most abundant hydrocarbons were C_{21} and C_{22} steranes (compounds called pregnanes and homopregnanes). These compounds were not found in the marls. The marls alone contained rearranged sterenes and spirosterenes as well as a C_{25} isoprenoid. Phytane and branched alkanes were relatively more abundant in the marl while hopanes (derived from eubacteria) were more abundant in gypsum. The distribution patterns of the C_{24} to C_{34} n-alkanes were also different between the less saline carbonates and the more hypersaline gypsum. C_{22} hydrocarbons were the most abundant class in both the marl and the gypsum, a feature also recognized in Gavish Sabkha microbial mats. Little is known about this potential biomarker of hypersaline environments or about the possible precursors of the pregnane and homopregnane compounds.

Stable carbon isotopes The distribution of stable carbon isotopes in hypersaline brines and sediments is subject to constraints due to hypersalinity and to biological fractionation. Stiller et al. (1985) described abiotic, $\delta^{13}C$ enrichment of inorganic carbon of up to +16.5‰ in solar evaporation ponds and up to +34.9‰ in brines evaporated in the laboratory. They attributed the isotopic fractionation values to non-equilibrium gas transfer from the atmosphere. Lazar and Erez (in preparation) found that the total CO_2 of the surficial brines of an oligotrophic solar saltern became slightly heavy ($\delta^{13}C$ up to ca. +3.5‰) as seawater was evaporated to 1.5 × concentration. The $\delta^{13}C$ fractionation decreased to minimum values in 6–8 × seawater concentration. Their lowest $\delta^{13}C$ values measured were near −8‰ in 16.0% salinity and near −9‰ in ca. 31.5% salinity. They estimated the minimum $\delta^{13}C$ value to be near −11‰ in 21–28% salinity. The accumulation of light inorganic carbon suggested that degradation exceeded productivity in the benthic microbial mats and associated sediments of the saltern, or that there were physical and chemical constraints to CO_2 equilibrium with the atmosphere. It is noteworthy that the minimum $\delta^{13}C$ values coincide with the maximum salinities that permit microbial mats to develop in salterns (see Chapter 13).

Schidlowski et al. (1984, 1985) found that organic carbon in microbial mats of Solar Lake and Gavish Sabkha was extremely heavy. The organic carbon in the surface of Solar Lake mats had a $\delta^{13}C$ value of −5.4‰ while the carbonates had a $\delta^{13}C$ value of +3.3‰. Some buried organic carbon was even heavier. The organic carbon in the surface of Gavish Sabkha mats had $\delta^{13}C$ values of −8.1 to −9.8‰ while the carbonates had $\delta^{13}C$ values of +2.0 to +4.7‰. The $\delta^{13}C$ values of the carbonates are typical of carbonates precipitated in closed environments. Organic carbon is typically much more depleted in ^{13}C due to the preference of the CO_2-fixing enzyme, RuBP carboxylase, for ^{12}C ($\delta^{13}C$ for CO_2 = −20 to −40‰). CO_2 diffusion from air into water only slightly discriminates against heavy carbon (−4.4‰). The authors attributed the occurrence of heavy organic carbon to a combination of CO_2 limitation (low CO_2 solubility and slow CO_2 diffusion from the atmosphere) and high rates of CO_2 fixation by the mats. The $\delta^{13}C$ values of sabkha sediments indicate whether they precipitated with carbon of a normal marine reservoir (+2‰) or carbon of a semi-restricted environment (−0.9‰) (Holser et al., 1981). A dolomite fraction was very light ($\delta^{13}C$ = −3.7 to −5.3‰), indicating a large contribution of biogenic CO_2 during dolomitization (see

Chapter 3). Dolomitization is further discussed with reference to the "sulfate problem" in Section 4.2.

The inorganic carbon of Dead Sea brines is light throughout the water column ($\delta^{13}C = -6.4$ to -15.4‰) as is the inorganic C of the inflowing Jordan River ($\delta^{13}C = -13$‰) (Nissenbaum and Kaplan, 1976). By comparing the stable isotopic signatures of the organic carbon and inorganic carbon (dissolved and particulate) in the Dead Sea, Nissenbaum and Kaplan concluded that biological processes (e.g., bacterial respiration) play only a minor role in the geochemistry of carbon in the Dead Sea.

In a study of stable-isotope mapping of carbonates in a late Miocene evaporite basin, McKenzie (1985) found that $\delta^{13}C$ values ranged from $+1.3$ to -49.0‰. Processes resulting in ^{13}C depletion are bacterial oxidation or sulfate reduction, both of which should result in formation of HCO_3^- with a stable isotope signature similar to the organic matter. Methane, which is strongly fractionated by methanogens ($\delta^{13}C = -40$ to -70‰) can be reoxidized to very light CO_2. The resulting CO_2 can be incorporated into carbonate rocks that are also isotopically light. Such regional mapping of the stable isotope signatures within a single basin of deposition can demonstrate local differences in organic matter accumulation and oxidation. McKenzie (1985) found that the lightest carbonates were associated with areas in Sicily with the most abundant sulfur mines. The sulfur in that region has been shown to be a by-product of the oxidation of sulfide produced by sulfate-reducing bacteria (see Section 4.2).

4.2 Sulfur cycle

The microbiology and ecology of sulfur cycling in a variety of aquatic environments, including hypersaline habitats, have been well documented. The description of sulfur cycling in the Solar Lake is exemplary (see Chapter 14). Sulfate-reducing bacteria are one of the most important agents for anaerobically oxidizing organic matter. They are probably the most important link between the carbon and sulfur cycles in evaporite environments. The basic formula for sulfate reduction is:

$$2\,(CH_2O) + SO_4^{2-} \rightarrow 2\,HCO_3^- + H_2S$$

where (CH_2O) represents organic matter at the oxidation level of carbohydrate. The generated bicarbonate increases the total alkalinity of the sediment porewaters and often leads to carbonate precipitation. The sulfide formed may eventually precipitate as pyrite or other metal sulfides. The typical organic substrates for sulfate-reducing bacteria are acetate and longer fatty acids, lactate, and aromatic compounds, in addition to H_2, CO_2, formate, ethanol, and other low molecular weight compounds. In hypersaline sediments, sulfate-reducers co-exist with methanogens which use methylamines preferentially in these environments (see Chapter 6). Other sulfide-generating bacteria in hypersaline environments include those that degrade proteins by fermentation, releasing sulfide from cysteine and methionine.

The review by ZoBell (1958) on sulfate-reducing bacteria in high salinities is the most comprehensive to date. Although sulfate reduction has been observed in NaCl-saturated sediments or enrichment cultures (see Chapter 5), the highest salinity so far found to be tolerated by sulfate-reducing bacteria in culture is 19% (Skyring et al., 1977). Maximum sulfate reduction rates measured in some hypersaline microbial mats are: 2–104 (mean = 21) $mmol \cdot m^{-2} \cdot d^{-1}$ in Spencer Gulf, Australia, where the salinity was 12.4–15.2% (Skyring et al., 1983); greater than 54.0 $mmol \cdot m^{-2} \cdot d^{-1}$ in Solar Lake, Sinai, where the salinity was about 18% (Jorgensen and Cohen, 1977); ca. 9 $mmol \cdot m^{-2} \cdot d^{-1}$ in 9.0% salinity and less than 2 $mmol \cdot m^{-2} \cdot d^{-1}$ in sediments of 15% and 30% salinity, respectively, in a solar saltern (Klug et al., 1985). Sulfate reduction rates in a saltern across a salinity gradient and with respect to sediment depth are shown in Figure 4.4. Although rates of sulfate reduction are greatly attenuated below the top centimeter of sediments, they continue to be low but measurable at depths in the hypersaline sediments where abundant sulfate is present and fermentation products are available as electron donors.

Sulfate reduction, alkalinity, and carbonate precipitation Sulfate reduction has long been linked with the generation of alkalinity and carbonate precipitation. The development of the extremely alkaline nature of the Wadi Natrun lakes in Egypt has been demonstrated to be essentially the result of bacterial sulfate reduction, although the bulk of the sulfate reduction takes place in the dilute groundwaters feeding the lakes (Abd-el-Malek and Rizk, 1963) (see Chapter 20). In alkaline Lake Magadi, direct evidence of the role of sulfate reduction was found in the precipitation of nahcolite and trona within the H_2S-rich muds (Eugster and Hardie, 1978). A similar mechanism was used to explain the occurrence of authigenic trona rosettes in the Green River Formation (Bradley and Eugster, 1969). Spiro (1977) described the appearance of pyrite and single crystals of calcite replacing gypsum in foraminifera found in bituminous shales. Sulfate reduction and carbonate precipitation have also been invoked to explain the deposition of caliche ($CaCO_3$) overlying gypsiferous sediments in some relatively arid soils (Lattman and Lauffenburger, 1974).

Unpublished data of Javor and Lazar show that sulfide and total alkalinity generation in hypersaline microbial mats and associated sediments from a marine-derived solar saltern (12.9–19.4% salinity) increased conservatively only when each concentration was relatively low (up to 4 mM and 12 $meq \cdot l^{-1}$, respectively) (Figure 4.5). To account for any differences due to salinity, the data are normalized to the degree of brine concentration above normal seawater salinity. When the top 5–10 cm of microbial mats and associated sed-

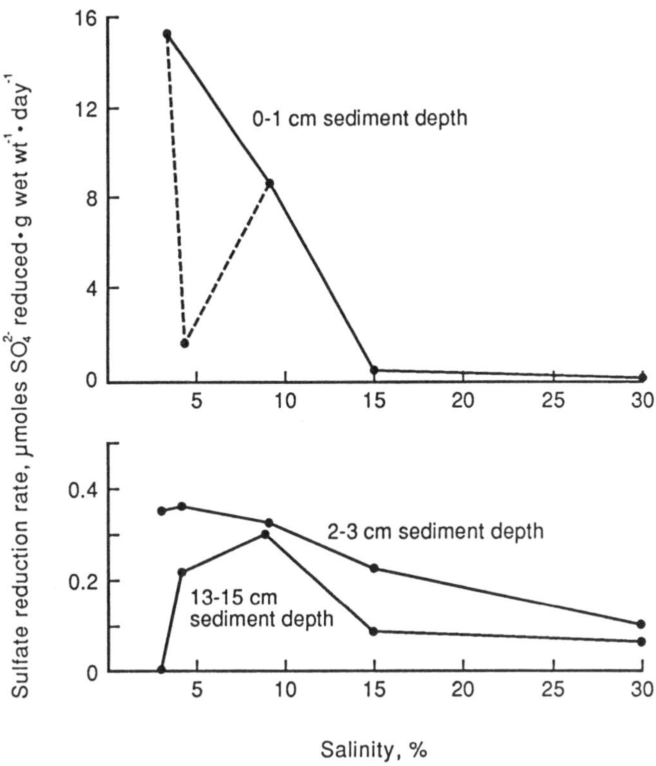

Figure 4.4 Sulfate reduction rates in relation to sediment depth and salinity in the San Francisco (California, U.S.A.) saltern, July–August, 1984 (from Klug et al., 1985).

iments were collected, mixed, and maintained 6 months in 10 cm × 40 cm, black-walled Winogradsky columns, exceptionally high concentrations of sulfide (up to 126 mM) and total alkalinity (up to 135 meq·l^{-1}) developed in the sediment porewaters. However, the ratios of total alkalinity to sulfide were less than two, which is that predicted by sulfate reduction. The "missing" alkalinity was precipitated as calcite (confirmed by the relative increase in calcite and Mg-calcite determined by x-ray diffraction). Even in sediments to which powdered aragonite was added, only calcite precipitated. These observations suggest that the organic acids and other DOC compounds in the organic-rich porewaters may have influenced the tendency for calcite precipitation. Low concentrations of dolomite were present in the original sediments making it difficult to determine whether any minor amount of dolomite has precipitated during the experiment.

The above experiment demonstrates the interrelationship between sulfide and total alkalinity in an extreme example, since in the Winogradsky columns a large volume of organic-rich microbial mat was "catastrophically" buried

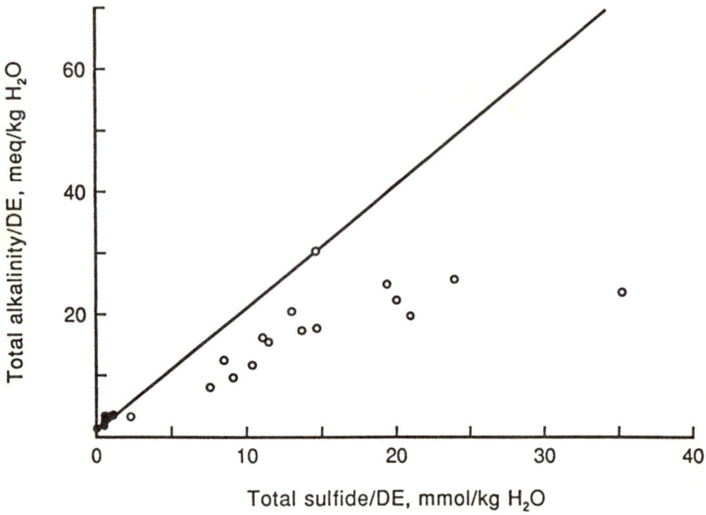

Figure 4.5 Total alkalinity vs. total sulfide concentration normalized to the degree of evaporation (DE) of porewaters from freshly collected cores of microbial mats in the Israel Salt Company (Eilat) saltern (closed circles) and from Winogradsky column sediments (open circles) described in text. The line indicates the predicted behavior of total alkalinity vs. total sulfide during sulfate reduction (slope = 2). The y-intercept is the total alkalinity of the overlying brines (Unpublished data of B. Javor and B. Lazar).

in the sediments. Analyses of sulfide and total alkalinity in the undisturbed microbial mats of samples taken at 2-mm intervals are presented in Figure 4.6. The photosynthetic zone (4 mm deep) of the microbial mat was dominated by the cyanobacterium *Aphanothece halophytica* and a *Chloroflexus*-like phototrophic bacterium. Below the photosynthetic zone was a 1-mm thick band of sand-sized calcite crystals underlain by black, reducing, organic-rich sediments. The porewater salinity was 15.2–17.6%. The total alkalinity (without the sulfide component) reached a maximum just below the photosynthetic zone where the calcite was precipitating. The maximum potential contribution to the total alkalinity by sulfide (two times the S^{2-} concentration) is presented in the same figure. This figure demonstrates that sulfide is either recycled, oxidized, or precipitated, or that much of the total alkalinity is generated from bicarbonate during photosynthesis or by ammonia accumulation near the photosynthetic zone. Clearly, the relationships between sulfate reduction and carbonate precipitation are complex near the sediment surface in hypersaline systems.

The fate of reduced sulfur The relationships between evaporites and stratiform metal deposits have been previously discussed (see Chapter 1).

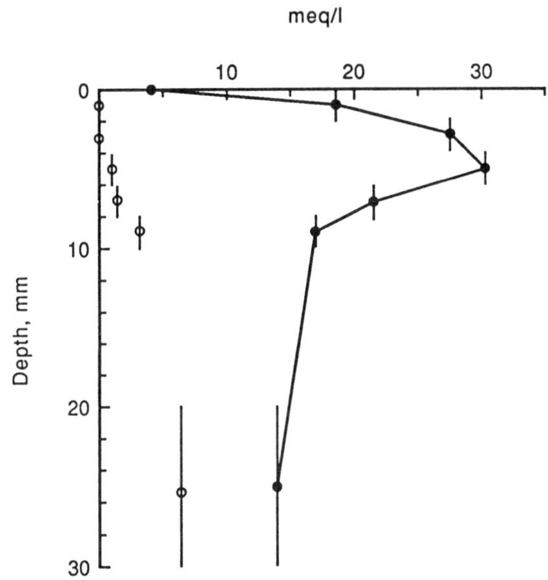

Figure 4.6 Total alkalinity (without sulfide), closed circles, and total sulfide, open circles (both expressed in meq·l^{-1}) in porewaters of *Aphanothece* mats of the Israel Salt Company (Eilat) (Unpublished data of B. Javor and B. Lazar).

Obviously, the mobilization of metals by the low redox potential of the reducing sediments and their precipitation with sulfide is a possible sink for generated sulfide. Some sulfide can escape as a volatile gas, but such gas fluxes have never been measured.

Sulfide can be oxidized by sulfur-oxidizing bacteria, producing elemental sulfur. Bacterial sulfide oxidation was proposed to explain the genesis of sulfur nodules in the Lake Eyre playa deposits (Baas-Becking and Kaplan, 1956) (see Chapter 17). The nodules contained plant and animal remains in a pellicle-lined cavity of the nodule core. Sulfur nodules have also been reported in Searles Lake, an alkaline playa deposit (Smith and Haines, 1964) and in the late Miocene marine evaporite deposit described by McKenzie (1985). Sulfide oxidation can apparently proceed all the way to sulfate, producing gypsum nodules at the sediment surface (West et al., 1979). Sulfide oxidation during tidal flushing by oxidized seawater may explain the occurrence of a thin band of gypsum crystals just below surficial and buried organic matter in the intertidal zone of a sabkha in Mexico (Javor, personal observations).

Elemental sulfur that forms the caprocks of salt domes is the principal commercial source of elemental sulfur in the western hemisphere (Davis and Kirkland, 1979). It is formed by the post-depositional bacterial reduction of anhydrite or gypsum to sulfide, which is then biologically or non-biologically oxidized to sulfur and becomes sealed by co-precipitating calcite and pyrite. Because the initial step of sulfur generation requires that halite be dissolved by seawater or groundwater in order to expose the anhydrite, it is not certain whether sulfate reduction takes place in dilute solutions, in marine salinities, or under hypersaline conditions. Such deposits are considered epigenetic. Some elemental sulfur deposits, including those mentioned by McKenzie (1985), may be syngenetic (formed at the time of the evaporite deposits). Davis and Kirkland state that biosyngenetic and bioepigenetic sulfur deposits cannot be differentiated by stable isotope analysis, but only by geological reconstruction of the environment of deposition.

Sulfide in the form of polysulfide can also react non-biologically with buried organic matter, producing sulfur-enriched organic residues. In Solar Lake microbial mats, only 1–2% of the organic matter near the surface had bound sulfur but 8% of the organic matter at 80 cm depth had bound sulfur (Cohen et al., 1980; Aizenshtat et al., 1983). The $\delta^{34}S$ of the protokerogen decreased with depth from −13‰ at the surface to as much as −26.5‰ at 45–80 cm depth, demonstrating the light isotope enrichment that would be predicted if the sulfide that was incorporated was generated by sulfate-reducing bacteria.

Stable sulfur isotopes Like the study of stable carbon isotopes, the field of stable sulfur isotopes is large and complex, but fortunately, well-studied. Stable sulfur isotopic analysis has proved to be an extremely helpful tool for understanding such concepts as the source of salts in the hypersaline lakes of Antarctica (see Chapter 16), the non-marine origin of a major halite deposit (Holser, 1979a), the source of sulfide in the hot brines of the Red Sea (Hartmann and Nielsen, 1966; Kaplan et al., 1969), major sedimentary trends over geologic time (discussed below), and the debate over the onset of bacterial sulfate reduction during the Precambrian (discussed below). The details of stable sulfur isotopic fractionation and its utility as a geochemical tool have been presented by Dean (1978), Chambers and Trudinger (1979), and Thode (1980).

Sulfate-reducing bacteria discriminate against ^{34}S by −40‰, but the differences between the sulfate and sulfide pools in sediments are often smaller. Thode (1980) stated that isotope discrimination due to sulfate reduction can result in differences of little as 0‰ to as much as about 60‰ between sulfate and sulfide. Recent seawater sulfate has a $\delta^{34}S$ value of about +20‰. Pierre (1985) found a dynamic redox cycle generated seasonally in a solar saltern. In the summer, porewater sulfide in a microbial mat was lighter ($\delta^{34}S$ = −15.0‰) than that in black mud ($\delta^{34}S$ = −2.4‰) with corresponding changes measured in the dissolved sulfate. Dissolution of minerals with fresh water from winter rains caused a shift in the distribution of ^{34}S. The average difference in $\delta^{34}S$ between oxidized and reduced S was 25‰.

Isotopic fractionation of sulfur can occur by processes other than bacterial sulfate reduction. Some sulfide enters porewaters by fermentative degradation of organic sulfur. Organic sulfur is typically light ($\delta^{34}S$ = −0.9 to −2.8‰) (Chambers and Trudinger, 1979). Evaporite precipitation of sulfates leads to slight sulfur fractionation, leaving behind heavier sulfate in the brines ($\delta^{34}S$ = +1.7‰) (Holser, 1979b). There is a slight isotopic effect associated with protonation of HS^- and volatilization of H_2S. At 22°C, H_2S is enriched in ^{34}S by +2.0 to +2.7‰ with respect to HS^-. Volatilized H_2S is 0.5‰ depleted in ^{34}S relative to H_2S remaining in solution (Fry et al., 1986).

Measuring the difference in the ^{34}S distribution between reduced and oxidized sulfur pools in hypersaline brines and sediments is sometimes a useful tool in interpreting the activity of sulfate-reducing bacteria when metabolic rate measurements or bacterial isolations have not been possible. In the Dead Sea, coexisting sulfate and sulfide have $\delta^{34}S$ values of +13.7 to +15.0‰ and −19.6 to −21.7‰, respectively, strongly suggesting isotopic fractionation by sulfate-reducers. In the hot brines of the Red Sea, no bacterial sulfate reduction could be demonstrated, although sulfide was present (see Chapter 12). Isotopic analyses of the sulfide and sulfate of the brines and the surrounding waters confirmed that sulfide was introduced into the brine by geothermal processes and not by *in situ* bacterial activity (Hartmann and Nielsen, 1966; Kaplan et al., 1969). The extremely heavy sulfate ($\delta^{34}S$ = +30.5‰ or as high as +37.5‰) in the Antarctic hypersaline Don Juan Pond occurs in the absence of any sulfides or obvious mechanism of ^{34}S enrichment (Tomiyama and Kitano, 1985). In this case, stable sulfur isotopes have not provided sufficient evidence to settle the controversy concerning the presence or absence of biological activity in this extremely saline brine (see Chapter 16).

Global trends in stable isotopic fractionations through Phanerozoic time
By estimating the size of recent reduced and oxidized reservoirs of carbon and sulfur, and using isotopic fractionation values of modern organisms as a key, a number of models based on the stable isotopes of evaporites have been constructed to describe variations in the atmospheric and tectonic history of the Phanerozoic era. Most of the more recent models are updates and elaborations of older models, which are not cited here. Both the models of Veizer et al. (1980) and Garrels and Lerman (1984) show that there is a broad negative correlation between the $\delta^{13}C$ values of sedimentary carbonates and the $\delta^{34}S$ values of sedimentary sulfates. Other models have shown correlations between the $\delta^{34}S$ and $\delta^{18}O$ values of sulfates (Holser et al., 1979; Zak et al., 1980). Garrels and Lerman used estimated modern pool sizes of sulfur present on the Earth's surface as 200×10^{18} moles each as gypsum and pyrite, and 42×10^{18} moles as dissolved sulfate in the oceans. Estimates by other investigators cited in that report ranged from 134×10^{18} moles reduced and 249×10^{18} moles oxidized sulfur, to 231×10^{18} moles reduced and 147×10^{18} moles oxidized sulfur. These estimates lead to the conclusion that some-

where between 30% and 50% of the sulfur on the surface of the Earth is currently found in evaporite rocks of hypersaline origin. Garrels and Lerman estimated the pool of carbonate-C to be 5200×10^{18} moles, and that of organic-C to be 1300×10^{18} moles. Other models they cite show that the pool size of carbonate-C is always about 4–5 times larger than that of reduced carbon. It is not stated how much of either pool of carbon was deposited under hypersaline conditions.

The following summary of $\delta^{34}S$ values in marine sulfates was taken from the models of Claypool et al. (1979), Holser (1979b), Veizer et al. (1980), and Garrels and Lerman (1984). In the late Proterozoic, from about 1000 to 700 million years ago, sedimentary marine sulfates had an average $\delta^{34}S$ value of +16 to +17‰. At the Precambrian-Cambrian boundary (about 550–600 million years ago), the sulfates rapidly became much heavier and had an average $\delta^{34}S$ value of +31‰, the heaviest ever in Earth history. Throughout the Paleozoic era there was a general trend of increasingly lighter sulfates, peaking with an average $\delta^{34}S$ value of +10.5‰ at the end of the Paleozoic era about 225 million years ago. Throughout the Mesozoic and Cenozoic eras there has been a temporal trend towards increasingly heavier sulfates to the current value of $\delta^{34}S$ = +20‰. Although these are broad temporal trends, there have been some dramatic, geologically rapid shifts in the average ^{34}S content of marine sulfates. The most dramatic example is the $\delta^{34}S$ shift from +27‰ to +18‰ to +28‰ to +17‰ in the course of about 80 million years during the Devonian and Mississippian. The broad variations over the last billion years are interpreted to reflect the variation in sulfate pool sizes as a result of erosion and deposition of sulfates, the reduction of sulfate, and the oxidation of sulfides (sedimentary, igneous, and metamorphic). The pool sizes in turn responded to major tectonic and atmospheric trends. The rapid excursions in the $\delta^{34}S$ values of sulfates probably do not reflect the sudden increase or decrease in bacterial activity, but they may indicate that only the ocean surface was mixed and that dense brines accumulated at depth were not necessarily recorded in the sedimentary record.

The "sulfate problem" in evaporite sequences According to both the predicted and analyzed chemical content of marine-derived brines, as evaporation progresses, bitterns brines should precipitate sulfate-bearing minerals along with the commonly found chlorides. The rarity of such sulfates in marine evaporite deposits constitutes the "sulfate problem." Sonnenfeld (1984) argued that bacterial sulfate reduction could effectively reduce the excess sulfate to sulfide, which would then escape as H_2S gas. However, Sonnenfeld's discussion on sulfate reduction has a number of inaccuracies and contradictions. For example, it is stated that the sulfate deficiency occurred very early in the concentration cycle, but few modern marine evaporite analogs show major sulfate deficiencies in moderately hypersaline environments. Sonnenfeld also states that gypsum is stable only in waters saturated with oxygen, but such a statement does not explain the existence of gypsum in Dead Sea sediments or in any other anaerobic, hypersaline environment where the activity of sulfate-reducing bacteria is insufficient to drive the total dissolution of gypsum or to prevent its saturation. It is also stated that H_2S in hypersaline brines escapes and does not attack ferrous hydroxides (Sonenfeld, 1984; page 105), but Fe^{2+} is an excellent scavenger of sulfide as demonstrated by the common occurrence of black, sulfidic muds (FeS and FeS_2) in hypersaline sediments.

Hite (1983) reviewed a variety of arguments used to explain the "sulfate problem." Model calculations based on known rates of organic productivity and sulfate reduction show that it is virtually impossible to lose sufficient volatile H_2S to explain the sulfate deficiency. The most convincing arguments show that dolomitization of $CaCO_3$ can drive the precipitation of gypsum with the excess sulfate during the process:

$$2\ CaCO_3 + Mg^{2+} + SO_4^{2-} \rightarrow CaMg(CO_3)_2 + CaSO_4$$

Because many ancient calcium carbonate deposits have dolomitized secondarily over geologic time, dolomitization and the co-precipitation of gypsum could remove sufficient sulfate to prevent large volumes of the extremely soluble K^+ or Mg^{2+} sulfates from precipitating. Detailed analysis of the middle Pennsylvanian evaporites of the Paradox basin indicate that dolomitization could account for the deficiency in sulfate minerals other than gypsum. The role of bacteria in such secondary dolomitization of $CaCO_3$ is not known with any certainty (see Chapter 3), but the carbon of dolomite formed in a sabkha in Mexico was found to be isotopically light due to biogenic influence (Holser et al., 1981). Thus, the "sulfate problem" in evaporites may be influenced by bacterial activity, but only through the mechanism of dolomitization.

4.3 Conclusion

It is clear from the above discussion that biological processes are preeminent in the carbon and sulfur cycles of hypersaline environments and that the carbon and sulfur chemistry of modern and ancient evaporites reflects the activities of living organisms. Because of the restricted biotas in hypersaline environments, there are a number of ways in which hypersaline environments differ significantly from conventional marine environments, and these differences are reflected in the carbon and sulfur chemistry. In the next part of the book we will discuss in detail the biology of hypersaline environments and emphasize the many ways in which this biology differs from that of marine environments.

References

Abd-el-Malek, Y., and **Rizk, S.G.** 1963. Bacterial sulphate reduction and the development of alkalinity. III. Experiments under natural conditions in the Wadi Natrun. *Journal of Applied Bacteriology* 26: 20–26.

Aizenshtat, Z., Stoler, A., Cohen, Y., and **Nielsen, H.** 1983. The geochemical sulphur enrichment of Recent organic matter by polysulfides in the Solar Lake, pp. 279–288 in Bjørnoy, M. et al. (editors), *Advances in Organic Geochemistry 1981*, John Wiley and Sons, New York.

Amit, O. and **Bein, A.** 1979. The genesis of the asphalt in the Dead Sea area. *Journal of Geochemical Exploration* 11: 211–225.

Anderson, R., Kates, M., Baedecker, M.J., Kaplan, I.R., and **Ackman, R.G.** 1977. The stereoisomeric composition of phytanyl chains in lipids of Dead Sea sediments. *Geochimica et Cosmochimica Acta* 41: 1381–1390.

Asselineau, J. and **Trüper, H.G.** 1982. Lipid composition of six species of the phototrophic bacterial genus *Ectothiorhodospira. Biochimica et Biophysica Acta* 712: 111–116.

Baas-Becking, L.G.M. and **Kaplan, I.R.** 1956. The microbiological origin of the sulphur nodules of Lake Eyre. *Transactions of the Royal Society of South Australia* 79: 52–65.

Ben-Amotz, A., Tornabene, T.G., and **Thomas, W.H.** 1985. Chemical profile of selected species of microalgae with emphasis on lipids. *Journal of Phycology* 21: 72–81.

Bivin, D.B. and **Stoeckenius, W.** 1986. Photoactive retinal pigments in haloalkaliphilic bacteria. *Journal of General Microbiology* 132: 2167–2177.

Borchert, H. and **Muir, R.O.** 1964. *Salt Deposits. The Origin, Metamorphism and Deformation of Evaporites.* D Van Nostrand Co., New York, 338 p.

Bradley, W.H. and **Eugster, H.P.** 1969. Geochemistry and paleolimnology of the trona deposits and associated authigenic minerals of the Green River Formation of Wyoming. *United States Geological Survey Professional Paper* 496-B. 71 pp.

Cardoso, J.N., Watts, C.D., Maxwell, J.R., Goodfellow, R., Eglinton, G., and **Golubic, S.** 1978. A biogeochemical study of the Abu Dhabi algal mats: a simplified ecosystem. *Chemical Geology* 23: 273–291.

Chambers, L.A. and **Trudinger, P.A.** 1979. Microbiological fractionation of stable sulfur isotopes: a review and critique. *Geomicrobiology Journal* 1: 249–293.

Claypool, G.E., Holser, W.T., Kaplan, I.R., Sakai, H., and **Zak, I.** 1980. The age curves of sulfur and oxygen isotopes in marine sulfate and their mutual interpretation. *Chemical Geology* 28: 199–260.

Cohen, Y., Aizenshtat, Z., Stoler, A., and **Jørgensen, B.B.** 1980. The microbial geochemistry of Solar Lake, Sinai, pp. 167–172 in Trudinger, P.A. and Walter, M.R. (editors), *Biogeochemistry of Ancient and Modern Environments*, Springer-Verlag, New York.

Colclasure, G.C., Yopp, J.H., and **Tindall, D.R.** 1974. Characterization of the fatty acids of the obligately halophilic blue-green alga, *Aphanothece halophytica. American Journal of Botany* 61 (Supplement): 21–22.

Collins, M.D. and **Tindall, B.J.** 1987. Occurrence of menaquinones and some novel methylated menaquinones in the alkaliphilic, extremely halophilic archaebacterium *Natronobacterium gregoryi. FEMS Microbiology Letters* 43: 307–312.

Collins, M.D., Ross, H.N.M., Tindall, B.J., and **Grant, W.D.** 1981. Distribution of isoprenoid quinones in halophilic bacteria. *Journal of Applied Bacteriology* 50: 559–565.

Davis, J.B. and **Kirkland, D.W.** 1979. Bioepigenetic sulfur deposits. *Economic Geology* 74: 462–468.

Dean, W.E. 1978. Some uses of stable isotopes of carbon, oxygen, and sulfur in solving problems related to evaporite deposits, pp. 124–143 in Dean, W.E. and Schreiber, B.C. (editors), *Marine Evaporites,* Society of Economic Paleontologists and Mineralogists Short Course no. 4, Tulsa.

Degens, E.T. and **Paluska, A.** 1979. Hypersaline solutions interact with organic detritus to produce oil. *Nature (London)* 281: 666–668.

Degens, E.T., Hunt, J.M., Reuter, J.M., and **Reed, W.E.** 1964. Data on the distribution of amino acids and oxygen isotopes in petroleum brine waters of various geologic ages. *Sedimentology* 3: 199–25.

de Leeuw, J.W., Sinninghe Damsté, J.S., Klok, J., Schenck, P.A., and **Boon, J.J.** 1985. Biogeochemistry of Gavish Sabkha sediments I. Studies on neutral reducing sugars and lipid moieties by gas chromatography-mass spectrometry, pp. 350–367 in Friedman, G.M. and Krumbein, W.E. (editors), *Hypersaline Ecosystems. The Gavish Sabkha,* Ecological Studies 53, Springer-Verlag, New York.

Dembicki, Jr., H., Meinschein, W.G., and **Hattin, D.E.** 1976. Possible ecological and environmental significance of the predominance of even-carbon number C_{20}–C_{30} n-alkanes. *Geochimica et Cosmochimica Acta* 40: 203–208.

De Rosa, M., Gambacorta, A., Nicolaus, B., Ross, H.N.M., Grant, W.D., and **Bu'Lock, J.D.** 1982. An asymmetric archaebacterial diether lipid from alkaliphilic halophiles. *Journal of General Microbiology* 128: 343–348.

De Rosa, M., Gambacorta, A., Nicolaus, B., and **Grant, W.D.** 1983. A C_{25}, C_{25} diether core lipid from archaebacterial haloalkalophilies. *Journal of General Microbiology* 129: 2333–2337.

Eugster, H.P. 1985. Oil shales, evaporites and ore deposits. *Geochimica et Cosmochimica Acta* 49: 619–635.

Eugster, H.P., and **Hardie, L.A.** 1975. Sedimentation in an ancient playa-lake complex: the Wilkins Peak Member of the Green River Formation of Wyoming. *Geological Society of America Bulletin* 86: 319–334.

Eugster, H.P. and **Hardie, L.A.** 1978. Saline lakes, pp. 237–293 in Lerman, A. (editor), *Lakes. Chemistry Geology Physics.* Springer-Verlag, New York.

Evans, R. and **Kirkland, D.W.** 1988. Evaporitic environments as a source of petroleum, pp. 256–299 in Schreiber, B.C. (editor), *Evaporites and Petroleum,* Columbia University Press, New York.

Evans, R.W., Kates, M., Ginzburg, M., and **Ginzburg, B.-Z.** 1982. Lipid composition of halotolerant algae, *Dunaliella parva* Lerche and *Dunaliella tertiolecta. Biochimica et Biophysica Acta* 712: 186–195.

Fried, A., Tietz, A., Ben-Amotz, A., and **Eichenberger, W.** 1982. Lipid composition of the halotolerant alga, *Dunaliella bardawil. Biochimica et Biophysica Acta* 713: 419–426.

Friedman, G.M., Sneh, A., and **Owen, R.W.** 1985. The Ras Muhammad Pool: implications for the Gavish Sabkha, pp. 218–237 in Friedman, G.M. and Krumbein, W.E. (editors), *Hypersaline Ecosystems. The Gavish Sabkha,* Ecological Studies 53, Springer-Verlag, New York.

Fry, B., Gest, H., and **Hayes, J.M.** 1986. Sulfur isotope effects associated with protonation of HS^- and volatilization of H_2S. *Chemical Geology* 58: 253–258.

Galinski, E.A., Pfeiffer, H.-P., and **Trüper, H.G.** 1985. 1, 4, 5, 6-Tetrahydro-2-methyl-4-pyrimidinecarboxylic acid: a novel cyclic amino acid from halophilic phototrophic bacteria of the genus *Ectothiorhodospira. European Journal of Biochemistry* 149: 135–140.

Garrels, R.M. and **Lerman, A.** 1984. Coupling of the sedimentary sulfur and carbon cycles—an improved model. *American Journal of Science* 284: 989–1007.

Grant, W.D., Pinch, G., Harris, J.E., De Rosa, M., and **Gambacorta, A.** 1985. Polar lipids in methanogen taxonomy. *Journal of General Microbiology* 131: 3277–3286.

Hahn, J. 1982. Geochemical fossils of a possibly archaebacterial origin in ancient sediments, pp. 40–52 in Kandler, O. (editor), *Archaebacteria*, Gustav Fischer, New York.
Hahn, J. and **Haug, P.** 1985. Sedimentary record and archaebacteria, pp. 215–253 in Woese, C.R. and Wolfe, R.S. (editors), *The Bacteria. A Treatise on Structure and Function*, Vol. VIII *Archaebacteria*, Academic Press, New York.
Hanna, K., Bengis-Garber, C., Kushner, D.J., Kogut, M., and **Kates, M.** 1984. The effect of salt concentrations on the phospholipid and fatty acid composition of the moderate halophile *Vibrio costicola*. *Canadian Journal of Microbiology* 30: 669–675.
Hartmann, M., and **Nielsen, H.** 1966. Sulfur isotopes in the hot brine and sediment of Atlantis II Deep (Red Sea). *Marine Geology* 4: 305–306.
Hite, R.J. 1983. The sulfate problem in marine evaporites, pp. 217–230 in Schreiber, B.C. and Harner, H.L. (editors), *Sixth International Symposium on Salt*, vol. 1, The Salt Institute, Toronto.
Hite, R.J., Anders, D.E., and **Ging, T.G.** 1984. Organic-rich source rocks of Pennsylvanian age in the Paradox basin of Utah and Colorado, p. 255–274 in Woodward, J., Meissner, F.F. and Clayton, J.L. (editors), *Hydrocarbon Source Rocks of the Greater Rocky Mountain Region*, Rocky Mountain Association of Geologists, Denver.
Holser, W.T. 1979a. Rotliegend evaporites, lower Permian of northwestern Europe. *Erdöl und Kohle, Erdgas, Petrochemistrie* 32: 159–162.
Holser, W.T. 1979b. Trace elements and isotopes in evaporites, pp. 295–346 in Burns, R.G. (editor), *Marine Minerals*, Mineralogical Society of America Short Course Notes, vol. 6, Washington, D.C.
Holser, W.T., Kaplan, I.R., Sakai, H., and **Zak, I.** 1979. Isotope geochemistry of oxygen in the sedimentary sulfate cycle. *Chemical Geology* 25: 1–17.
Holser, W.T., Javor, B., and **Pierre, C.** 1981. Geochemistry and ecology of salt pans at Guerrero Negro, Baja California. *Geological Society of America Cordilleran Section Field Trip*, no. 1, 56 pp.
Imhoff, J.F. 1986. Osmoregulation and compatible solutes in eubacteria. *FEMS Microbiology Reviews* 39: 57–66.
Javor, B.J. 1983. Nutrients and ecology of the Western Salt and Exportadora de Sal saltern brines, pp. 195–205 in Schreiber, B.C. and Harner, H.L. (editors), *Sixth International Symposium on Salt*, vol. 1, The Salt Institute, Toronto.
Jørgensen, B.B. and **Cohen, Y.** 1977. Solar Lake (Sinai). 5. The sulfur cycle of the benthic cyanobacterial mats. *Limnology and Oceanography* 22: 657–666.
Kaplan, I.R. and **Baedecker, M.J.** 1970. Biological productivity in the Dead Sea Part II. Evidence for phosphatidyl glycerophosphate lipid in sediment. *Israel Journal of Chemistry* 8: 529–533.
Kaplan, I.R., Sweeney, R.E., and **Nissenbaum, A.** 1969. Sulfur-isotope studies on Red Sea geothermal brines and sediments, pp. 474–498 in Degens, E.T. and Ross, D.A. (editors), *Hot Brines and Recent Heavy Metal Deposits in the Red Sea*, Springer-Verlag, New York.
Kirkland, D.W. and **Evans, R.** 1981. Source-rock potential of evaporitic environment. *Bulletin of the American Association of Petroleum Geologists* 65: 181–190.
Klug, M., Boston, P., Francois, R., Gyure, R., Javor, B., Tribble, G., and **Vairavamurthy, A.** 1985. Sulfur reduction in sediments of marine and evaporite environments, pp. 128–157 in Sagen, D. (editor), *The Global Sulfur Cycle*, National Aeronautics and Space Administration Technical Memorandum 87570, Washington, D.C.
Kushner, D.J. 1985. The Halobacteriaceae, pp. 171–214 in Woese, C.R. and Wolfe, R.S. (editors), *The Bacteria. A Treatise on Structure and Function*, Vol. VIII *Archaebacteria*, Academic Press, New York.
Kushwaha, S.C., Gochnauer, M.B., Kushner, D.J., and **Kates, M.** 1974. Pigments and isoprenoid compounds in extremely and moderately halophilic bacteria. *Canadian Journal of Microbiology* 20: 241–245.

Lattman, L.H. and **Lauffenburger, S.K.** 1974. Proposed role of gypsum in the formation of caliche. *Zeitschrift für Geomorphologie Supplementband* 20: 169–170.

Matsumoto, G., Torii, T., and **Hanya, T.** 1984. Vertical distribution of organic constituents in an Antarctic lake: Lake Vanda. *Hydrobiologia* 111: 119–126.

McKenzie, J.A. 1985. Stable-isotope mapping in Messinian evaporative carbonates of central Sicily. *Geology* 13: 851–854.

McKirdy, D.M. and **Kantsler, A.J.** 1980. Oil geochemistry and potential source rocks of the Officer Basin, South Australia. *Journal of the Australian Petroleum Exploration Association* 20: 68–86.

Nissenbaum, A. and **Kaplan, I.R.** 1976. Sulfur and carbon isotopic evidence for biogeochemical processes in the Dead Sea ecosystem, pp. 309–325 in Nriagu, J.O. (editor), *Environmental Biogeochemistry,* Vol. 1, *Carbon, Nitrogen, Phosphorus, Sulfur and Selenium Cycles,* Ann Arbor Science, Ann Arbor.

Nissenbaum, A. and **Goldberg, M.** 1980. Asphalts, heavy oils, ozocerite and gases in the Dead Sea basin. *Organic Geochemistry* 2: 167–180.

Nissenbaum, A., Baedecker, M.J., and **Kaplan, I.R.** 1972. Organic geochemistry of Dead Sea sediments. *Geochimica et Cosmochimica Acta* 36: 709–727.

Ohno, Y., Yano, I., and **Masui, M.** 1979. Effect of NaCl concentration and temperature on the phospholipid and fatty acid compositon of a moderately halophilic bacterium, *Pseudomonas halosaccharolytica. Journal of Biochemistry (Japan)* 85: 413–422.

Oren, A., Fattom, A., Padan, E., and **Tietz, A.** 1985. Unsaturated fatty acid composition and biosynthesis in *Oscillatoria limnetica* and other cyanobacteria. *Archives of Microbiology* 141: 138–142.

Peleg, E. and **Tietz, A.** 1971. Glycolipids of a halotolerant moderately halophilic bacterium. *FEBS Letters* 15: 309–312.

Pierre, C. 1985. Isotopic evidence for the dynamic redox cycles of dissolved sulphur compounds between free and interstitial solutions in marine salt pans. *Chemical Geology* 53: 191–196.

Robinson, W.E. 1979. The origin, deposition, and alteration of the organic material in Green River shale. *Organic Geochemistry* 1: 205–218.

Schidlowski, M, Matzigkeit, U., and **Krumbein, W.E.** 1984. Superheavy organic carbon from hypersaline microbial mats. *Naturwissenschaften* 71: 303–308.

Schidlowski, M., Matzigkeit, U., Mook, W.G., and **Krumbein, W.E.** 1985. Carbon isotope geochemistry and ^{14}C ages of microbial mats from Gavish Sabkha and the Solar Lake, pp. 381–401 in Friedman, G.M. and Krumbein, W.E. (editors), *Hypersaline Ecosystems. The Gavish Sabkha,* Ecological Studies 53, Springer-Verlag, New York.

Schreiber, B.C. (editor). 1988. *Evaporites and Petroleum.* Columbia University Press, New York.

Skyring, G.W., Jones, H.E., and **Goodchild, D.** 1977. The taxonomy of some new isolates of dissimilatory sulfate-reducing bacteria. *Canadian Journal of Microbiology* 23: 1415–1425.

Skyring, G.W., Chambers, L.A., and **Bauld, J.** 1983. Sulfate reduction in sediments colonized by cyanobacteria, Spencer Gulf, South Australia. *Australian Journal of Marine and Freshwater Research* 34: 359–374.

Smith, G.I. and **Haines, D.V.** 1964. Character and distribution of nonclastic minerals in the Searles Lake evaporite deposit, California. *United States Geological Survey Bulletin* 1181-P, 58 pp.

Sonnenfeld, P. 1984. *Brines and Evaporites.* Academic Press. New York. 613 pp.

Spiro, B. 1977. Bacterial sulphate reduction and calcite precipitation in hypersaline deposition of bituminous shales. *Nature (London)* 269: 235–237.

Spiro, B., Welte, D.H., Rullkotter, J., and **Schaefer, R.G.** 1983. Asphalts, oils, and bituminous rocks from the Dead Sea area—a geochemical correlation study. *Bulletin of the American Association of Petroleum Geologists* 67: 1163–1175.

Stiller, M., Rounick, J.S., and **Shasha, S.** 1985. Extreme carbon-isotope enrichments in evaporating brines. *Nature (London)* 316: 434–435.

Surdam, R.C. and **Wolfbauer, C.A.** 1975. Green River Formation: a playa-lake complex. *Geological Society of America Bulletin* 86: 335–345.

ten Haven, H.L., de Leeuw, J.W., and **Schenck, P.A.** 1985. Organic geochemical studies of a Messinian evaporite basin, northern Apennines (Italy) I: Hydrocarbon biological markers for a hypersaline environment. *Geochimica et Cosmochimica Acta* 49: 2181–2191.

Thode, H. 1980. Sulphur isotope ratios in late and early Precambrian sediments and their implications regarding early environments and early life. *Origins of Life* 10: 127–136.

Tindall, B.J. 1985. Qualitative and quantitative distribution of diether lipids in haloalkliphilic archaebacteria. *Systematic and Applied Microbiology* 6: 243–246.

Tomiyama, C. and **Kitano, Y.** 1985. Salt origin in the Wright Valley, Antarctica. *Antarctic Record* 86: 17–27.

Tornabene, T.G., Holzer, G., and **Peterson, S.L.** 1980. Lipid profile of the halophilic alga, *Dunaliella salina. Biochemical and Biophysical Research Communications* 96: 1349–1356.

Veizer, J., Holser, W.T., and **Wilgus, C.K.** 1980. Correlation of $^{13}C/^{12}C$ and $^{34}S/^{32}S$ secular variations. *Geochimica et Cosmochimica Acta* 44: 579–587.

Waples, D.W., Haug, P., and **Welte, D.H.** 1974. Occurrence of a regular C_{25} isoprenoid hydrocarbon in Tertiary sediments representing a lagoonal-type, saline environment. *Geochimica et Cosmochimica Acta* 38: 381–387.

Ward, D.M., and **Brock, T.D.** 1978. Hydrocarbon biodegradation in hypersaline environments. *Applied and Environmental Microbiology* 35: 353–359.

West, I.M., Ali, Y.A., and **Hilmy, M.E.** 1979. Primary gypsum nodules in a modern sabkha on the Mediterranean coast of Egypt. *Geology* 7: 354–358.

Wright, J.L.C. 1979. The occurrence of ergosterol and (22*E*, 24*R*)-24-ethylcholesta-5, 7, 22-trien-3β-ol in the unicellular chlorophyte *Dunaliella tertiolecta. Canadian Journal of Chemistry* 57: 2569–2571.

Zak, I., Sakai, H., and **Kaplan, I.R.** 1980. Factors controlling the $^{18}O/^{16}O$ and $^{34}S/^{32}S$ isotope ratios of ocean sulfates, evaporites, and interstitial sulfates from modern deep sea sediments, pp. 339–373 in Goldberg, E.D., Horibe, Y. and Saruhashi, K. (editors), *Isotope Marine Chemistry,* Geochemistry Research Association, Uchida Rokakuho Publishing Company, Tokyo.

ZoBell, C.E. 1958. Ecology of sulfate reducing bacteria. *Producers' Monthly* 22: 12–29.

5

Halophilic and Halotolerant Non-phototrophic Eubacteria

Halophilic, heterotrophic eubacteria share in common their general cell organization, but show many differences in modes of metabolism, salinity tolerances, and ranges of environmental conditions that permit growth. While numerous eubacteria have been isolated from hypersaline environments, relatively little is known about their activities and roles *in situ*. This gap in knowledge arises from the fact that little is known about the nature and concentrations of the primary organic matter produced in the euphotic zones of hypersaline environments. With the wealth of techniques used to study bacterial processes in freshwater and marine environments, microbiologists should eventually be able to describe more precisely how rapidly and to what extent halophilic eubacteria recycle organic matter in their natural habitats. The following descriptions of the ecology, physiology, and biochemistry of isolated, heterotrophic eubacteria indicate the potential role these bacteria may play in the degradation of organic matter in evaporite environments. The descriptions include the kinds of lipids they produce that might eventually become part of the hydrocarbons remaining in hypersaline sediments.

5.1 General physiological and ecological characteristics

Studies on halophilic and halotolerant, non-phototrophic eubacteria can be broadly divided into three types: ecological and enrichment studies to describe the variety of metabolic types and rates of activity in hypersaline habitats; taxonomic research on bacterial isolates describing general growth requirements and salt tolerances; and investigations detailing specific aspects of metabolism, osmoregulation, and biochemistry. Reviews of some of these aspects of the biology of moderately halophilic and halotolerant bacteria have been published by Kushner (1968, 1978), Morishita and Masui (1980), Starr et al. (1981), Brown (1983), Oren (1986b), Kushner and Kamekura (1988), Rodriguez-Valera (1988), and Ventosa (1988).

Table 5.1 lists the salt tolerances and optima of various metabolic activities detected in enrichments and isolates of halotolerant and halophilic bacteria (both eubacteria and archaebacteria). These summaries indicate that some microbial processes are largely absent or have not been studied in extremely hypersaline environments (e.g., nitrification and N_2 fixation), some processes are relatively attenuated in high salt concentration (for example, sulfate reduction and methanogenesis), and other microbial activities are relatively unaffected by hypersalinity (for example, aerobic carbohydrate and amino acid degradation). Table 5.2 indicates that little is known about the activity of halophiles in other extreme environments. Some of these "extreme" conditions could result from burial of evaporites and diagenesis. These tables are far from being complete as some physiological types have not been assayed or successfully demonstrated in hypersaline habitats. These data indicate that "classical" microbial nitrogen, sulfur, and carbon cycles found in freshwater and normal marine environments may be at least partially attenuated or interrupted in extremely hypersaline milieux.

Moderately halophilic bacteria grow well in 0.5–2.5 M NaCl and extreme halophiles grow best in 2.5–5 M NaCl (Kushner and Kamekura, 1988). Halotolerant bacteria are those that can tolerate a broad range of NaCl concentrations. While the populations and metabolic activities of eubacteria (moderate halophiles and halotolerant bacteria) and archaebacteria (most of which are extreme halophiles) overlap in hypersaline habitats, salinity and temperature regimes impose limits to the distribution of each.

The study of solar saltern brines by Rodriguez-Valera et al. (1981) evaluates some of these differences. Using colonies selected from agar plates from direct streaks of the brines (from 10% total dissolved solids to NaCl saturation), 150 strains of bacteria were studied to determine growth characteristics and nutrient requirements. The red-pink halobacteria were separated from eubacteria by color, and eubacteria were further distinguished by morphology. Table 5.3 summarizes some of their data. Not only did the total number of moderate halophiles decrease in the stronger brines, but the relative numbers of spiral-shaped cells decreased and the number of non-motile rods increased. The eubacteria in general grew best in 10–15% salt while the halobacteria preferred 25–30% salt in culture. These growth experiments were performed at 38°C. Growth tests at 30°C showed that the halobacteria could not grow at this temperature but that the eubacteria could. The halobacteria all required complex medium. The nutritional requirements of the moderately halophilic eubacteria changed with increased salinity. They required more complex media in higher salts.

Further taxonomic attributes of these bacteria were presented by Rod-

Table 5.1 Salt tolerances and optima of various physiological groups of halototolerant and halophilic bacteria

	Salinity, %		
Group	Range	Optimum	References
Sulfur oxidizers (non-phototrophic)			
Thiobacillus sp.	2–22	6	Saslowsky (1927) *in* Hof (1935)
3 species	up to 24		Issatchenko and Salimoska (1929) *in* Hof (1935)
2 species	up to 30	0, 12	Hof (1935)
Sulfate reducers	0–30	0–3	Saslowsky (1928)
	up to 20	8	Rubentschik (1928) *in* Benecke (1933)
	up to 30		Hof (1935), Rubentschik (1946), Klug et al. (1985)
	5–25	10–20	Baas-Becking and Kaplan (1956)
	up to 30	10–12	ZoBell (1958)
	up to 19		Skyring et al. (1977)
Nitrate reducers	5–20		Rubentschik et al. (1937)
	0.5–30	12	Elazari-Volcani (1940)
	15–30[a]	25[a]	Hochstein and Tomlinson (1985)
Nitrifiers	1–15	3–7	Rubentschik (1929)
N_2 fixers	up to 16[b]		Potts (1980)
Urea oxidizers	0–19	0–5	Rubentschik (1926a, 1926b)
Amino acid oxidizers	3–28	3–12	Ward and Brock (1978)
	15–30[a]	25[a]	Kushner (1985)
Hydrocarbon oxidizers			
Hexadecane	3–20		Ward and Brock (1978)
Tweens	15–30[a]	25[a]	Colwell et al. (1979)
Fermenters			
H_2 production	5–25	20	Baas-Becking and Kaplan (1956)
Cellulose decomposition	5–25		Baas-Becking and Kaplan (1956)
	up to $\geq$15		Rubentschik (1933)
Glucose	4–15	8–10	Oren(1983)
	2–30	13	Zeikus et al. (1983)
	8–16	9–15	Oren et al. (1984b)
	15–30[a]	25[a]	Javor (1984)
Amino acids	2–30	13	Zeikus et al. (1983)
	15–30[a]	25[a]	Javor (1984)
Methanogens	5–15		Baas-Becking and Kaplan (1956)
	5–20	12	Mathrani and Boone (1985); Paterek and Smith (1985)
	up to 30	25	Zhilina (1986)

[a]Salinity tolerances and optima of most halobacteria.
[b]Measured in cyanobacterial populations.
Included in Table 5.1 are studies on both pure cultures and enrichment cultures. No studies on ammonia oxidizers have been done.

riguez-Valera et al. (1985), Rodriguez-Valera (1986), and Ventosa (1988). It was concluded that aerobic halophilic eubacteria from hypersaline brines belong to genera that predominate in normal seawater (*Vibrio, Pseudomonas,* and *Flavobacterium*). Halophilic eubacteria in hypersaline soils belong to genera that are typical of soils (*Pseudomonas, Bacillus,* and Gram-positive cocci).

Table 5.2 Tolerances to other environmental extremes of halophilic and halotolerant, non-phototrophic bacteria

Physiological type	Environmental limit	Organism	Reference
Acid-tolerant	pH 4.4	*Pediococcus halophilus*	Noda et al. (1980)
Thermotolerant	50°C	*Bacillus*-like	Pfiffner et al. (1986)
	100°C	Methanogen	Zhilina (1986)
Psychrotolerant	0°–5°C	*Planococcus* sp.	Miller and Leschine (1984)
	−5°C	*Halomonas subglaciescola*	Franzmann et al. (1987)
Obligate anaerobes	?	Halophilic methanogens	Mathrani and Boone (1985);
	?	*Methanohalophilus mahii*	Paterek and Smith (1985, 1988)
	?	*M. zhilinae*	Mathrani et al. (1988)
	?	*Methanococcus halophilus*	Zhilina (1986)
	?	*Haloanaerobium praevalens*	Zeikus et al. (1983);
	?	*Halobacteroides halobius*	Oren (1984b)
	?	*Sporohalobacter lortetii*	Oren (1983)
	?	*Sporohalobacter marismortui*	Oren et al. (1987)

Acidophilic, psychrophilic, thermophilic, barophilic, and barotolerant halophiles have not been reported.

Table 5.3 Distribution of eubacteria and archaebacteria (halobacteria) in solar saltern brines[a]

	Colonies from 10–20% TDS brine	Colonies from 20–30% TDS brine	Colonies from >30% TDS brine
Eubacteria			
Total colonies	186	119	79
Spirals	56%	31%	7%
Cocci	4%	5%	20%
Non-motile rods	6%	24%	30%
Motile rods	34%	40%	43%
Halobacteria			
Total colonies	27	64	55
Rods	70%	61%	67%
Pleomorphic rods	18%	0%	4%
Cocci	12%	39%	29%

TDS, total dissolved solids.
[a]From Rodriguez-Valera et al. (1981). The % TDS of sea salt was determined by dry weight and compared with Cl^- content by argentometric titration: 25% TDS=17% NaCl; 33–43% TDS=22.5–29% NaCl.

A similar taxonomic study of the heterotrophic bacteria from other salterns showed the same genera of bacteria predominated (Marquez et al., 1987).

Javor (1984) found that saltern brines amended with NaCl inhibited moderate halophiles (*Vibrio*) at lower salt concentrations than extreme halophiles. Within the range of Mg^{2+}, K^+, Cl^-, and SO_4^{2-} concentrations found in solar saltern brines up to and including NaCl saturation, the moderate halophiles were relatively intolerant of Mg^{2+} and Cl^-, but relatively tolerant of K^+ and SO_4^{2-}. The halobacteria showed the opposite behavior with respect to Mg^{2+}, Cl^-, and SO_4^{2-}. In brines derived from seawater by evaporation, Mg^{2+} reaches high concentrations (0.5 M) in brines that are saturated with respect to NaCl. It was suggested that Mg^{2+} intolerance may be an important factor for limiting the success of moderate halophiles in such strong, marine-derived brines.

These microbial activities in hypersaline environments must be viewed together with geochemical observations to account for the very large standing crops of active populations and the great accumulations of organic matter that are often associated with moderately hypersaline environments. The accumulation of organic-rich sediments indicates that primary productivity greatly exceeds decomposition at salinities below gypsum saturation (at about 4× seawater concentration or approximately 15% salinity), especially in marine-derived systems. Such accumulations occur even though many moderately halophilic heterotrophs grow optimally in about 10–15% salt. Environments associated with massive gypsum or halite precipitation often lack such large accumulations of organic-rich sediments. The inability of microbial mats to develop and accumulate in such environments may result from 1) the relatively slow growth or the inability of many eubacteria to develop in salinities greater than about 15–20% TDS; and 2) the rapid overgrowth and "dilution" of the organic matter in and on the sediments by chemical precipitates. Apparently the relative decrease in the organic content of evaporite-rich sediments with rising salinity also inhibits the development of tightly-coupled microbial processes that dominate in organic-rich sediments and microbial mats of lower salinities. A more complete comparison and assessment of microbial activities and organic accumulation in hypersaline habitats in which there is little salinity fluctuation in individual basins (e.g., solar salterns and some salt lakes where gypsum and halite can accumulate) and those in which broad fluctuations occur (e.g., the limans studied by Rubentschik and associates where gypsum and halite do not accumulate) would indicate the role evaporite precipitation plays in decoupling potential microbial degradation processes.

5.2 Pure cultures of halophilic eubacteria

The eubacterial isolates described in Table 5.4 are all chemoheterotrophs. Most of the isolates are aerobes or facultative anaerobes (growth by nitrate reduction or fermentation). The preponderance of non-specialized heterotrophs listed in Table 5.4 may not reflect their dominance in brines and associated sediments, but rather the relative ease of culturing bacteria on nonspecific media by classical techniques. The enrichment and pure culture studies of Hirsch (1980) in the Solar Lake, Sinai, demonstrated that 104 distinct morphotypes of heterotrophic bacteria could be recognized microscopically

Table 5.4 Isolates of non-phototrophic, halotolerant, and moderately halophilic eubacteria

Organism	M NaCl optimum	M NaCl tolerance	Atmosphere	Reference[a]
Gram-negative				
Acinetobacter sp.	0.9–1.7	up to 4.3	Strict aerobe	6, 25, 26
Alcaligenes sp.	0.9–1.7	up to 4.3	Strict aerobe	6, 25, 26
Alteromonas sp.		≤0.5–≥3.5	Strict aerobe	6, 26
"Chromobacterium maris-mortui"	2.1	0.1–5.0	Facultative anaerobe	1
Deleya halophila	1.3	0.3–5.0	Strict aerobe	27
Desulfovibrio sp.		0.9–3.3	Strict anaerobe	29
D. salexigens	0.5–0.9	0.5–1.7	Strict anaerobe	29
Flavobacterium sp.	0.9–1.7	up to 4.3	Facultative anaerobe	25, 26
F. halmephilum		0.5–5.0	Facultative anaerobe	1
Haloanaerobium praevalens	2.2	0.3–5.0	Strict anaerobe	33
Halobacteroides halobius	1.5–2.5	1.4–2.8	Strict anaerobe	22
Halomonas elongata	0.4–1.4	0.05–4.5	Facultative anaerobe	32
H. halodurans		0.3–2.65	Strict aerobe?	28, 35
H. subglaciescola	0.4–3.0	0.1–3.4	Facultative anaerobe	34
Pseudomonas sp.	0.9–1.7	up to 4.3	Strict aerobe?	6, 25, 26
P. halestorgus		up to 5.0	Facultative anaerobe	1
P. halosaccharolytica		0.5–4.25	Strict aerobe?	5, 17
Sporohalobacter lortetii[b]	1.4–1.7	0.7–2.5	Strict anaerobe	19
Sporohalobacter marismortui	0.5–2.0		Strict anaerobe	20, 21, 23
Vibrio costicola	1.0	0.5–3.5	Facultative anaerobe	4, 12
Gram-positive				
Actinopolyspora halophila	3.4	2.0–5.0	Strict aerobe	3, 7, 8, 9
Arthrobacter sp.	0.9–2.6	up to 4.3	Strict aerobe	25
Bacillus sp.	0.9–1.7	up to 4.3	Facultative anaerobe	25
Bacillus-like	0.7–1.7	0–2.1	Facultative anaerobe	24
Brevibacterium sp.	0.9–2.6	up to 4.3		25
Corynebacterium sp.	0.9–1.7	up to 4.3	Facultative anaerobe	25
Micrococcus halobius	1.0–2.0	0.5–4.0	Strict aerobe	13, 18
M. varians		≤1.0–4.3	Strict aerobe	10
Nocardia sp.	0.3–0.9	up to 4.3	Strict aerobe	25
Paracoccus halodenitrificans	1.0–2.0	0.6–4.0	Facultative anaerobe	2, 13
Pediococcus halophilus	1.1–1.7	>0–≥3.1	Microaerophile	15, 33
Planococcus sp.		0–2.0	Strict aerobe	14
Planococcus sp.	1.7–2.6	up to 4.3	Strict aerobe	25
P. halophilus	1–2	0–5.5	Strict aerobe	16
Sporosarcina halophila	1.7–2.6	ca. 0.3–5	Strict aerobe	30
Staphylococcus epidermidis	0	0–4.2	Facultative anaerobe	11
Staphylococcus sp.	1.7–2.6	up to 4.2	Facultative anaerobe	25

but few could be identified as representatives of known genera. Some identifiable sulfur-oxidizing bacteria, such as *Achromatium, Beggiatoa,* and related taxa, are important organisms in hypersaline microbial mats in the Solar Lake and in solar salt ponds in up to about 15% salinity (Krumbein et al., 1977; Jørgensen and Des Marais, 1986), but they have never been cultured. However, such bacteria have eluded attempts of cultivation from non-hypersaline environments as well (La Riviere and Schmidt, 1981). Enrichments for sulfate-

Table 5.4 References

[a]*References*	
(1) Elazari-Volcani (1940)	(18) Onishi and Kamekura (1972)
(2) Gibbons (1958)	(19) Oren (1983)
(3) Gochnauer et al. (1975)	(20) Oren (1986b)
(4) Hipkiss et al. (1980)	(21) Oren (1987)
(5) Hiramatsu et al. (1980)	(22) Oren et al. (1984b)
(6) Imhoff and Rodriguez-Valera (1984)	(23) Oren et al. (1987)
(7) Johnson et al. (1986a)	(24) Pfiffner et al. (1986)
(8) Johnson et al. (1986b)	(25) Quesada et al. (1982)
(9) Johnson et al. (1986c)	(26) Quesada et al. (1983)
(10) Kamekura and Onishi (1974)	(27) Quesada et al. (1984)
(11) Komaratat and Kates (1975)	(28) Rosenberg (1983)
(12) Kushner (1968)	(29) Skyring et al. (1977)
(13) Kushner (1978)	(30) Ventosa et al. (1983)
(14) Miller and Leschine (1984)	(31) Villar et al. (1985)
(15) Noda et al. (1980)	(32) Vreeland et al. (1980)
(16) Novitsky and Kushner (1976)	(33) Zeikus et al. (1983)
(17) Ohno et al. (1979)	(34) Franzmann et al. (1987)
	(35) Hebert and Vreeland (1988)

[b]Formerly *Clostridium* (Oren et al., 1987).

reducing bacteria in high salt media sometimes have yielded positive results, but no halophilic sulfate-reducer has ever been described. Skyring et al. (1977) reported isolates of *Desulfovibrio* capable of growth in 19% salt (see Chapter 4). In the next two sections, the characteristics of these halophilic eubacteria are presented.

5.3 Gram-negative isolates

Vibrio *V. costicola* is perhaps the best known moderately halophilic bacterium because it has served as a model in studies on the effects of salt on nutrition, transport, protein turnover, lipids, and osmoregulation in eubacteria. *Vibrio* is a common isolate from >10% salinities in salterns (Rodriguez-Valera et al., 1985), but it constituted only 3% of the isolates from hypersaline soils (Rodriguez-Valera, 1986). Studies on its physiology are presented in Section 6.5.

Pseudomonas, Alteromonas, and Alcaligenes These three genera comprise the most common eubacteria isolated from solar salt ponds. In many cases they may represent nearly 60% of the aerobic heterotrophic eubacteria found in plate counts in culture media made with brine of 25% TDS (Ventosa, et al., 1982; Rodriguez-Valera et al., 1985). Strains of *Pseudomonas* and *Alcaligenes* are also found in hypersaline soils (Quesada et al., 1982, 1983; Rodriguez-Valera, 1986). These strains all tolerate ≥20% salt, while *P. halosaccharolytica* tolerates 25% salt (Hiramatsu et al., 1980). "*Ps. halestorgus,*" a Dead Sea isolate that is no longer available in culture, tolerated 24% NaCl. In high salt agar medium it produced bright brown colonies while at lower salt concentrations it produced grey-white colonies (Elazari-Volcani, 1940). A moderately halophilic *Pseudomonas* was found to secrete a soluble protease that demonstrated maximum enzymatic activity in 18% salt (Van Qua et al., 1981).

Acinetobacter Members of this genus have been isolated from hypersaline soils (Quesada et al., 1983) and solar salt ponds (Rodriguez-Valera et al., 1985) where it constitutes a minor component in all salinities above 10% TDS. A moderately halophilic strain isolated from sea sands produced two soluble amylases with maximum enzyme production in 1–2 M NaCl (Onishi and Hidaka, 1978).

Deleya *Deleya halophila* is a slightly halophilic rod isolated from hypersaline soils (Quesada et al., 1984). It grows optimally in 7.5% marine salts but tolerates 2–30% salinity. It is a strict aerobe that can grow on a variety of carbon sources. It grows optimally at 30°–37°C, although a few strains show some growth at 4°C.

Chromobacterium An organism called *Chromobacterium maris-mortui* was observed as a motile rod in isolations from the Dead Sea (Elazari-Volcani, 1940), but its classification as a species of this genus is uncertain (Ventosa, 1988). It grew optimally in 12% salt and produced blue-brown colonies. The colonies grew in concentric rings on agar and left a deep blue print but in >13% salt the colonies were colorless. Strains of *Chromobacterium* were isolated only from >25% TDS brines in a saltern (Rodriguez-Valera et al., 1985).

Flavobacterium *Flavobacterium halmephilum*, a Dead Sea isolate, is a non-motile, halotolerant rod that formed yellow colonies (Elazari-Volcani, 1940). Moderately halophilic strains of *Flavobacterium* have also been isolated from hypersaline soils (Quesada et al., 1983) and from solar salt ponds in >10% TDS brines (Rodriguez-Valera et al., 1985), where they constituted 8–12% of the total heterotrophic eubacterial isolates.

Quesada et al. (1987) studied 33 strains of *Flavobacterium*-like moderate halophiles and 22 strains of *Acinetobacter*-like strains of moderate halophiles and found none that showed any close relationships to *F. halmephilum* or "*C. marismortui*," and their DNA base composition indicated that they were not closely related to non-halophilic strains of those genera. Quesada et al. concluded that the moderate halophiles probably belonged to yet unnamed genera.

Halomonas *H. elongata* is a broadly halotolerant rod isolated from a solar saltern (Vreeland and Martin, 1980; Vreeland et al., 1980). It can grow as a facultative anaerobe by reducing nitrate or fermenting glucose. It grows at the expense of a variety of carbohydrates and amino acids. Although it is primarily an aerobe, assays for the presence of cytochrome oxidase were negative. Growth between 20° and 40°C was recorded, but maximal salt tolerance was noted at 30°C. *H. halodurans* (formerly *Pseudomonas halodurans*), another halotolerant species, was isolated from seawater (Rosenberg, 1983; Hebert and Vreeland, 1987). *H. subglaciescola*, an isolate from a hypersaline Antarctic lake, grows at 0°C and 25°C, while some strains can grow at −5°C (Franzmann et al., 1987).

Four obligately anaerobic, heterotrophic eubacteria are known to be moderately halophilic. They are all Gram-negative rods that thrive by fermentation. Their 16S rRNA sequences show they are all related to each other but are unrelated to any other subgroup among the eubacteria (Oren, 1986b).

Halobacteroides halobius, a novel genus with but one species, was characterized from isolations from Dead Sea sediments (Oren et al., 1984b). It requires 1.4–2.8 M NaCl. The cells are long rods that can be induced to produce spores (Oren, 1987). Glucose is fermented to ethanol, acetate, H_2, and CO_2. Several other sugars are also fermented as well as pyruvate.

Haloanaerobium praevalens, also a novel genus with only one species, was isolated from the Great Salt Lake, Utah (Zeikus et al., 1983). It grows optimally in 2.2 M salt but can tolerate 5 M NaCl. It is a non-motile, non-sporing rod that was detected in all the enrichments for obligately fermentative bacteria in the Great Salt Lake. It thrives on carbohydrates and amino acids. Carbohydrate fermentation leads to the production of H_2, CO_2, acetate, propionate, and butyrate but not ethanol or lactate. Methionine degradation results in methylmercaptan production. Based on 16S rRNA oligonucleotide cataloging, *H. praevalens* and *Halobacteroides* have been placed in a new family of moderately halophilic, obligately anaerobic bacteria, the Haloanaerobiaceae (Oren et al., 1984a).

Sporohalobacter lortetii (formerly *Clostridium lortetii*), an isolate from Dead Sea sediments, requires 1–2 M NaCl (Oren, 1983; Oren et al., 1987). It is a rod that produces gas vacuoles in sporulating cells only near the terminal endospore. Similar rods were observed in sediments of the Great Salt Lake. It ferments glucose and several other carbohydrates, producing H_2 and low-molecular-weight fatty acids. Lactate was not detected. Up to 1 mM sulfide was produced in medium with casamino acids, probably as a result of cysteine degradation.

Sporohalobacter marismortui was also isolated from Dead Sea sediments. (Oren et al., 1987). It is a motile, sporulating rod that ferments glucose and other sugars, but not glycerol or pyruvate. Fermentation products from glucose include ethanol, acetate, butyrate, formate, H_2, and CO_2. It grows optimally in 0.5–2 M NaCl at 36–45°C, but tolerates as much as 3 M NaCl.

5.4 Gram-positive isolates

Micrococcus Moderately halophilic species of *Micrococcus* include *M. varians, M. morrhuae,* and *M. halobius. M. varians* var. *halophilus,* an isolate from soy sauce mash, produces extracellular nuclease and amylase in medium with 2.5–3.5 M NaCl (Kamekura and Onishi, 1974, 1976, 1978). In some media, the cells formed in clumps and enzymatic activity was absent in the presence of >40 mM Mg^{2+} or Ca^{2+}. The extracellular nuclease from *M. varians* had DNase and RNase activity. Maximal activity occurred in 2.9 M NaCl or 2.1 M KCl at 40°C. Extracellular amylase activity has also been reported in cultures of other unidentified strains of *Micrococcus* (Onishi, 1972; Onishi and Sonoda, 1979).

M. morrhuae K-17 grows in at least 2 M NaCl and utilizes a wide variety of carbon sources. Like many other moderate halophiles, the salt range of this strain can be extended by growth in complex medium that provides the substrates the cells cannot synthesize under stressful growth conditions (Chan and Leung, 1979).

Paracoccus *P. halodenitrificans,* as its name implies, can grow by anaerobic nitrate reduction, reducing nitrate to N_2 (Robinson, 1952; Robinson and Gibbons, 1952). The strain of Gibbons was isolated from meat-curing brines. It grew optimally in 4.4–8.8% NaCl but tolerated 23.4% NaCl. The presence of low concentrations of Ca^{2+} in the medium permitted cells to grow in 1% salt (Gibbons, 1958; Takahashi and Gibbons, 1959).

Marinococcus and Planococcus *M. halophilus* (previously identified as *Planococcus halophilus* and *Paracoccus haloxanthus*) is a Gram-variable halophilic coccus that produces yellow colonies. It is motile and strictly aerobic (Novitsky and Kushner, 1975, 1976). At lower temperatures it can grow without NaCl, but requires 0.5 M NaCl at 25°C. *M. halophilus* and *Planococcus* sp. are frequently isolated from hypersaline soils and salt ponds (Ventosa et al., 1983). A halotolerant strain of *Planococcus* was isolated from soil from the Antarctic Dry Valley (Miller and Leschine, 1984). It grows in 0–2.0 M NaCl between 0° and 40°C, although growth in all salt concentrations is poor at 0°–5°C. The medium used contained a high concentration of Mg^{2+} (200 mM) that apparently did not hinder growth.

Pediococcus *P. halophilus* is a facultative anaerobe that apparently prefers organic-rich hypersaline habitats since it is associated with soy sauce production (Noda et al., 1980) and salted fish (Villar et al., 1985). *P. halophilus* ferments glucose to lactic acid and acetic acid in developing soy sauce (18% salt). These metabolites, as well as the resulting low pH (4.7–4.8), strongly inhibit osmophilic shoyu yeasts during the early stage of fermentation. Villar et al. (1985) reported their strains to be microaerophilic and catalase negative. The bacteria could not reduce nitrate to nitrite. They produce acid via glucose fermentation, and are tolerant of very low pH (4.4). Optimal temperatures for growth are 25°–30°C, and up to 40°C is tolerated. *P. halophilus* is unique among the isolated halophilic and halotolerant bacteria in its apparent tolerance of acidic medium that results from sugar fermentation.

Actinomycetes *Actinopolyspora halophila* was described as a contaminant in unsterilized medium used to grow extreme halophiles (Gochnauer et al., 1975). Studies by Johnson and co-workers (Johnson and Lanthier, 1986; Johnson et al., 1986a, 1986b) showed that *A. halophila* is essentially an extreme halophile since it requires >10% NaCl for growth. It grows on casein and a variety of sugars. It has an optimal temperature of 37°C, tolerates up to 43°C, and shows slight growth at 10°C. The wild-type strain grows optimally in 12% salt while an erythromycin-resistant strain grows optimally in 20% salt. The antibiotic-resistant strain, however, also tolerates lower salt (6%) than the wild-type. *A. halophila* produces a variety of extracellular enzymes, including cell wall lytic enzymes, carbohydrate-degrading enzymes, β-lactamases (which degrade antibiotics in the penicillin family), and protease. Enzyme activities were higher in 15% salt than in 0% salt.

Other halophilic actinomycetes have been briefly described. Rubentschik (1946) noted that Zavialov (no reference given) isolated an anaerobe named by him as *Actinomyces pliogenes* from the silt of a salt lake. On lactose medium with asparagine it formed colonies surrounded by a black zone of iron sulfide. The cells appeared as rods or branching filaments. A similar report was published by Benecke (1933) describing the investigator as Sawialoff and the bacterium as *A. pelogenes*. Of 724 strains of heterorophic, halophilic eubacteria isolated from a saltern, only two were actinomycetes (Rodriguez-Valera et al., 1985). They both came from brines with 10–15% TDS.

Other bacteria Many other moderately halophilic bacteria have been well studied but not described further than by strain designation. Still others have been described and named but once. The latter case is especially true in the older literature from a time when bacterial nomenclature was not subject to the rigorous physiological and biochemical testing used in modern bacterial taxonomy. Some examples of these lesser known halophiles are given here.

Staphylococcus epidermidis, a halotolerant strain, was originally isolated as a contaminant in medium containg 25% salt. It grows best in medium without NaCl (Komaratat and Kates, 1975).

A *Bacillis*-like, Gram-positive, spore-forming rod was isolated from a landfill associated with an oil well, brine injection water, and anaerobic sewage sludge (Pfiffner et al., 1986). It grows anaerobically at up to 50°C in medium with 0% to 10–12% salt. These bacteria produce extracellular heteropolysaccharides when grown on sugars. A marine *Bacillus* isolated from rotting wood produces extracellular nuclease when the cells are grown aerobically in 1–2 M NaCl (Onishi et al., 1983). Weisser and Trüper (1985) described an alkaliphilic, halotolerant *Bacillus* isolated from the Wadi Natrun lakes that tolerated up to 20% NaCl.

Lactobacillus casei, which is used to make cheese, grows in up to 9–12% NaCl (Hegazi, 1984). Staphylococci were also found in those salt concentrations. *Staphylococcus* and *Micrococcus* grow in up to 16% salt in broth and up to 19% salt on agar (Hill and White, 1929). Some other pathogens, including *Bacillus anthracis* and *Corynebacterium pseudodiptheriticum*, grow slowly in 7–10% salt agar but survive in 17% to ≥20% NaCl.

Sporosarcina halophila, a Gram-positive coccus, has been isolated from several hypersaline soils and salt ponds (Ventosa et al., 1983). Unidentified Gram-positive cocci and Gram-positive rods constituted 8–14% and 3–5%, respectively, of the heterotrophic, eubacterial isolates from solar saltern brines (Rodriguez-Valera et al., 1985).

A variety of other successful enrichments for halotolerant and halophilic bacteria, including eubacteria, have been described: enrichments from garden soil and salted beans (Hof, 1935); seawater and mackerel (Venkataraman and Sreenivasan, 1954); and enrichments from intertidal marine habitats (Forsyth et al., 1971). *Achromobacter venenosum*, isolated from the English channel, could grow in ≥10% salt (Vargues, 1962). Enrichments in 4 M NaCl for bacteria in dried and salted fish, solar salt, soy sauce mash, marine sands, and seaweeds are described by Onishi and Kamekura (1980) and Onishi et al., (1980). A large (3–6 μm), colorless, Gram-positive sarcina, *Sarcina gigantea*, was isolated from salted fish (Petter, 1931). It grew in 3–35% salt at 25°C. It was an aerobe but was shown to be catalase-negative.

Strain Ba_1, an obligately aerobic, Gram-negative rod isolated from the Dead Sea, can grow in ≥2 M NaCl although it respires optimally in 0.2–0.8 M NaCl. Addition of choline or betaine (compatible solutes used in control of osmotic pressure) increases its resistance to higher salinities (Rafaeli-Eshkol, 1968; Rafaeli-Eshkol and Avi-Dor, 1968; Shkedy-Vinkler and Avi-Dor, 1975). Choline not only stimulates growth, but it could serve as the carbon source for unidentified vibrios isolated from solar salt ponds (Javor, 1984). Choline is presumably converted to betaine by the cells.

More detailed taxonomic discussions of halophilic and halotolerant eubacteria are given by Ventosa (1988) and Kushner and Kamekura (1988).

5.5 Physiology and biochemistry

Bacteria and wall-less eucaryotes such as *Dunaliella* are in osmotic equilibrium with their environment. Because high internal salt concentrations interfere with biochemical functions within the cells of most organisms, the organisms must find a way to maintain osmotic balance without interfering with cell function. A variety of intracellular solutes are used for this purpose. The internal solute used must be non-toxic; such compounds are called compatible solutes. Examples of compatible solutes used by various halophilic or halotolerant organisms include glutamate, proline, choline, betaine, polyamines, and glycerol. The internal concentration of the compatible solute will vary with the external salt concentration.

In addition, several other changes occur in halotolerant and moderately halophilic eubacteria in response to changing salinities. 1) Most moderate halophiles have more demanding nutritional requirements at higher salinities. Temperature optima or tolerances also increase in high salt and requirements for divalent cations change with salinity. 2) Ion and amino acid transport are affected by changes in salinity. 3) The lipid composition often differs when cells are grown at high and low salinities. 4) Most moderate halophiles produce glycine, betaine or other compatible organic solutes in response to increased salinity rather than increase their intracellular inorganic ion content. The following discussion highlights these aspects of physiology and biochemistry that might be useful in understanding in which environments these bacteria might be detected and where they would be metabolically active. A more thorough review of the biology of halotolerant and moderately halophilic eubacteria can be found in Kushner and Kamekura (1988).

The ability to tolerate high salt concentrations in defined medium has been described in several moderately halophilic and halotolerant eubacteria (Keller and Henis, 1967; Vreeland and Martin, 1980; Forsyth and Kushner, 1970; Rodriguez-Valera et al., 1981). It is generally believed that the salt range of growth is extended when cultures are grown in complex media which provide growth factors that cannot be synthesized at high salinities (Chan and Leung, 1979). Such growth factors could possibly include compatible solutes or other compounds synthesized more slowly under high salt conditions when the metabolic machinery is operating at a lower rate due to salt stress. For example, Kamekura et al. (1985) successfully extended the salt tolerance of *V. costicola* in chemically defined medium by including a high concentration of glutamate (2% w/v), which may be important for activating protein synthesis in the presence of high concentrations of Cl^-.

Salt requirements and tolerances are also temperature-dependent. A strain of *Planococcus* grew in 0–5.5 M NaCl, but it thrived in dilute medium (0.1 M NaCl) only at <30°C. In 1.0 M NaCl its optimal temperature was 35°C (Novitsky and Kushner, 1975, 1976). Two strains of *Micrococcus* tolerated as little as 0.5 M NaCl at 30°C, but could only grow in 0.2 M NaCl when the temperature was lowered to 25°C (Chan and Leung, 1979). The ability to tolerate higher salt concentrations at higher temperatures was also noted in *P. halosaccharolytica* (Ohno et al., 1979).

Both Ca^{2+} and Mg^{2+} affect growth and survival of moderate halophiles. As little as 5 mM $CaCl_2$ permitted the salt range of growth of *P. halodenitri-*

ficans to be lowered from 0.65 M NaCl to 0.3 M NaCl (Gibbons, 1958; Takahashi and Gibbons, 1959). The composition of the medium also apparently affects the tolerance of cells to divalent cations. In some media *M. varians* var. *halophilus* tolerates as much as 2 M $MgSO_4$ while in other media as little as 81 mM Mg^{2+} causes cell flocculation (Kamekura and Onishi, 1976).

Salt affects the ability of moderate halophiles to take up exogenous compounds and control the rate of protein synthesis. The ability of *V. costicola* to transport an amino acid analog (α-aminoisobutyric acid, AIB) decreases when cells are transferred from high salt (4 M NaCl) to low salt (0.5 M NaCl) or when cells are transferred from 0.5 M NaCl to 4 M NaCl (Kushner et al, 1983; Hamaide et al., 1984). Protein turnover rate in *V. costicola* also increases in very low salt (Hipkiss et al., 1980). The authors concluded that higher turnover rates in lower salt may be caused by changes in protein conformation and increased susceptibility to proteolysis.

In some halotolerant or halophilic bacteria, it is possible that salinity-dependent transport may involve changes in the conformation of pre-made receptors or transport proteins. There is evidence that the protein species of the outer membrane of moderate halophiles change in response to salt concentration of the culture medium (Hiramatsu et al., 1980). Alternatively, changes in polar lipid composition in the membranes may affect active transport (Hanna et al., 1984).

Moderate halophiles and halotolerant eubacteria typically produce phosphatidylglycerol (PG), phosphatidylglycerophosphate (PGP), phosphatidylethanolamine (PE), cardiolipin, and sometimes several other lipids. The effect of culture conditions on lipid composition has been investigated in several moderately halophilic bacteria. In general, an increase in salt concentration causes an increase in negatively charged polar lipids. In *P. halosaccharolytica*, an increase in salt concentration in the medium or an increase in growth temperature resulted in a relative decrease in PE (Ohno et al., 1979; Hiramatsu et al., 1980; Hara, 1982; Hara and Masui, 1985). Growth in medium with glucose resulted in an increase in proportions of glucosylphosphatidylglycerol and diphosphatidylglycerol (Ohno et al., 1979). Similar changes in the proportion of PE as a result of changes in salt concentration have been observed in several other moderate halophiles (Kogut and Russell, 1984; Hanna et al., 1984; Vreeland et al., 1984; Russell et al., 1985; Miller, 1985), but not in a moderate halophile from the Dead Sea (strain Ba_1) grown in 2 and 4 M NaCl (Peleg and Tietz, 1971).

The fatty acid composition of membrane lipids can also be affected by culture conditions in moderate halophiles. Both high temperature and high salt cause a relative increase in saturated and cyclopropanoic fatty acids and acidic phospholipids in *Ps. halosaccharolytica* (Ohno et al., 1979; Hara, 1982). The major fatty acids were $C_{16:0}$, $C_{16:1}$, $C_{18:0}$, $C_{18:1}$, and cyclopropanoic C_{17} and C_{19} fatty acids (Ohno et al., 1976). A $C_{19:0}$ cyclopropanoic fatty acid was also a major component in the lipids of Dead Sea strain Ba_1 grown in 2 M NaCl

+ 0.5 M KCl (Peleg and Tietz, 1971). High salt also affected the proportion of certain fatty acids in *V. costicola*, but this bacterium does not produce cyclopropanoic C_{17} or C_{19} fatty acids (Hanna et al., 1984).

In halotolerant *Planococcus* strain A4a, low temperatures induced the formation of monounsaturated, branched-chain fatty acids while high NaCl (1.5 M) caused an increase in the relative amount of $C_{15:0}$ fatty acid (Miller, 1985). In another halotolerant bacterium, *Staphylococcus epidermidis*, the major lipids all had essentially the same fatty acid concentration in cells grown in 0–15% salt (Komaratat and Kates, 1975). *S. epidermidis* produced $C_{15:0}$ fatty acid as the major constituent of the total fatty acids followed by $C_{17:0}$ fatty acids.

Culture age apparently affects lipid composition in some moderate halophiles. As the cells of *P. halosaccharolytica* changed from logarithmic to stationary phase, there was a relative decrease in PG and PE while the relative content of cardiolipin and a phosphoglycolipid increased (Ohno et al., 1976). In strain Ba_1, 48-hr cultures contained less PG and cardiolipin than 18-hr cultures (Stern and Tietz, 1973). $C_{16:1}$ and $C_{18:1}$ constituted 12% and 25%, respectively, of the fatty acids of PE in 18-hr cultures, but these fatty acids were absent in 48-hr cultures in which only $C_{16:0}$ and cyclopropanoic fatty acids C_{17} and C_{19} were found. However, in *V. costicola* grown in 1 M NaCl, little change in fatty acid composition or proportion of different lipids was observed in cultures of different growth phases (Hanna et al., 1984).

Diverse pigments and isoprenoid compounds are produced among moderately halophilic, non-phototrophic bacteria. Most of the bacteria surveyed by Kushwaha et al. (1974) produced squalene, dihydrosqualene, and tetrahydrosqualene. All of the surveyed moderate halophiles produced menaquinone, except a rod identified as $A_{31}C$ that contained only ubiquinone. *Marinococcus halophilus* produced phytoene but rod $A_{31}C$ did not. Phytoene is also found in halobacteria. β-carotene and bacterioruberin were absent in the moderate halophiles (although it should be noted that the moderately halophilic phototrophic bacteria produce a variety of carotenoids). In another survey by Collins (1981), some moderate halophiles were shown to contain large amounts of unsaturated menaquinones with 7 and 8 isoprene units (*Brevibacterium halotolerans* and *M. halobius*, respectively). *Actinopolyspora halophila* produced a mixture of partially saturated quinones with 9 isoprene units. While *V. costicola* produced both menaquinones and ubiquinones with 8 isoprene units, several other species produced only ubiquinones with 8 or 9 isoprene units.

In studies of the nature and concentrations of compatible solutes in moderate halophiles and halotolerant bacteria, betaine has been cited as the dominant intracellular solute (Imhoff and Rodriguez-Valera, 1984; Imhoff, 1986). Table 5.5 shows the concentrations of betaine and other nitrogenous substances accumulated at different salt concentrations by several moderate halophiles. It is noteworthy that the increase in betaine concentration correlates

Table 5.5 Nitrogenous compatible solutes in halophilic eubacteria[a]

Organism	External M NaCl	Internal compatible solutes: Amino acids	Betaine	Polyamines	Reference[b]
Micrococcus halobius	0.52	0.384[c]	0.290		3
	1.72	0.113[c]	0.851		3
	3.45	0.049[c]	1.10		3
Micrococcus sp.	0.52	0.052[c]	0.158		3
	1.72	0.133[c]	0.684		3
	3.45	0.076[c]	1.07		3
M. varians subsp. *halophilus*	3			1.24[d]	2
P. halodenitrificans	1.0	0.244			1
Vibrio costicola	0.52	0.308[c]	0.219		3
	1.72	0.428[c]	0.636		3
	3.45	0.389[c]	1.20		3
	1.0	0.334			1
	0.5–3.0			0.300[e]	4
Halomonas elongata	0.05	0.005			5
	1.37	0.102			5
	3.41	0.371			5

[a]All values are molar concentrations unless otherwise stated.
[b]References: (1) Christian and Waltho (1962); (2) Hamana et al. (1985); (3) Imhoff and Rodriguez-Valera (1984); (4) Kamekura et al. (1986); (5) Vreeland et al. (1983).
[c]Glutamic acid.
[d]Micromoles per g wet weight.
[e]Micromoles per mg protein.

well with the external salt concentration up to 10% salt, which is approximately the optimal salt concentration for growth. The inability to accumulate sufficient betaine or glutamate to counterbalance the external osmolarity at salt concentrations higher than 10% may well be why these bacteria grow poorly in highly concentrated brines. Other nitrogenous compatible solutes include ectoine in the phototrophic bacterium *Ectothiorhodospira* (Galinski et al., 1985) and *Micrococcus* (Imhoff, 1986), amino acids (Christian and Waltho, 1962; Measures, 1975; Unemoto and Hayashi, 1979; Vreeland et al., 1983), and polyamines (Hamana et al., 1985; Kamekura et al., 1986). While ectoine and amino acids may accumulate in millimolar quantities, total polyamines never accumulate to more than micromolar concentrations in these bacteria.

In addition to biosynthesis of a compatible solute, some of the osmotic balance is maintained by the accumulation of Na^+ and K^+ in most moderate halophiles (Christian and Waltho, 1962; Masui and Wada, 1973; Shindler et al., 1977; Matheson et al., 1976; Sadler et al., 1980; Vreeland et al., 1983) (Table 5.6). Some of the variations found between different studies of the same organisms may be due to differences in techniques and inherent problems in measurement. Internal anion concentrations in the bacteria were not reported in most of the studies.

In contrast to the aerobic moderate halophiles, the obligate anaerobes *Haloanaerobium praevalens* and *Halobacteroides halobius* were found to contain high levels of intracellular K^+ (0.76–2.05 M) and Na^+ (0.28–2.6 M) (Oren,

Table 5.6 Inorganic ion concentrations in halophilic eubacteria

	External ions, M			Internal ions, M			
Organism	Na^+	K^+	Cl^-	Na^+	K^+	Cl^-	Reference[a]
Paracoccus halodenitrificans	1.0	0.004	1.0	0.311	0.474	0.055	1
Vibrio costicola	1.0	0.004	1.0	0.684	0.221	0.139	1
	0.6	0.01		0.51–0.65	0.72		6
	1.0	0.01		0.58–0.89	0.82		6
	1.6	0.01		1.09–1.39	0.57		6
	2.0	0.01		0.90–1.29	0.55		6
	3.1	0.01		1.78	0.37		6
(log phase)	0.8		0.8	0.51	0.52		3
Unidentified (log)	0.5		0.5	0.05	0.34		3
(stationary)	0.5		0.5	0.29	0.32		3
(log)	4.3		4.3	0.62	0.58		3
(stationary)	4.3		4.3	1.01	0.66		3
Pseudomonas halodenitrificans (log)	1.0	0.01	1.0	0.50	0.10		5
(stationary)	3.0	0.01	3.0	1.1	0.10		5
Pseudomonas sp.	1–3	0.055	1–3	0.9–1.15	0.67–0.89		2
Halomonas elongata	0.64	0.017		0.042	0.002		7
	1.38	0.016		0.312	0.016		7
	3.41	0.016		0.630	0.018		7
Haloanaerobium praevalens	0.1	0.01		0.46	0.95		4
	2.22	0.01	2.3	1.52	1.59	2.24	4
	3.08	0.01	3.16	2.63	2.05	3.28	4
Halobacteroides halobius	1.56	0.013		0.28–0.67	0.76–1.21		4

[a]References: (1) Christian and Waltho (1962); (2) Masui and Wada (1973); (3) Matheson et al. (1976); (4) Oren (1986a); (5) Sadler et al. (1980); (6) Shindler et al. (1977); (7) Vreeland et al. (1983).

1986a). While the internal Na^+ concentrations correlated with increasing external salinity, internal K^+ concentrations did not. The sum of these two internal cations were approximately equal to the total cation concentration outside the cells. Potential intracellular organic osmolytes (betaine, glycerol, and amino acids) were insignificant. The ability to accumulate such high internal salt concentrations was previously believed to be a property of halobacteria only.

5.6 Summary

Halotolerant and moderately halophilic, non-phototrophic eubacteria are a diverse group of microbes that share in common their eubacterial classification and their ability to live in high salt. Physiologically and biochemically, the bacteria are as diverse as their taxonomy indicates. The lack of halophilic isolates of nitrogen-fixers or sulfate-reducers may reflect their actual absence in nature, or it may reflect the complex nature of their natural niches which have not been duplicated under laboratory conditions. Although the halobacteria discussed in the next chapter are preeminent in many hypersaline environments, it is clear that the diversity of bacteria at high salt concentrations often can be quite high.

References

Baas-Becking, L.G.M. and **Kaplan, I.R.** 1956. The microbiological origin of the sulphur nodules of Lake Eyre. *Transactions of the Royal Society of South Australia* 79: 52–65.

Benecke, W. 1933. Bacteriology of the sea, pp. 717–854 in Benecke, W., *Abderhalden's Handbuch der biologischen Arbeitsmethoden*, vol. 9, Westfalia-Wilhelms University, Munster.

Brown, A.D. 1983. Halophilic prokaryotes, pp. 137–162 in Lange, O.L., Nobel, P.S., Osmond, C.B. and Ziegler, H. (editors), *Physiological Plant Ecology* III, Springer-Verlag, New York.

Chan, K.-Y. and **Leung, O.C.** 1979. Nutrition and growth of the moderately halophilic bacteria *Micrococcus morrhuae* K-17 and *Micrococcus luteus* K-15. *Microbios* 25: 71–84.

Christian, J.H.B. and **Waltho, J.A.** 1962. Solute concentrations within cells of halophilic and non-halophilic bacteria. *Biochimica et Biophysica Acta* 65: 506–508.

Collins, M.D., Ross, H.N.M., Tindall, B.J., and **Grant, W.D.** 1981. Distribution of isoprenoid quinones in halophilic bacteria. *Journal of Applied Bacteriology* 50: 559–565.

Colwell, R.R., Litchfield, C.D., Vreeland, R.H., Kiefer, L.A., and **Gibbons, N.E.** 1979. Taxonomic studies of red halophilic bacteria. *International Journal of Systematic Bacteriology* 29: 379–399.

Elazari-Volcani, B. 1940. *Studies on the microflora of the Dead Sea*. Ph.D. Thesis. Hebrew University of Jerusalem. 119 pp.

Forsyth, M.P. and **Kushner, D.J.** 1970. Nutrition and distribution of salt response in populations of moderately halophilic bacteria. *Canadian Journal of Microbiology* 16: 253–261.

Forsyth, M.P., Schindler, D.B., Gochnauer, M.B., and **Kushner, D.J.** 1971. Salt tolerance of intertidal marine bacteria. *Canadian Journal of Microbiology* 17: 825–828.

Franzmann, P.D., Burton, H.R., and **McMeekin, T.A.** 1987. *Halomonas subglaciescola*, a new species of halotolerant bacteria isolated from Antarctica. *International Journal of Systematic Bacteriology* 37: 27–34.

Galinski, E.A., Pfeiffer, H.-P., and **Trüper, H.G.** 1985. 1, 4, 5, 6-Tetrahydro-2-methyl-4-pyrimidinecarboxylic acid: a novel cyclic amino acid from halophilic phototrophic bacteria of the genus *Ectothiorhodospira*. *European Journal of Biochemistry* 149: 135–140.

Gibbons, N.E. 1958. The effect of salt on the metabolism of halophilic bacteria, pp. 69–76 in Eddy, B.P. (editor), *Proceedings of the 2nd International Symposium on Food Microbiology*, H.M. Stationery Office, London.

Gochnauer, M.B., Leppard, G.G., Komaratat, P., Kates, M., Novitsky, T., and **Kushner, D.J.** 1975. Isolation and characterization of *Actinopolyspora halophila* gen. et sp. nov., an extremely halophilic actinomycete. *Canadian Journal of Microbiology* 21: 1500–1511.

Hamaide, F., Sprott, G.D., and **Kushner, D.J.** 1984. Energetic basis of development of salt-tolerant transport in a moderately halophilic bacterium, *Vibrio costicola*. *Archives of Microbiology* 140: 231–235.

Hamana, K., Kamekura, M., Onishi, H., Akazawa, T., and **Matsuzaki, S.** 1985. Polyamines in photosynthetic eubacteria and extreme-halophilic archaebacteria. *Journal of Biochemistry (Tokyo)* 97: 1653–1658.

Hanna, K., Bengis-Garber, C., Kushner, D.J., Kogut, M., and **Kates, M.** 1984. The effect of salt concentration on the phospholipid and fatty acid conposition of the moderate halophile *Vibrio costicola*. *Canadian Journal of Microbiology* 30: 669–675.

Hara, H. 1982. Effect of extreme growth conditions on lipid composition and liposome properties in a moderately halophilic bacterium. *Japanese Journal of Bacteriology* 37: 851–862.

Hara, H. and **Masui, M.** 1985. Effect of sodium chloride concentration on the synthesis of membrane phospholipid in a halophilic bacterium. *FEMS Microbiology Letters* 31: 279–282.

Hebert, A.M. and **Vreeland, R.H.** 1987. Phenotypic comparison of halotolerant bacteria. *Halomonas halodurans* sp. nov., nom. rev., comb. nov. *International Journal of Systematic Bacteriology* 37: 347–350.

Hegazi, F.Z. 1984. The microbial flora of salted raw milk. *Systematic and Applied Microbiology* 5: 527–533.

Hill, J.H. and **White, E.C.** 1929. Sodium chloride media for the separation of certain Gram-positive cocci from Gram-negative bacilli. *Journal of Bacteriology* 18: 43–57.

Hipkiss, A.R., Armstrong, D.W., and **Kushner, D.J.** 1980. Protein turnover in a moderately halophilic bacterium. *Canadian Journal of Microbiology* 26: 196–203.

Hiramatsu, T., Yano, I., and **Masui, M.** 1980. Effect of NaCl concentration on the protein species and phospholipid composition of the outer membrane in a moderately halophilic bacterium. *FEMS Microbiology Letters* 7: 289–292.

Hirsch, P. 1980. Distribution and pure culture studies of morphologically distinct Solar Lake microorganisms, pp. 41–60 in Nissenbaum, A. (editor), *Hypersaline Brines and Evaporitic Environments*, Developments in Sedimentology 28, Elsevier Scientific, New York.

Hochstein, L.I. and **Tomlinson, G.A.** 1985. Denitrification by extremely halophilic bacteria. *FEMS Microbiology Letters* 27: 329–331.

Hof, T. 1935. Investigations concerning bacterial life in strong brines. *Extrait du Recueil des Travaux Botaniques Néerlandais* 32: 92–173.

Imhoff, J.F. 1986. Osmoregulation and compatible solutes in eubacteria. *FEMS Microbiology Reviews* 39: 57–66.

Imhoff, J.F. and **Rodriguez-Valera, F.** 1984. Betaine is the main compatible solute of halophilic eubacteria. *Journal of Bacteriology* 160: 478–479.

Javor, B.J. 1984. Growth potential of halophilic bacteria isolated from solar salt environments: carbon sources and salt requirements. *Applied and Environmental Microbiology* 48: 352–360.

Johnson, K.G. and **Lanthier, P.H.** 1986. β-Lactamases from *Actinopolyspora halophila*, an extremely halophilic actinomycete. *Archives of Microbiology* 143: 379–386.

Johnson, K.G., Lanthier, P.H., and **Gochnauer, M.B.** 1986a. Cell walls from *Actinopolyspora halophila*, an extremely halophilic actinomycete. *Archives of Microbiology* 143: 365–369.

Johnson, K.G., Lanthier, P.H., and **Gochnauer, M.B.** 1986b. Studies of two strains of *Actinopolyspora halophila*, an extremely halophilic actinomycete. *Archives of Microbiology* 143: 370–378.

Jørgensen, B.B. and **Des Marais, D.J.** 1986. Competition for sulfide among colorless and purple sulfur bacteria in cyanobacterial mats. *FEMS Microbiology Ecology* 38: 179–186.

Kamekura, M. and **Onishi, H.** 1974. Halophilic nuclease from a moderately halophilic *Micrococcus varians. Journal of Bacteriology* 119: 339–344.

Kamekura, M. and **Onishi, H.** 1976. Effect of magnesium and some nutrients on the growth and nuclease formation of a moderate halophile, *Micrococcus varians* var. *halophilus. Canadian Journal of Microbiology* 22: 1567–1576.

Kamekura, M. and **Onishi, H.** 1978. Flocculation and adsorption of enzymes during growth of a moderate halophile, *Micrococcus varians* var. *halophilus. Canadian Journal of Microbiology* 24: 703–709.

Kamekura, M., Wallace, R., Hipkiss, A.R., and **Kushner, D.J.** 1985. Growth of *Vibrio costicola* and other moderate halophiles in a chemically defined mineral medium. *Canadian Journal of Microbiology* 31: 870–872.

Kamekura, M., Bardocz, S., Anderson, P., Wallace, R., and **Kushner, D.J.** 1986. Polyamines in moderately and extremely halophilic bacteria. *Biochimica et Biophysica Acta* 880: 204–208.

Keller, P. and **Henis, Y.** 1967. The effect of yeast extract and casitone on the response to salt of the microflora of a highly saline soil. *Canadian Journal of Microbiology* 13: 1427–1432.

Klug, M., Boston, P., Francois, R., Gyure, R., Javor, B., Tribble, G., and **Vairavamurthy, A.** 1985. Sulfur reduction in sediments of marine and evaporite environments, pp. 128–157 in Sagen, D. (editor), *The Global Sulfur Cycle,* National Aeronautics and Space Administration Technical Memorandum 87570, Washington, D.C.

Kogut, M. and **Russell, N.J.** 1984. The growth and phospholipid composition of a moderately halophilic bacterium (*Vibrio costicola*) during adaptation to changes in salinity. *Current Microbiology* 10: 95–98.

Komaratat, P. and **Kates, M.** 1975. The lipid composition of a halotolerant species of *Staphylococcus epidermidis. Biochimica et Biophysica Acta* 398: 464–484.

Krumbein, W.E., Cohen, Y., and **Shilo, M.** 1977. Solar Lake (Sinai). 4. Stromatolitic cyanobacterial mats. *Limnology and Oceanography* 22: 635–656.

Kushner, D.J. 1968. Halophilic bacteria. *Advances in Applied Microbiology* 10: 73–97.

Kushner, D.J. 1978. Life in high salt and solute concentrations, pp. 317– 368 in Kushner, D.J. (editor), *Microbial Life in Extreme Environments,* Academic Press, London.

Kushner, D.J. and **Kamekura, M.** 1988. Physiology of halophilic eubacteria, p. 109–138 in Rodriguez-Valera, F. (editor), *Halophilic Bacteria,* Vol. I. CRC Press, Boca Raton.

Kushner, D.J., Hamaide, F., and **Macleod, R.A.** 1983. Development of salt-resistant active transport in a moderately halophilic bacterium. *Journal of Bacteriology* 153: 1163–1171.

Kushwaha, S.C., Gochnauer, M.B., Kushner, D.J., and **Kates, M.** 1974. Pigments and isoprenoid compounds in extremely and moderately halophilic bacteria. *Canadian Journal of Microbiology* 20: 241–245.

La Riviere, J.W.M. and **Schmidt, K.** 1981. Morphologically conspicuous sulfur-oxidizing eubacteria, pp. 1037–1048 in Starr, M.P., Stolp, H., Trüper, H.G., Balows, A. and Schlegel, H.G. (editors), *The Prokaryotes. A Handbook on Habitats, Isolation, and Identification of Bacteria,* vol. 1, Springer-Verlag, New York.

Marquez, M.C., Ventosa, A., and **Ruiz-Berraquero, F.** 1987. A taxonomic study of heterotrophic halophilic and non-halophilic bacteria in a solar saltern. *Journal of General Microbiology* 133: 45–56.

Masui, M. and **Wada, S.** 1973. Intracellular concentrations of Na^+, K^+, and Cl^- of a moderately halophilic bacterium. *Canadian Journal of Microbiology* 19: 1181–1186.

Matheson, A.T., Sprott, G.D., McDonald, I.J., and **Tessier, H.** 1976. Some properties of an unidentified halophile: growth characteristics, internal salt concentration, and morphology. *Canadian Journal of Microbiology* 22: 780–786.

Mathrani, I.M. and **Boone, D.R.** 1985. Isolation and characterization of a moderately halophilic methanogen from a solar saltern. *Applied and Environmental Microbiology* 50: 140–143.

Mathrani, I.M., Boone, D.R., Mah, R.A., Fox, G.E., and **Lau, P.P.** 1988. *Methanohalophilus zhilinae* sp. nov., an alkaliphilic, halophilic, methlotrophic methanogen. *International Journal of Systematic Bacteriology* 38: 139–142.

Measures, J.C. 1975. Role of amino acids in osmoregulation of non-halophilic bacteria. *Nature (London)* 257: 398–400.

Miller, K.J. 1985. Effects of temperature and sodium chloride concentration on the phospholipid and fatty acid compositions of a halotolerant *Planococcus* sp. *Journal of Bacteriology* 162: 263–270.

Miller, K.J. and **Leschine, S.B.** 1984. A halotolerant *Planococcus* from Antarctic Dry Valley soil. *Current Microbiology* 11: 205–210.

Morishita, H. and **Masui, M. (editors).** 1980. *Saline Environment. Physiological and Biochemical Adaptation in Halophilic Microorganisms.* Japanese Conference on Halophilic Microbiology. Osaka. 200 pp.

Noda, F., Hayashi, K., and **Mizunuma, T.** 1980. Antagonism between osmophilic lactic acid bacteria and yeasts in brine fermentation of soy sauce. *Applied and Environmental Microbiology* 40: 452–457.

Novitsky, T.J. and **Kushner, D.J.** 1975. Influence of temperature and salt concentration on the growth of a facultatively halophilic *"Micrococcus"* sp. *Canadian Journal of Microbiology* 21: 107–110.

Novitsky, T.J. and **Kushner, D.J.** 1976. *Planococcus halophilus* sp. nov., a facultatively halophilic coccus. *International Journal of Systematic Bacteriology* 26: 53–57.

Ohno, Y., Yano, I., Hiramatsu, T., and **Masui, M.** 1976. Lipids and fatty acids of a moderately halophilic bacterium, no. 101. *Biochimica et Biophysica Acta* 424: 337–350.

Ohno, Y., Yano, I., and **Masui, M.** 1979. Effect of NaCl concentration and temperature on the phospholipid and fatty acid composition of a moderately halophilic bacterium, *Pseudomonas halosaccharolytica. Journal of Biochemistry (Japan)* 85: 413–422.

Onishi, H. 1972. Halophilic amylase from a moderately halophilic *Micrococcus. Journal of Bacteriology* 109: 570–574.

Onishi, H. and **Kamekura, M.** 1972. *Micrococcus halobius* sp. n. *International Journal of Systematic Bacteriology* 22: 233–236.

Onishi, H. and **Hidaka, O.** 1978. Purification and properties of amylase produced by a moderately halophilic *Acinetobacter* sp. *Canadian Journal of Microbiology* 24: 1017–1023.

Onishi, H. and **Sonoda, K.** 1979. Purification and properties of an extracellular amylase from a moderate halophile, *Micrococcus halobius. Applied and Environmental Microbiology* 38: 616–620.

Onishi, H. and **Kamekura, M.** 1980. Characterization of salt response, lysis and flocculation of halophilic bacteria, pp. 111–121 in Morishita, H. and Masui, M. (editors), *Saline Environment. Physiological and Biochemical Adaptation in Halophilic Microorganisms,* Japanese Conference on Halophilic Microbiology, Osaka.

Onishi, H., Fuchi, H., Konomi, K., Hidaka, O., and **Kamekura, M.** 1980. Isolation and distribution of a variety of halophilic bacteria and their classification by salt-response. *Agricultural and Biological Chemistry* 44: 1253–1258.

Onishi, H., Mori, T., Takeuchi, S., Tani, K., Kobayashi, T., and **Kamekura, M.** 1983. Halophilic nuclease of a moderately halophilic *Bacillus* sp.: production, purification and characteristics. *Applied and Environmental Microbiology* 45: 24–30.

Oren, A. 1983. *Clostridium lortetii* sp. nov., a halophilic obligatory anaerobic bacterium producing endospores with attached vacuoles. *Archives of Microbiology* 136: 42–48.

Oren, A. 1986a. Intracellular salt concentrations of the anaerobic halophilic eubacteria

Haloanaerobium praevalens and *Halobacteroides halobius*. *Canadian Journal of Microbiology* 32: 4–9.

Oren, A. 1986b. The ecology and taxonomy of anaerobic halophilic eubacteria. *FEMS Microbiology Reviews* 39: 23–29.

Oren, A. 1987. A procedure for the selective enrichment of *Halobacteroides halobius* and related bacteria from anaerobic hypersaline sediments. *FEMS Microbiology Letters* 42: 201–204.

Oren, A., Paster, B.J., and **Woese, C.R.** 1984a. *Haloanoerobiaceae*: a new family of moderately halophilic, obligatory anaerobic bacteria. *Systematic and Applied Microbiology* 5: 71–80.

Oren, A., Weisburg, W.G., Kessel, M., and **Woese, C.R.** 1984b. *Halobacteroides halobius* gen. nov., sp. nov., a moderately halophilic anaerobic bacterium from the bottom sediments of the Dead Sea. *Systematic and Applied Microbiology* 5: 58–70.

Oren, A., Pohla, H., and **Stackebrandt, E.** 1987. Transfer of *Clostridium lortetii* to a new genus *Sporohalobacter* gen nov. as *Sporohalobacter lortetii* comb. nov., and description of *Sporohalobacter marismortui* sp. nov. *Systematic and Applied Microbiology* 9: 239–246.

Paterek, J.R. and **Smith, P.H.** 1985. Isolation and characterization of a halophilic methanogen from Great Salt Lake. *Applied and Environmental Microbiology* 50: 877–881.

Paterek, J.R. and **Smith, P.H.** 1988. *Methanohalophilus mahii* gen. nov., sp. nov., a methylotrophic halophilic methanogen. *International Journal of Systematic Bacteriology* 38: 122–123.

Peleg, E. and **Tietz, A.** 1971. Glycolipids of a halotolerant moderately halophilic bacterium. *FEBS Letters* 15: 309–312.

Petter, H.F.M. 1931. On bacteria of salted fish. *Koninklijke Akademie van Wetenschappen te Amsterdam. Proceedings,* Series B 58: 160–181.

Pfiffner, S.M., McInerney, M.J., Jenneman, G.E., and **Knapp, R.M.** 1986. Isolation of halotolerant, thermotolerant, facultative polymer-producing bacteria and characterization of the exopolymer. *Applied and Environmental Microbiology* 51: 1224–1229.

Potts, M. 1980. Blue-green algae (Cyanophyta) in marine coastal environments of the Sinai Peninsula; distribution, zonation, stratification and taxonomic diversity. *Phycologia* 19: 60–73.

Quesada, E., Ventosa, A., Rodriguez-Valera, F., and **Ramos-Cormenzana, A.** 1982. Types and properties of some bacteria isolated from hypersaline soils. *Journal of Applied Bacteriology* 53: 155–162.

Quesada, E., Ventosa, A., Rodriguez-Valera, F., Megias, L., and **Ramos-Cormenzana, A.** 1983. Numerical taxonomy of moderately halophilic Gram-negative bacteria from hypersaline soils. *Journal of General Microbiology* 129: 2649–2658.

Quesada, E., Ventosa, A., Ruiz-Berraquero, F., and **Ramos-Cormenzana, A.** 1984. *Deleya halophila,* a new species of moderately halophilic bacteria. *International Journal of Systematic Bacteriology* 34: 287–292.

Quesada, E., Valderrama, M.J., Bejar, V., Ventosa, A., Ruiz-Berraquero, F., and **Ramos-Cormenzana, A.** 1987. Numerical taxonomy of moderately halophilic Gram-negative nonmotile eubacteria. *Systematic and Applied Microbiology* 9: 132–137.

Rafaeli-Eshkol, D. 1968. Studies on halotolerance in a moderately halophilic bacterium. *Biochemical Journal* 109: 679–685.

Rafaeli-Eshkol, D. and **Avi-Dor, Y.** 1968. Studies on halotolerance in a moderately halophilic bacterium. *Biochemical Journal* 109: 687–691.

Robinson, J. 1952. The effects of salts on the nitritase and lactic dehydrogenase activity of *Micrococcus halodenitrificans*. *Canadian Journal of Botany* 30: 155–163.

Robinson, J. and **Gibbons, N.** 1952. The effect of salts on the growth of *Micrococcus halodenitrificans* n. sp. *Canadian Journal of Botany* 30: 147- 154.

Rodriguez-Valera, F. 1986. The ecology and taxonomy of aerobic chemoorganotrophic halophilic eubacteria. *FEMS Microbiology Reviews* 39: 17–22.

Rodriguez-Valera, F. 1988. Characteristics and microbial ecology of hypersaline environments, p. 3–30 in Rodriguez-Valera, F. (editor), *Halophilic Bacteria,* Vol. I. CRC Press, Boca Raton.

Rodriguez-Valera, F., Ruiz-Berraquero, F., and **Ramos-Cormenzana, A.** 1981. Characteristics of the heterotrophic bacterial populations in hypersaline environments of different salt concentrations. *Microbial Ecology* 7: 235–243.

Rodriguez-Valera, F., Ventosa, A., Juez, G., and **Imhoff, J.F.** 1985. Variation of environmental features and microbial populations with salt concentration in a multi-pond saltern. *Microbial Ecology* 11: 107–115.

Rosenberg, A. 1983. *Pseudomonas halodurans* sp. nov., a halotolerant bacterium. *Archives of Microbiology* 136: 117–123.

Rubentschik, L. 1926a. Über die Einwirkung von Salzen auf die Lebenstatigkeit der Urobakterien. *Zentralblatt für Bakteriologie II* 67: 167–194.

Rubentschik, L. 1926b. Über einige neue Urobakterienarten. *Zentralblatt für Bakteriologie II* 66: 161–180.

Rubentschik, L. 1929. Zur Nitrifikation bei hohen Salzkonzentrationen. *Zentralblatt für Bakteriologie II* 77: 1–18.

Rubentschik, L. 1933. Zur anaeroben Zelluosezersetzung in Salzseen. *Zentralblatt für Bakteriologie II* 88: 182–186.

Rubentschik, L. 1946. Sulfate reducing bacteria. *Microbiology* (English translation) 15: 443–456.

Rubentschik, L.J., Roisen, M.B., Bilansky, F.M., and **Shamis, D.L.** 1937. Denitrifying bacteria of the Odessa limans and of Odessa Bay. *Works of the Odessa State University. Biology* 2: 171–207.

Russell, N.J., Kogut, M., and **Kates, M.** 1985. Phospholipid biosynthesis in the moderately halophilic bacterium *Vibrio costicola* during adaptation to changing salt concentrations. *Journal of General Microbiology* 131: 781–790.

Sadler, M., McAninch, M., Alico, R., and **Hochstein, L.I.** 1980. The intracellular sodium and potassium composition of the moderately halophilic bacterium, *Paracoccus halodenitrificans. Canadian Journal of Microbiology* 26: 496–502.

Saslawsky, A.S. 1928. Zur Frage der Wirkung hoher Salzkonzentrationen auf die biochemischen Prozesse im Limanschlamm. *Zentralblatt für Bakteriologie II* 73: 18–28.

Shindler, D.B., Wydro, R.M., and **Kushner, D.J.** 1977. Cell-bound cations of the moderately halophilic bacterium *Vibrio costicola. Journal of Bacteriology* 130: 698–703.

Shkedy-Vinkler, C. and **Avi-Dor, Y.** 1975. Betaine-induced stimulation of respiration at high osmolarities in a halotolerant bacterium. *Biochemical Journal* 150: 219–226.

Skyring, G.W., Jones, H.E., and **Goodchild, D.** 1977. The taxonomy of some new isolates of dissimulatory sulfate-reducing bacteria. *Canadian Journal of Microbiology* 23: 1415–1425.

Starr, M.P. Stolp, H., Trüper, H.G., Balows, A., and **Schlegel, H.G.** (editors). 1981. *The Prokaryotes. A Handbook on Habitats, Isolation, and Identification of Bacteria.* Springer-Verlag. New York. 2284 pp.

Stern, N. and **Tietz, A.** 1973. Glycolipids of a halotolerant, moderately halophilic bacterium. I. The effect of growth medium and age of culture on lipid composition. *Biochimica et Biophysica Acta* 296: 130–135.

Takahashi, I. and **Gibbons, G.E.** 1959. Effect of salt concentration on the morphology and chemical composition of *Micrococcus halodenitrificans. Canadian Journal of Microbiology* 5: 25–35.

Unemoto, T. and **Hayashi, M.** 1979. Regulation of internal solute concentrations of marine *Vibrio alginolyticus* in response to external NaCl concentration. *Canadian Journal of Microbiology* 25: 922–926.

Van Qua, D., Simidu, U., and **Taga, N.** 1981. Purification and some properties of halophilic protease produced by a moderately halophilic marine *Pseudomonas* sp. *Canadian Journal of Microbiology* 27: 505–510.

Vargues, H. 1962. Contribution à l'étude du caractère halophile chez les bacteries isolées due milieu marin. *Bulletin de l'Institution Océanographique* 1231, 176 pp.

Venkataraman, R. and **Sreenivasan, A.** 1954. Salt tolerance of marine bacteria. *Food Research* 19: 311–313.

Ventosa, A. 1988. Taxonomy of moderately halophilic heterotrophic eubacteria, p. 71–84 in Rodriguez-Valera, F. (editor), *Halophilic Bacteria,* Vol. I. CRC Press, Boca Raton.

Ventosa, A., Quesada, E., Rodriguez-Valera, F., Ruiz-Berraquero, F., and **Ramos-Cormenzana, A.** 1982. Numerical taxonomy of moderately halophilic Gram-negative rods. *Journal of General Microbiology* 128: 1959–1968.

Ventosa, A., Ramos-Cormenzana, A., and **Kocur, M.** 1983. Moderately halophilic Gram-positive cocci from hypersaline environments. *Systematic and Applied Microbiology* 4: 564–570.

Villar, M., de Ruiz Holgado, A.P., Sanchez, J.J., Trucco, R.E., and **Oliver, G.** 1985. Isolation and characterization of *Pediococcus halophilus* from salted anchovies (*Engraulis anchoita*). *Applied and Environmental Microbiology* 49: 664–666.

Vreeland, R.H. and **Martin, E.L.** 1980. Growth characteristics, effects of temperature, and ion specificity of the halotolerant bacterium *Halomonas elongata. Canadian Journal of Microbiology* 26: 746–752.

Vreeland, R.H., Litchfield, C.D., Martin, E.L., and **Elliot, E.** 1980. *Halomonas elongata,* a new genus and species of extremely salt-tolerant bacteria. *International Journal of Systematic Bacteriology* 30: 485–495.

Vreeland, R.H., Mierau, B.D., Litchfield, C.D., and **Martin, E.** 1983. Relationship of the internal solute composition to the salt tolerance of *Halomonas elongata. Canadian Journal of Microbiology* 29: 407–414.

Vreeland, R.H., Anderson, R., and **Murray, R.G.E.** 1984. Cell wall and phospholipid composition and their contribution to the salt tolerance of *Halomonas elongata. Journal of Bacteriology* 160: 879–883.

Ward, D.M. and **Brock, T.D.** 1978. Hydrocarbon biodegradation in hypersaline environments. *Applied and Environmental Microbiology* 35: 353–359.

Weisser, J. and **Trüper, H.** 1985. Osmoregulation in a new haloalkaliphilic *Bacillus* from the Wadi Natrun (Egypt). *Systematic and Applied Microbiology* 6: 7–11.

Zeikus, J.F., Hegge, P.W., Thompson, T.E., Phelps, T.J., and **Langworthy, T.A.** 1983. Isolation and description of *Haloanaerobium praevalens* gen. nov. and sp. nov., an obligately anaerobic halophile common to Great Salt Lake sediments. *Current Microbiology* 9: 225–234.

Zhilina, T.N. 1986. Methanogenic bacteria from hypersaline environments. *Systematic and Applied Microbiology* 7: 216–222.

ZoBell, C.E. 1958. Ecology of sulfate reducing bacteria. *Producers' Monthly* 22: 12–29.

6

Halophilic Archaebacteria

6.1 General physiology and biochemistry of *Halobacterium*

The bacteria to be discussed in this chapter are so widespread in high salt environments that they virtually define the hypersaline niche. Often highly pigmented, they are mainly responsible for the intense reddish color of salterns and salt lakes. Indeed, solar salt crystals are themselves frequently pink in color due to entrapped halobacteria, and cultures can often be prepared from such sources. Fish preserved in brines frequently exhibit extensive halobacterial growth and the first pure cultures were isolated from these sources.

Halophilic archaebacteria include heterotrophic halobacteria and methanogens. In terms of ecophysiology, halobacteria probably perform many of the same metabolic functions as halophilic eubacteria. Factors involved in the competition between the two major groups may involve different tolerances to salinity, oxygen, concentration of dissolved organic carbon, temperature, and other chemical and physical parameters of natural hypersaline environments. However, *in situ* studies on the ecophysiology of these organisms are lacking. As obligate anaerobes, methanogens probably also face competitive challenges from anaerobic eubacteria, but relatively little is known about either group of anaerobes. The lipids produced by archaebacteria may prove to be the best evidence demonstrating the presence of halobacteria and methanogens in both ancient hypersaline environments and in modern evaporite environments in which the microbiology has not been studied. The following discussion highlights the distinctive nature of the archaebacteria and the physiological and biochemical attributes important for understanding how they thrive and survive under extremely hypersaline conditions in nature.

Halobacteria are generally considered to be extreme halophiles, in most cases requiring at least 12–15% salt for growth. A wealth of studies have demonstrated the unique adaptations as well as the variety of strategies for growth and survival in hypersaline habitats by halobacteria. The present chapter concentrates on those studies that are of ecological or biogeochemical interest. Much of the literature has already been compiled in general reviews and articles addressing specific aspects of halobacteria, the most recent including those of Kushner (1978, 1985), Dundas (1977), Larsen (1981), Brown (1983), and Hochstein (1988). Reviews of more specific topics include those on taxonomy (Larsen, 1984; Juez, 1988), tolerance to alkaline conditions (Grant and Tindall, 1986; Tindall, 1988), ecophysiology (Tindall and Trüper, 1986; Rodriguez-Valera, 1988), cell walls (Kandler, 1982; Wieland, 1988), salt tolerance (Lanyi, 1976, 1980), genetics and information transfer (Bayley, 1976; Pfeifer, 1988), lipids (Langworthy et al., 1982; De Rosa et al., 1986; Kamekura and Kates, 1988), bacteriorhodopsin (Stoeckenius et al., 1979, Stoeckenius and Bogomolni, 1982; Wagner and Linhart, 1988), and phototrophic growth (Oesterhelt and Krippahl, 1983). A brief general description of halobacteria will be included here, with more extensive discussions addressing the interrelationships of halobacteria and their environments.

Halobacteria are archaebacteria, a classification they share with methanogens and certain thermophiles. Archaebacteria share certain cellular features that separate them from the majority of bacteria, the eubacteria. Archaebacterial lipids are glycerol diethers linked to isoprenoid alcohols containing 20, 25, or 40 carbon atoms. Halobacteria produce C_{50} carotenoid pigments (bacterioruberins). In contrast, eubacterial lipids are based predominantly on ester linkages formed by the condensation of alcohols and fatty acids. One of the best known exceptions is a thermophilic, anaerobic, sulfate-reducing bacterium that produces glycerol diethers (Langworthy et al., 1983). Halobacteria have C_{30} squalenes and C_{20} diether isoprenoids whereas methanogens produce C_{14} to C_{30} isoprenoids. Haloalkalophilic archaebacteria (see Section 6.3) produce asymmetric isoprenoid ethers. These membrane features are important both as taxonomic and biogeochemical indicators providing evidence that archaebacteria thrive or thrived in hypersaline environments. Halobacterial lipids have been identified in Dead Sea sediments (Kaplan and Baedecker, 1970b; Nissenbaum et al., 1972; Anderson et al., 1977).

Archaebacteria also differ from eubacteria on the level of information transfer, with differences in ribosomal composition, transfer RNAs, RNA polymerases, and translation systems. They have been found to have a unique 7S non-ribosomal RNA (Luehrsen et al., 1985). Unlike eubacteria, archaebacteria lack peptidoglycan in their cell walls. In some respects they resemble eucaryotes, having histone-like proteins and comparable ribosomal protein sequences as well as exhibiting a sensitivity to certain antibiotics such as aphidicolin (Forterre et al., 1984; Schinzel and Burger, 1984). Although most halobacteria are sensitive to many of the antibiotics that inhibit other bacteria such as bacitracin, chloramphenicol, anisomycin, and rifampicin, halobacteria and other archaebacteria are insensitive to some antibiotics that inhibit eubacteria, including the peptidoglycan cell wall inhibitor penicillin (Pecher and Böck, 1981; Bonelo et al., 1984). Selective use of cell-wall-active antibiotics in enrichment cultures can be a valuable tool for the specific enrichment of halophilic archaebacteria in the presence of eubacteria.

All known archaebacteria thrive only in "extreme" environments: high salt, high temperature, or low redox potential. They are thought to belong to a branch of bacterial evolution that diverged early from the mainstream eubacterial line (Woese, 1987). The positions of the various branches of the archaebacterial kingdom with relation to the evolution of procaryotes and eucaryotes are questions of debate and will not be discussed here.

Halobacteria differ from other archaebacteria by the presence in some members of a unique, retinal-based protein pigment, bacteriorhodopsin. The purple pigment (absorption maximum of 568 nm in the unexcited state) mediates a light-driven proton pump that drives ATP synthesis. In cells that synthesize bacteriorhodopsin, photophosphorylation can augment growth. Unlike the phototrophic eubacteria, the pigment in the halobacteria is located in the membrane of the cell envelope rather than on intracytoplasmic membranes. Other rhodopsin pigments in some halobacteria include halorhodopsin (a chloride pump) and sensory rhodopsin (involved in phototaxis). The red color typical of most halobacteria is not due to bacteriorhodopsin, but rather to the carotenoids β-carotene and bacterioruberin.

Some halobacteria are motile by means of one or more flagella. The flagellins contain sulfate oligosaccharides (Wieland et al., 1985). Some halobacteria produce gas vacuoles to regulate their buoyancy. Natural brines in which halobacteria thrive are often of such high specific gravity that cells remain buoyant without gas vacuoles (Javor, personal observation). Optimal temperature for growth is usually about 40°–50°C.

The membranes and proteins of halobacteria are well adapted for the highly ionic environments in which halobacteria are found. Polar lipids always exceed non-polar lipids and acidic amino acids always exceed neutral and basic amino acids. The negatively charged residues are required for ionic shielding to maintain the protein stability. The low proportion of non-polar amino acids in the highly ionic environment is believed to be necessary to induce hydrophobic bond formation within the proteins. Many halobacterial proteins dissociate upon dilution. The dissolution of the cell envelope (lipids plus proteins) in diluted brines, in addition to the inactivation and dissociation of enzymes and ribosomes, accounts for the inability of many halobacteria to thrive in habitats of widely varying salinity.

Halobacteria maintain osmotic balance in brines by accumulating inorganic ions rather than forming an organic compatible solute. Christian and Waltho (1962) found that *Halobacterium salinarium* accumulated 4.6 M K^+. Many soluble and membrane-bound enzymes of extreme halophiles show maximal activity at high salt concentrations, and are irreversibly inactivated by exposure to low salt. However, glycine betaine, a compatible solute, has been detected in both *H. salinarium* and in alkalophilic halobacteria (De Rosa et al., 1988), but is not present in sufficient concentration to balance the external osmotic pressure. Halobacteria synthesize little or no polyamines (Chen and Martinowicz, 1984; Hanana et al., 1985; Carteni-Farina et al., 1985; Kamekura et al., 1986).

Halobacteria and halococci have been isolated from a variety of highly saline environments, including natural salinas and solar salt ponds of marine origin, rock salt, hypersaline soils, inland salt lakes, and salted fish, hides, bacon, and sausage. There are no published reports of halobacteria isolated from deep sea brines, Antarctic hypersaline lakes, and certain highly salted food products (cheese, pickles, and soy sauce mash). Three described genera of non-alkalophilic halobacterial rods or pleomorphic forms are in the current literature: *Halobacterium, Haloarcula,* and *Haloferax* (Table 6.1). *Halococcus,* which is more tolerant of brine dilution that *Halobacterium,* has also been isolated from seawater (Rodriguez-Valera et al., 1979). In solar salt ponds, both halobacterial and eubacterial populations are present (Rodriguez-Valera et al., 1981, 1985) (see Chapter 5).

Table 6.1 Growth attributes of various halobacteria[a]

Species	Carbon sources for growth		bR[b]	Source	Reference
	Amino acids	Carbohydrates			
Halobacterium					
H. salinarium[c]	+	−	+	Salt ponds	1, 9
H. saccharovorum	−	+	−	Salt ponds	13
H. sodomense	+	+	+	Dead Sea	11
"H. marismortui"	+	+	−	Dead Sea	2, 4
H. denitrificans	−	+	ND	Salt ponds	14
Haloferax					
H. gibbonsii	+	+	ND	Salt ponds	6
H. volcanii	+	+	−	Dead Sea	4, 10
H. mediterranei	−	+	ND	Salt ponds	12
Haloarcula					
H. hispanica	+	+	ND	Salt ponds	6
H. vallismortis	−	+	−	Inland brine	3, 15
"H. californiae"	+	+	+	Salt ponds	4, 5
"H. sinaiiensis"	+	−	+	Gavish Sabkha	5, 7
Halococcus					
H. morrhuae	+	+	−	Various	4, 8

[a]ND = not determined. References noted by numbers: (1) Colwell et al., 1979; (2) Ginzburg et al., 1970; (3) Gonzalez et al., 1978; (4) Javor, 1984; (5) Javor et al., 1982; (6) Juez et al., 1986; (7) Kessel et al., 1985; (8) Kocur and Hodgkiss, 1973; (9) Larsen, 1984; (10) Mullakhanbhai and Larsen, 1975; (11) Oren, 1983c; (12) Rodriguez-Valera et al., 1985; (13) Tomlinson and Hochstein, 1976; (14) Tomlinson et al., 1986; (15) Juez, 1988.
[b]bR, Bacteriorhodopsin.
[c]Includes strains previously described as *H. halobium* and *H. cutirubrum*.

Before about 1960, many of the reports describing the isolation of red halophiles described general growth and physiology. Some of these reports are particularly interesting because they demonstrated the degree of nutritional versatility and the ability for anaerobic growth or metabolism in a variety of extreme halophiles. Schoop (1934a, 1934b, and 1935) conducted studies on obligate halophilism in fungi and red and colorless bacteria. He noted that obligate halophilism was probably first recognized by Le Dantec in 1906, who called such miroorganisms "chlorurophiles." Schoop's red isolates from salted fish were cocci (called *"Micrococcus litoralis"*). Many strains were reported to produce acid from sugars (indicative of incomplete oxidation of sugars or anaerobic fermentation) and they showed the ability to denitrify. Elazari-Volcani (1940) reported the isolation of two red halophiles from the Dead Sea that could denitrify, and one of those isolates also produced acid from sugars. Petter (1931) described red and orange rods and a red sarcina (coccus) from salted fish and solar salt. The bacteria grew optimally at 37°C. They could not ferment sugars but they did reduce nitrate to nitrite.

Extreme halophiles that could grow on glucose and chitin were isolated from salted hides (Stuart 1935, 1936). On solid glucose medium the cells were spindle-shaped rods that piled up as they grew and formed sessile fruiting bodies or cysts. In some cases, the formation of cyst-like colonies was preceded by a peculiar type of growth in which no well-defined bacterial cells could be discerned. Similar morphogenesis was noted in an isolate from a salina (Wais, 1985) but it has not been further studied in halobacteria.

The interrelated effects of Eh, pH, NaCl concentration, and protein concentration were examined in a red coccus, *"Sarcina littoralis"* (probably *Halococcus morrhuae*) by Stuart (1941a, 1941b) and Stuart and James (1938a, 1938b). In these studies it was found that the combination of high pH and high salt caused a lowering of the redox potential (oxygen solubility is lower in high salt). Pigment production only occurred in high salt or relatively high redox potential. None of the isolates grew anaerobically. The bacteria grew luxuriantly in 3–3.5 M NaCl in the presence of 5–10% dissolved protein, but only at lower protein concentrations in 4–5 M NaCl. The poor solubility of proteins (salting-out effect) was believed to have removed sufficient salt to inhibit growth in high-protein, high-salt medium.

Venkataraman and Sreenivasan (1954, 1956) described several red halophilic bacteria isolated from salted fish. Several strains were found to denitrify and three strains (called *"Sarcina"* sp.) definitely showed the ability to ferment sugars. Aerobically most of the isolated cocci raised the pH when growing on sugars but none of the isolated rods produced acid. The ability to respire various sugars was investigated in two extremely halophilic rods and in *Sarcina littoralis* by Katznelson and Robinson (1956). All cultures respired glycerol but only *S. littoralis* showed the ability to respire sugars. Anaerobic respiration was not discussed.

The ability of halobacteria to ferment and to grow anaerobically has been further elucidated for nitrate reduction (Werber and Mevarech, 1978; Hochstein and Tomlinson, 1985; Mancinelli and Hochstein, 1986; Tomlinson et al., 1986), dark fermentative growth (Javor, 1984), and light-stimulated fermentative growth (Hartmann et al., 1980; Oesterhelt, 1982; Rodriguez- Valera et al., 1983b). In spite of a growing body of evidence that halobacteria as a group are facultative anaerobes, they still are reported to be strict aerobes (Brown, 1983) or that only some strains can grown anaerobically (by respiration of nitrate) (Larsen, 1981). Fermentation as an alternative way of life is not generally recognized. This misconception probably arose because *Halobacterium salinarium* and related strains (*H. halobium* and *H. cutirubrum*) have been the focus of most modern studies of halobacteria, and these strains have limited or no ability to grow anaerobically in the dark. No obligately anaerobic halobacteria have ever been described.

Another popular misconception is that halobacteria grow primarily on amino acids and have a limited ability to attack carbohydrates. This is certainly true for *H. halobium* and related strains. In addition to the other isolates listed above, many well-characterized strains show exceptional ability to grow at the expense of carbohydrates: *Halobacterium saccharovorum* (Tomlinson and Hochstein, 1972, 1976; Tomlinson et al., 1974), "*H. marismortui*" (Javor, 1984), *H. denitrificans* (Tomlinson et al., 1986), *H. sodomense* (Oren, 1983c); *Haloarcula vallismortis* (Gonzalez et al., 1978), strains of *Haloarcula* (Javor, 1984, Juez et al., 1986); *Haloferax mediterranei* (Rodriguez-Valera et al., 1983a), *H. volcanii* (Javor, 1984), and *H. gibbonsii* (Juez et al., 1986) (Table 6.1). Javor (1984) showed that of 18 strains of rods, cocci, and square extreme halophiles isolated from four environments, 13 grew on sugars. Of those 13 strains, 4 strains grew poorly or not at all on amino acids (peptone or casamino acids plus yeast extract). Nearly all strains grew on glycerol or pyruvate, and some strains grew on acetate, citrate, succinate, or lactate.

Halobacteria have been shown to fix CO_2 in the presence of propionate, and in some cases CO_2 assimilation is stimulated in the light (Danon and Caplan, 1979; Oren, 1983b; Javor, 1988). CO_2 assimilation in a strain of *Haloarcula* that lacked bacteriorhodopsin was inhibited by light (Javor, 1988). There is no evidence that halobacteria fix CO_2 by a reductive pathway. The study by Javor indicates that propionate-stimulated CO_2 fixation is an anaplerotic process that is probably linked to amino acid synthesis.

Studies of carbon metabolism in halobacteria indicate that these bacteria are largely similar to aerobic eubacteria. Halobacteria have all the enzymes of the tricarboxylic acid cycle (Aitken and Brown, 1969; Danson et al., 1985). Extracellular protease (Norberg and Hofsten, 1969), amylase (Good and Hartmann, 1970), and amyloglucosidase (Oren, 1983a) have been partially characterized and numerous other enzymatic activities have been noted in the course of the biochemical classification of all published species. Two carbohydrate-utilizing halobacteria have been shown to accumulate β-hydroxybutyrate under conditions of nitrogen limitation and abundant organic carbon

(Fernandez-Castillo, et al., 1986). However, novel or modified pathways of metabolism have been noted during carbohydrate catabolism (Tomlinson et al., 1974), cysteine degradation (Newton and Javor, 1985), and CO_2 fixation (Javor, 1988).

Halobacteria are generally believed to be heterotrophs that require undefined complex media (Larsen, 1981) or defined media with many required amino acids and vitamins (Onishi et al., 1965). However, *H. mediterranei* has been shown to grow well on single carbon sources in rather simple inorganic media (Rodriguez-Valera et al., 1980). In both this study and others that have demonstrated carbohydrate utilization, the inclusion of high concentrations of buffer in the medium has been required to prevent bacterial acidification. It is possible that growth tests on carbohydrates were negative in many other reports because the bacteria rapidly acidified the medium before growth could be detected. When amino acids are present in the medium, deamination and release of ammonia would prevent this acidification.

Salt requirements for halobacteria have been shown in numerous studies. Among the non-alkalophilic extreme halophilies, all require both the sodium and the chloride ion. Halobacteria usually grow optimally in 3–4 M NaCl. Most strains can grow in $\geq$5 M (saturated) salt. The ability to grown in $\leq$2.5 M NaCl depends on the strain, the temperature, and other salts in solution. Halobacterial rods tend to round up into spheres or otherwise show pleomorphy in less concentrated brines. This phenomenon makes even gross identification of field material very difficult.

In natural, marine-derived brines, most extreme halophiles studied by Javor (1984) were inhibited by 500 mM Mg^{2+} introduced as Cl^- and especially as SO_4^{2-} salts. Such natural brines are saturated with respect to NaCl when concentrations of Mg^{2+} are $\geq$500 mM. In contrast, isolates from the Dead Sea (*H. volcanii* and *H. sodomense*) have high requirements and a broad tolerance of Mg^{2+} (0.075–1.5 M and 0.6–1.2 M Mg^{2+}, respectively) (Mullakhanbhai and Larsen, 1975; Cohen et al., 1983; Oren, 1983c). Optimal Mg^{2+} concentration for *H. volcanii* is 100–500 mM. The inhibition of growth caused by low Mg^{2+} can be partially relieved by as little as 1 mM Ca^{2+}. Mg^{2+} has been shown to be essential for motility (Baryshev, 1982) and for maintaining cell membrane and envelope integrity (Rayman et al., 1967; Matsuoka et al., 1981), but the mechanism causing the negative effects of high extracellular concentrations of Mg^{2+} has not been described. Likewise, the mechanism of inhibition by high concentrations of SO_4^{2-} is not known. Because Mg^{2+} and SO_4^{2-} are divalent ions, their presence may upset the charge balance in or on the cells.

Edgerton and Brimblecombe (1981) constructed a thermodynamic model of halobacterial environments to take into account charge concentrations and mole fractions of monovalent ions. By plotting the growth areas of halobacterial isolates with respect to these factors, they found that the environmental limits for the isolates were not easily defined by either Na^+ or Cl^- activity.

The model is instructive for determining the activity of specific ions at different salt concentrations and as a tool for understanding the effects of components of natural brines on cell growth. Other ecological aspects of halobacteria are discussed in the chapters concerning the environments in which they are found.

Antibiotic activity possibly produced by some strains of red halophiles against other strains was noted by Dussault (1954). Halocins, proteins with antibiotic activity secreted by halobacteria, are produced by *Haloferax mediterranei* (Rodriguez-Valera et al., 1982a; Meseguer and Rodriguez-Valera, 1985). One halocin inhibited other halobacterial rods, but not halococci or eubacterial halophiles. The protein is adsorbed by the target cell and induces lysis.

Environmental conditions influence pigment synthesis in halobacteria. *H. mediterranei* produced much more bacterioruberin and β-carotene in 15% salt than in 25% or 35% salt (Kushwaha et al., 1982). In contrast, *H. cutirubrum* produced bacterioruberin and β-carotenes only when salt concentrations exceeded 15%. In a study of 47 strains of red extreme halophiles, Gibbons (1958) noted that increasing salt concentrations caused some bacteria to lose color and in others it caused them to be redder. In *H. halobium* and *H. cutirubrum* grown in defined medium, the presence of 0.1% glycerol stimulated growth but caused the cells to lose their carotenoids (Gochnauer et al., 1972). In complex medium, growth in the presence of 0.1% glycerol did not change the carotenoid content of the cells but growth with glucose did. In *H. cutirubrum,* Kushwaha and Kates (1979) found that the concentration of bacterioruberin decreased in the presence of 0.1–0.5% glycerol but the content of C_{40} carotenoids (β-carotene and lycopene) increased.

Carotenogenesis requires aerobic conditions (Tornabene, 1978; Tomlinson et al., 1986). In *H. cutirubrum,* the decrease in isoprenoid pigments with decreases in ambient O_2 was accompanied by decreases in total squalene, dihydrosqualene, and tetrahydrosqualene content of the cells. However, the relative cellular concentrations of tetrahydrosqualene to total squalene increased with anaerobiosis, and the relative cellular concentration of dihydrosqualene was maximal under microaerophilic conditions. Similarly, bacteriorhodopsin requires aerobic conditions for the synthesis of precursors, since oxygen is necessary for the formation of retinal from β-carotene (Hartmann et al., 1980).

Bacteriorhodopsin has been found only in several strains or species of *Halobacterium* and *Haloarcula*. Apart from the numerous purely biochemical, biophysical, and genetic studies of bacteriorhodopsin, some investigations have addressed the physiological and ecological function of the pigment. Oesterhelt and Krippahl (1973) found that 560 nm light inhibited respiration in cells with the pigment. Light was found to be an energy source for cultures of bacteriorhodopsin-containing *H. halobium* (Hartmann et al., 1980; Rodriguez-Valera et al., 1983b). Light was also found to stimulate CO_2 fixation in bacteriodopsin-containing cells (Danon and Caplan, 1979; Oren, 1983b; Javor,

1988). When bacteriorhodopsin-containing cells were maintained under starvation conditions, light in combination with O_2 accelerated cell death (Brock and Peterson, 1976). Under anaerobic conditions, cell death was retarded in the light but cells died rapidly in the dark. Rhodopsin pigments of halobacteria have also been shown to be involved in photosensory and phototactic responses toward yellow-green light and away from ultraviolet and blue light (Dencher and Hildebrand, 1979, 1982; Wagner, 1984; Spudich, 1985; Plotkin et al., 1985; Takahashi et al., 1985a, 1985b). Bacteriorhodopsin has been found in natural populations of halobacteria in the Dead Sea (Oren and Shilo, 1981) and in solar salterns (Javor, 1983).

The number of genetic studies of halobacteria has greatly increased since Moore and McCarthy (1969a, 1969b) found that extrachromosomal DNA constitutes up to 11–36% of the total DNA in halobacteria. While the chromosomal DNA has a G+C base composition of 66–68 mol%, the extrachromosomal DNA has a G+C composition of 57–60 mol%. Much of the extrachromosomal DNA is believed to be found in plasmids in the cells. Gutierrez et al. (1986) found megaplasmids (100 mDal to over 300 mDal) in halobacteria. Many of the studies related to both chromosomal and plasmid sequences and activity have been summarized by Kushner (1985) and Pfeifer (1988). Of particular physiological or ecological significance is the finding of high rates of spontaneous mutations and that these mutations occur at "hot spots" in the DNA (Pfeifer et al., 1981; Sapienza and Doolittle, 1982).

Phages may prove to be important vehicles for information transfer in halobacteria. A variety of phages specific for halobacteria have been isolated from various sources: as a contaminant in a maintained culture (Torsvik and Dundas, 1974, 1980; Schnabel et al., 1982a, 1982b; Schnabel and Zillig, 1982, 1984; Schnabel, 1984a, 1984b), from fermented fish sauce (Pauling, 1982; Rohrmann and Cheney, 1983), and from solar salt pond sediments (Wais et al., 1975; Wais and Daniels, 1985). These phages have linear, double-stranded DNA and they all require high salt concentrations for stability. The studies by Schnabel and co-workers have provided a detailed molecular characterization of phage ϕH of *H. halobium*. It has a molecular weight of 39 kdal, a G+C content of 65 mol%, and shows a partial homology with the AT-rich regions of the host extrachromosomal DNA. Variants in the phages occurred spontaneously by different insertions, deletions, or inversions in its DNA.

Wais and Daniels (1985) found that natural populations of halobacterial phages increased dramatically after heavy rainfall caused a brief period of brine dilution in solar salt ponds. The dilution was accompanied by mass death of the halobacterial population. They found that the phage isolates failed to produce plaques in media with high salt. High degrees of virulence were found only when the phage was introduced in cultures at a relatively high multiplicity of infection. The authors concluded that in nature the phages probably change from prophage to lytic form when halobacterial populations are subjected to brine dilution. Once the population has lysed (either due to phage activity or to envelope dissolution in dilute solution), the phages may act as DNA "reservoirs" for the ensuing halobacterial populations when brines have again concentrated enough to support halobacteria.

6.2 Taxonomy of the Halobacteriaceae

Non-alkalophilic and alkalophilic halobacteria both belong to the family Halobacteriaceae. The review by Juez (1988) on the taxonomy of the Halobacteriaceae includes a compilation of accepted generic and specific names of this family. Four genera of non-alkalophilic halobacteria are currently accepted: *Halobacterium, Halococcus,* and *Haloarcula* and *Haloferax. Haloferax* has been introduced as a genus to describe rods that are different from *Halobacterium* based on polar lipid composition, nucleic acid differences, tolerance of lower salt concentrations, and numerical taxonomy (Torreblanca et al., 1986; Juez et al., 1986; Juez, 1988). Immunologic differences have also been noted (Conway de Macario et al., 1986). Table 6.1 lists some of the attributes of recognized species and strains of halobacteria. Tindall and Trüper (1986) also classified most of these species according to their ecophysiology. The non-alkalophilic halobacteria fell into seven distinct groups.

It should be noted that numerous isolates of halophiles referred to only by strain number are often used for microbiological studies. In addition, rather broad phenotypic differences are often noted between strains of the same species. "*Halobacterium marismortui*" (also called "*Halobacterium* of the Dead Sea)," a well-studied microorganism, has been given the name of an isolate previously described by Elazari-Volcani (1940) but which was subsequently lost.

Among the rod-shaped halobacteria, a broad variety of physiological types are known (see Table 6.1). Some are strictly proteolytic, some are facultatively proteolytic with ability to grow on glycerol and pyruvate as well, some are strict carbohydrate-utilizers, and some can grow on a broad variety of substrates. An unnamed isolate was noted to grow only on acetate and pyruvate (Javor, 1984). Bacteriorhodopsin has been found in only some species and only some strains produce gas vacuoles. Both strict aerobes and facultative anaerobes (growth by fermentation or nitrate respiration) have been described. Anaerobic growth may be important for these bacteria in hypersaline environments because the solubility of oxygen is very low in strong brines.

Square-shaped bacteria were noted by Walsby (1980) in a brine pool of the Gavish Sabkha, Sinai. The bacteria appeared to divide in two dimensions, giving rise to floating, gas-vacuolated, "postage stamp"-like sheets of cells (Figure 6.1). Similar microorganisms were described from a salt lake (Romanenko, 1981). The ultrastructure of the Sinai material was described by Parkes and Walsby (1981), Stoeckenius (1981), and Kessel and Cohen (1982).

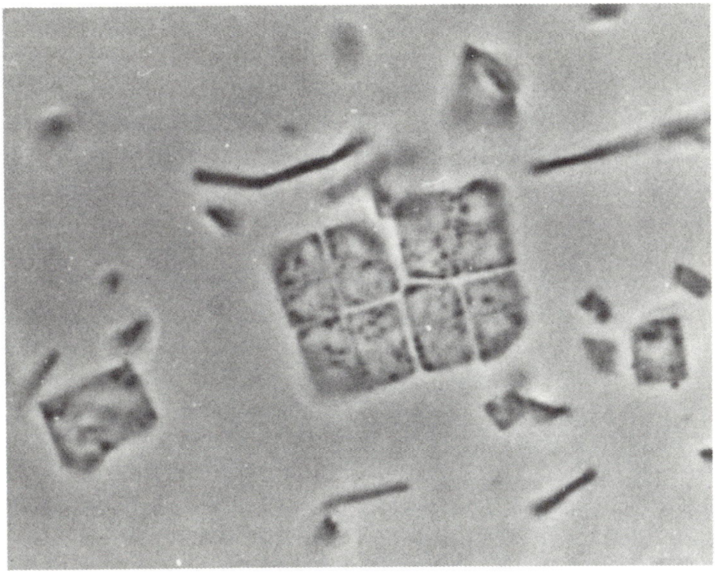

Figure 6.1 Square-shaped bacteria. Phase-contrast photomicrograph. Cells in the center are about 2 μm wide.

Electron micrographs showed the cells to be flat, rectangular boxes, 0.25 μm thick and up to 4 μm or more on a side. A bacteriorhodopsin-like pigment was detected in the natural population and an isolated strain was found to have patches of purple membrane (Stoeckenius et al., 1985). A strain from Mexico had a similar pigment apparently dispersed in the membrane (Javor, et al., 1982). Motility and flagellar composition in a Gavish Sabkha isolate were described by Alam et al. (1984). While motility appeared similar to that of other halobacteria, protein composition appeared markedly different.

Square-shaped or box-shaped bacteria were isolated by Javor et al. (1982) but they were somewhat smaller and more variable than the natural populations. Gas vacuoles could not be induced. The generic name *Haloarcula* was suggested for these isolates. Taxonomic studies by Torreblanca et al. (1986) and Juez et al. (1986) described the genus *Haloarcula* including an additional species from a Spanish solar saltern, *H. hispanica*.

Although halococci have long been noted among the extreme halophiles, they are the least studied. Besides their morphology, they differ from rod- and box-shaped halobacteria in several respects: resistance to lysis as well as relatively high metabolic rates in diluted medium (Rodriguez-Valera et al., 1982b); the nature of the complex amino sugar composition of their relatively thick cell walls (Reistad, 1975; Steber and Schleifer, 1975; Hunter and Millar, 1980); and a somewhat lower G+C composition of the major component of the DNA (60.5–65.8 mol %) (Kocur and Bohacek, 1972) than that of *Halo-*

bacterium (66–68 mol %) (Larsen, 1984). Bacteriorhodopsin has never been found in the halococci.

All strains of halococci have been lumped into one taxon, *Halococcus morrhuae* (Kocur and Hodgkiss, 1973), an accepted classification in the 9th edition of Bergey's manual (Larsen, 1984). This classification includes all strains previously referred to as species of red halophilic "*Sarcina*" and red halophilic "*Micrococcus.*" The differentiation between these strains originally was based on phenotypic differences in colony formation: "*Micrococcus*"-type colonies were small, round, and smooth, being composed of single or paired cells, or cells in tetrads. "*Sarcina*"-type colonies were larger, rugose, and coarsely granular, being composed of large bundles or packets of cells (Elazari-Volcani, 1940).

In spite of clear morphological differences, there are physiological and biochemical similarities between many strains of the two types of halococci (Kocur and Bohacek, 1972; Kocur and Hodgkiss, 1973; Montero et al., 1988). DNA-16S rRNA hybridization studies by Ross and Grant (1985) showed that all halococci were distinct from rod-shaped extreme halophiles, but no comparisons were made between the morphologically distinct halococci (only two strains were compared). Phylogenetic relationships determined by 5S rRNA "fingerprinting" also showed halococci to be different from the rods, but further differences between different strains of *Halococcus* were not assayed (Nicholson and Fox, 1983).

Kocur and Hodgkiss (1973) and Montero et al. (1988) stated that all strains of halococci were strict aerobes, but the latter investigators found evidence of anaerobic metabolism in many strains (reduction of nitrate to nitrite or gas, and reduction of cysteine to H_2S). Javor (1984) and Montero et al. (1988) found that some non-alkalophilic halococci could grow on a variety of organic carbon compounds, including carbohydrates and low-molecular-weight carboxylic acids. In addition, Javor (1984) found some strains that could grow as facultative anaerobes. These data support similar findings of Venkataraman and Sreenivasan (1954, 1956). Of the 96 strains of halococci studied by Montero et al. (1988), four had markedly different physiological and biochemical attributes, leading the authors to suggest that these strains probably belong to a separate taxon.

6.3 Alkalophilic Halobacteriaceae

Extremely halophilic, alkalophilic archaebacteria have been enriched for and isolated from hypersaline soda lakes. Grant and Tindall (1980) described complex media for enrichments from Lake Magadi, Kenya, using 16% NaCl and 16% Na_2CO_3 (w/v) at pH 10.6–10.8. Tindall et al. (1980) described the first isolates of these halobacteria from Lake Magadi. The isolates required high pH for optimum growth (pH 9.0–10.0) and they had a very low Mg^{2+}

requirement (0.1–2.0 mM). The organisms are similar in size to halobacteria (0.7 x 1.5–3.0 μm), they grow optimally in 4 M NaCl, and they lyse in <1 M NaCl. Like non-alkalophilic halobacteria, they synthesize bacterioruberins which give them an orange-red color. Other similarities to halobacteria include similar antibiotic sensitivities and similar amino acid content of the bulk proteins (acidic amino acids exceed basic amino acids).

Soliman and Trüper (1982) described an obligately aerobic, haloalkalophilic archaebacterium isolated from the Wadi Natrun, Egypt. It was given the name *Halobacterium pharaonis*, but the genus was subsequently changed to *Natronobacterium* based on lipid and nucleic acid differences between the alkalophilic strains and *Halobacterium* strains (Tindall et al., 1984). Like the strain from Lake Magadi, *N. pharaonis* grows well at alkaline pH (7.7–9.3), requires high salt (>2 M NaCl), and is inhibited by relatively low Mg^{2+} concentrations (10 mM).

N. pharaonis can grow on a broad range of carbon sources only in the presence of low concentrations of glutamic acid or casamino acids. It grows on single carbon sources such as formate, pyruvate, butyrate, and fumarate, but not on sugars, acetate, or simple alcohols. Glucose enhances bacterioruberin synthesis but not growth.

In addition to *N. pharaonis*, which has been isolated from both Wadi Natrun and Lake Magadi, *N. magadii*, *N. gregoryi*, and an alkalophilic halococcus, *Natronococcus occultus*, have been described from the latter alkaline lake (Tindall et al., 1984). Other unnamed haloalkalophiles have been isolated from Owens Lake, California, U.S.A. (Tindall, 1985), where they apparently coexist with halophilic phototrophic bacteria *Ectothiorhodospira* (Tew, 1980). In agitated enrichment cultures of such brines, alkalophilic halobacteria dominate whereas in stationary (non-oxygenated) enrichments, the phototrophic bacteria dominate (Grant and Tindall, 1986).

Like many non-alkalophilic halobacteria, many strains of alkalophilic halobacteria from Kenyan soda lakes and Wadi Natrun (including *N. pharaonis*) produce photoactive retinal pigments (Bivin and Stoeckenius, 1986). Two pigments have been noted with absorption maxima near 580 nm and 500 nm, respectively. The P_{580} pigment behaves like halorhodopsin while the P_{500} pigment shows some similarities to sensory rhodopsin. However, the latter has an absorption maximum of 587 nm in halobacteria. Bacteriorhodopsin was not identified in the 51 strains tested.

Perhaps the best studied attributes of the alkalophilic halobacteria are their lipids. Like all archaebacteria, their lipids have ether-linked side chains. They synthesize asymmetric C_{20},C_{25} diethers unlike halobacteria that produce only C_{20},C_{20} diethers (De Rosa et al., 1982). C_{20},C_{25} diether lipids have also been found in some methanogens (Grant et al., 1985).

Further studies of a Lake Magadi isolate demonstrated that of the total isoprenoid diether fraction, 9% (w/w) was C_{20},C_{20} lipid, 6% was a novel C_{25},C_{25} lipid, and 85% was C_{20},C_{25} lipid (de Rosa et al., 1983). In a survey of

12 strains of haloalkalophiles, Tindall (1985) found that C_{20},C_{20} diether lipids comprised from as little as 11% of the total lipids (*N. pharaonis*) to as much as 99.9% of the total lipids (unnamed strains from Wadi Natrun). Only two isolates synthesized C_{25},C_{25} diether lipids (which constituted $\leq$1% of the total lipids), and both strains were from Lake Magadi (*N. pharaonis* and *N. gregoryi*). C_{20},C_{25} diether lipids therefore constitute anywhere from the bulk to only a minor fraction of diether lipids in natronobacteria as a group. Increasing the salinity from 3 M to 4.5 M NaCl resulted in the relative increase of C_{20},C_{25} lipids in most strains (Morth and Tindall, 1985a).

Polar lipids were also investigated by Morth and Tindall (1985b). Glycolipids and nitrogen-containing lipids are absent in the alkalophilic halobacteria. Phosphatidylglycerophosphate (which constituted >50% of the total) and phosphatidylglycerol (which constituted >10% of the total) were found in all the 10 strains assayed. Three unidentified polar lipids were found in various distributions among the strains. *Natronococcus occultus* was unique in carrying the first and second unknown polar lipids while the rods carried the first and/or the third unknown.

Like other halobacteria, the alkalophiles synthesize menaquinone and dihydrogenated menaquinones as their major respiratory quinones (Collins and Tindall, 1987). Unlike non-alkalophilic halobacteria, *Natronobacterium gregoryi* was found to synthesize minor amounts of novel menaquinones that are methylated at the eighth isoprenyl unit. Some eubacteria have methylated menaquinones only at the sixth or seventh isoprenyl unit. The unique halobacterial menaquinones could serve as biogeochemical markers.

6.4 Halophilic methanogens

Methane has been noted in various hypersaline environments, including Solar Lake, Sinai (Giani et al., 1984), solar saltern ponds (A. Oren, personal communication), and hypersaline alkaline lakes (Oremland et al., 1982; Tindall, 1988). Isolates of halophilic methanogenic bacteria have been obtained from hypersaline marine stromatolites and microbial mats (Zhilina, 1983; Giani et al., 1984), solar salt ponds (Mathrani and Boone, 1985; Yu et al., 1985), Great Salt Lake, Utah, U.S.A. (Paterek and Smith, 1985), hypersaline Crimean lagoons (Zhilina, 1986), and hypersaline alkaline lakes (Oremland et al., 1982; Mathrani et al., 1988). Table 6.2 summarizes some of the characteristics of these isolates.

The isolates range from moderate halophiles to extreme halophiles. None of the isolates can grow on $CO_2 + H_2$ or on acetate, substrates typically used by freshwater and normal marine methanogens and sulfate-reducing bacteria. They all thrive on methylamines and usually methanol. Enrichment studies show that methanogens and sulfate-reducing bacteria do not compete for

Table 6.2 Halophilic methanogens known in pure culture

	M NaCl		
Isolate	Optimum	Range	Reference
Methanococcus halophilus	1.2	0.3–2.6	Zhilina, 1983; Zhilina & Kevbrin, 1985
Methanococcus halophilus-like	ND	ND	Zhilina, 1986
Large, flat cells	4.3	≤5	Zhilina, 1986
Methanosarcina sp.	ND	ND	Giani et al., 1984
Coccus SF1	2.1	0.9–3.4	Mathrani & Boone, 1985
Coccus	2.7	≥1.8	Yu et al., 1985
Methanohalophilus mahii	1.0–2.0	0.5–3.5	Paterek & Smith, 1985, 1988
M. zhilinae	0.7	0.2–2.1	Mathrani et al., 1988

their carbon substrates in hypersaline sediments (Giani et al., 1984; Zhilina, 1986).

All the isolates listed demonstrate typical archaebacterial antibiotic sensitivities. The isolate of Mathrani and Boone (1985) tolerated exposure to air. Most of the isolates are mesophilic, with temperature optima near 35°–37°C although temperatures of ca. 20°C to 40°–45°C support growth. The isolate that formed large flat cells (Zhilina, 1986) grew optimally at 50°–55°C and could withstand 20 min exposure to 100°C. The pH range for growth of all the listed isolates except *Methanohalophilus mahii* is near 6 to 8. *M. mahii,* an isolate from a Wadi Natrun lake, grows optimally at pH 9.2 (Mathrani et al., 1988).

The biochemistry of the halophilic methanogens is poorly known. Zhilina (1986) described bulk composition of proteins, carbohydrates, and nucleic acids as well as the amino acid composition of the bulk proteins in *M. halophilus.* The lipids are similar to those of *Methanosarcina.*

6.5 Other halophilic archaebacteria?

Certain possible "missing links" have not been found among the halophilic archaebacteria. No obligately anaerobic halobacterium has been found, although halophilic obligate anaerobes (methanogens) are known. It should be possible to use the immunologic distinctiveness between halobacteria and methanogens (Conway de Macario et al., 1986) to screen natural populations and enrichment cultures. The discovery of marine, thermophilic, sulfate-reducing archaebacteria in a novel branch of this bacterial kingdom (Stetter et al., 1987) may mean that halophilic sulfate reducers could be found within the archaebacteria. The existence of thermophilic archaebacteria in marine hot springs (Zillig et al., 1987) also suggests that halophilic extreme thermophiles may also be found among the archaebacteria. Because organic matter buried with evaporites is subjected to microbial degradation, barotolerance and barophily may also be important physiological features to investigate in the halophilic archaebacteria.

References

Aitken, D.M. and **Brown, A.D.** 1969. Citrate and glyoxylate cycles in the halophil, *Halobacterium salinarium. Biochimica et Biophysica Acta* 177: 351–354.

Alam, M., Claviez, M., Oesterhelt, D., and **Kessel, M.** 1984. Flagella and motility behaviour of square bacteria. *EMBO Journal* 3: 2899-2903.

Anderson, R., Kates, M., Baedecker, M.J., Kaplan, I.R., and **Ackman, R.G.** 1977. The stereoisomeric composition of phytanyl chains in lipids of Dead Sea sediments. *Geochimica et Cosmochimica Acta* 41: 1381–1390.

Baryshev, V.A. 1982. Regulation of *Halobacterium halobium* motility by the $Mg^{2+}$and Ca^{2+} ions. *FEMS Microbiology Letters* 14: 139–143.

Bayley, S.T. 1976. Information transfer: salt effects, pp. 119–136 in Heinrich, M.R. (editor), *Extreme Environments. Mechanisms of Microbial Adaptation,* Academic Press, New York.

Bivin, D.B. and **Stoeckenius, W.** 1986. Photoactive retinal pigments in haloalkaliphilic bacteria. *Journal of General Microbiology* 132: 2167–2177.

Bonelo, G., Ventosa, A., Megias, M., and **Ruiz-Berraquero, F.** 1984. The sensitivity of halobacteria to antibiotics. *FEMS Microbiology Letters* 21: 341–345.

Brock, T.D. and **Peterson, S.** 1976. Some effects of light on the viability of rhodopsin-containing halobacteria. *Archives of Microbiology* 109: 199–200.

Brown, A.D. 1983. Halophilic prokaryotes, pp. 137–162 in Lange, O.L., Nobel, P.S., Osmond, C.B. and Ziegler, H. (editors), *Physiological Plant Ecology* III, Springer-Verlag, New York.

Carteni-Farina, M., Porcelli, M., Cacciapuoti, G., de Rosa, M., Gambacorta, A., Grant, W.D., and **Ross, H.N.M.** 1985. Polyamines in halophilic archaebacteria. *FEMS Microbiology Letters* 28: 323–327.

Chen, K.Y. and **Martinowicz, H.** 1984. Lack of detectable polyamines in an extremely halophilic bacterium. *Biochemical and Biophysical Research Communications* 124: 423–429.

Christian, J.H.B. and **Waltho, J.A.** 1962. Solute concentrations within cells of halophilic and non-halophilic bacteria. *Biochimica et Biophysica Acta* 65: 506–508.

Cohen, S., Oren, A., and **Shilo, M.** 1983. The divalent cation requirement of Dead Sea halobacteria. *Archives of Microbiology* 136: 184–190.

Collins, M.D. and **Tindall, B.J.** 1987. Occurrence of menaquinones and some novel methylated menaquinones in the alkaliphilic, extremely halophilic archaebacterium *Natronobacterium gregoryi. FEMS Microbiology Letters* 43: 307–312.

Colwell, R.R., Litchfield, C.D., Vreeland, R.H., Kiefer, L.A., and **Gibbons, N.E.** 1979. Taxonomic studies of red halophilic bacteria. *International Journal of Systematic Bacteriology* 29: 379–399.

Conway de Macario, E., Konig, H., and **Macario, A.J.L.** 1986. Immunologic distinctiveness of archaebacteria that grow in high salt. *Journal of Bacteriology* 168: 425–427.

Danon, A. and **Caplan, S.R.** 1979. CO_2 fixation by *Halobacterium halobium. FEBS Letters* 74: 255–258.

Danson, M.J., Black, S.C., Woodland, D.L., and **Wood, P.A.** 1985. Citric acid cycle enzymes of the archaebacteria: citrate synthase and succinate thiokinase. *FEBS Letters* 179: 120–124.

Dencher, N.A. and **Hildebrand, E.** 1979. Sensory transduction in *Halobacterium halobium*: retinal protein pigment controls UV-induced behavioral response. *Zeitschrift für Naturforschungen* 34c: 841–847.

Dencher, N.A. and **Hildebrand, E.** 1982. Photobehavior of *Halobacterium halobium*. *Methods in Enzymology* 88: 420–426.

De Rosa, M., Gambacorta, A., Nicolaus, B., Ross, H.N.M., Grant, W.D., and **Bu'Lock, J.D.** 1982. An asymmetric archaebacterial diether lipid from alkaliphilic halophiles. *Journal of General Microbiology* 128: 343–348.

De Rosa, M., Gambacorta, A., Nicolaus, B., and **Grant, W.D.** 1983. A C_{25},C_{25} diether core lipid from archaebacterial haloalkalophiles. *Journal of General Microbiology* 129: 2333–2337.

De Rosa, M., Gambacorta, A., and **Gliozzi, A.** 1986. Structure, biosynthesis, and physiocochemical properties of archaebacterial lipids. *FEMS Microbiology Reviews* 50: 70–80.

De Rosa, M., Gambacorta, A., Grant, W.D., Lanzotti, V., and **Nicolaus, B.** 1988. Polar lipids and glycine betaine from haloalkaliphilic bacteria. *Journal of General Microbiology* 134: 205–211.

Dundas, I.E.D. 1977. Physiology of *Halobacteriaceae*. *Advances in Microbial Physiology* 15: 85–120.

Dussault, H.P. 1954. Destruction of "red halophiles' by antagonistic bacteria. *Gaspé Fisheries Experimental Station Note* 34: 3–5.

Edgerton, M.E. and **Brimblecombe, P.** 1981. Thermodynamics of halobacterial environments. *Canadian Journal of Microbiology* 27: 899–909.

Elazari-Volcani, B. 1940. Studies on the Microflora of the Dead Sea. Ph.D. Thesis. Hebrew University of Jerusalem. 119 pp.

Fernandez-Castillo, R., Rodriguez-Valera, F., Gonzalez-Ramos, J., and **Ruiz-Berraquero, F.** 1986. Accumulation of poly(β-hydroxybutyrate) by halobacteria. *Applied and Environmental Microbiology* 51: 214–216.

Forterre, P., Elie, C., and **Kohiyama, M.** 1984. Aphidicolin inhibits growth and DNA synthesis in halophilic archaebacteria. *Journal of Bacteriology* 159: 800–802.

Giani, D., Giani, L., Cohen, Y., and **Krumbein, W.E.** 1984. Methanogenesis in the hypersaline Solar Lake (Sinai). *FEMS Microbiology Letters* 25: 219–224.

Gibbons, N.W. 1958. The effect of salt on the metabolism of halophilic bacteria, pp. 69–76 in Eddy, B.P. (editor), *Proceedings of the Second International Symposium on Food Microbiology*, H.M. Stationery Office, London.

Ginzburg, M., Sachs, L., and **Ginzburg, B.A.** 1970. Ion metabolism in a *Halobacterium*. I. Influence of age of culture on intracellular concentrations. *Journal of General Physiology* 55: 187–207.

Gochnauer, M.B., Kushwaha, S.C., Kates, M., and **Kushner, D.J.** 1972. Nutritional control of pigment and isoprenoid compound formation in extremely halophilic bacteria. *Archiv für Mikrobiologie* 84: 339–349.

Gonzalez, C., Gutierrez, C., and **Ramirez, C.** 1978. *Halobacterium vallismortis* sp. nov. An amylolytic and carbohydrate-metabolizing, extremely halophilic bacterium. *Canadian Journal of Microbiology* 24: 710– 715.

Good, W.A. and **Hartman, P.A.** 1970. Properties of the amylase from *Halobacterium halobium*. *Journal of Bacteriology* 104: 601–603.

Grant, W.D. and **Tindall, B.J.** 1980. The isolation of alkalophilic bacteria, pp. 27–38 in Gould, G.W. and Corry, J.E.C. (editors), *Microbial Growth and Survival in Extremes of Environment*, Society of Applied Bacteriology Technical Service 15, Academic Press, New York.

Grant, W.D. and **Tindall, B.J.** 1986. The alkaline saline environment, p. 25–54 in Herbert,R.A. and Codd, G.A. (editors), *Microbes in Extreme Environments*. Society of General Microbiology Special Publication 17, Academic Press, London.

Grant, W.D., Pinch, G., Harris, J.E., De Rosa, M., and **Gambacorta, A.** 1985. Polar lipids in methanogen taxonomy. *Journal of General Microbiology* 131: 3277–3286.

Gutierrez, M.C., Garcia, M.T., Ventosa, A., Nieto, J.J., and **Ruiz-Berraquero, F.** 1986. Occurrence of megaplasmids in halobacteria. *Journal of Applied Bacteriology* 61: 67–71.

Hamana, K., Kamekura, M., Onishi, H., Akazawa, T., and **Matsuzaki, S.** 1985. Polyamines in photosynthetic eubacteria and extreme-halophilic archaebacteria. *Journal of Biochemistry (Japan)* 97: 1653–1658.

Hartmann, R., Sickinger, H.-D., and **Oesterhelt, D.** 1980. Anaerobic growth of halobacteria. *Proceedings of the National Academy of Science U.S.A.* 77: 3821–3825.

Hochstein, L.I. 1988. The physiology and metabolism of the extremely halophilic bacteria, p. 67–83 in Rodriguez-Valera, F. (editor), *Halophilic Bacteria,* Vol. II. CRC Press, Boca Raton.

Hochstein, L.I. and **Tomlinson, G.A.** 1985. Denitrification by extremely halophilic bacteia. *FEMS Microbiology Letters* 27: 329–331.

Hunter, M.I.S. and **Millar, S.J.W.** 1980. Effect of wall antibiotics on the growth of the extremely halophilic coccus, *Sarcina marina* NCMB 778. *Journal of General Microbiology* 120: 255–258.

Javor, B.J. 1983. Planktonic standing crop and nutrients in a saltern ecosystem. *Limnology and Oceanography* 28: 153–159.

Javor, B.J. 1984. Growth potential of halophilic bacteria isolated from solar salt environments: carbon sources and salt requirements. *Applied and Environmental Microbiology* 48: 352–360.

Javor, B.J. 1988. CO_2 fixation in halobacteria. *Archives of Microbiology* 149: 433–440.

Javor, B., Requadt, C., and **Stoeckenius, W.** 1982. Box-shaped halophilic bacteria. *Journal of Bacteriology* 151: 1532–1542.

Juez, G. 1988. Taxonomy of extremely halophilic archaebacteria, p. 3–24 in Rodriguez-Valera, F. (editor), *Halophilic Bacteria,* Vol. II. CRC Press, Boca Raton.

Juez, G., Rodriguez-Valera, F., Ventosa, A., and **Kushner, D.J.** 1986. *Haloarcula hispanica* spec. nov. and *Haloferax gibbonsii* spec. nov., two new species of extremely halophilic archaebacteria. *Systematic and Applied Microbiology* 8: 75–79.

Kamekura, M. and **Kates, M.** 1988. Lipids of halophilic archaebacteria, p. 25–54 in Rodriguez-Valera, F. (editor), *Halophilic Bacteria,* Vol. II. CRC Press, Boca Raton.

Kamekura, M., Bardocz, S., Anderson, P., Wallace, R., and **Kushner, D.J.** 1986. Polyamines in moderately and extremely halophilic bacteria. *Biochimica et Biophysica Acta* 880: 204–208.

Kandler, O. 1982. Cell wall structures and their phylogenetic implications, pp. 149–160 in Kandler, O. (editor), *Archaebacteria,* Gustav Fischer, New York.

Kaplan, I.R. and **M.J. Baedecker.** 1970. Biological productivity in the Dead Sea. Part II. Evidence for phosphatidyl glycerophosphate lipid in sediment. *Israel Journal of Chemistry* 8: 529–533.

Katznelson, H. and **Robinson, J.** 1956. Observations on the respiratory activity of certain obligately halophilic bacteria with high salt requirements. *Journal of Bacteriology* 71: 244–249.

Kessel, M. and **Cohen, Y.** 1982. Ultrastructure of square bacteria from a brine pool in southern Sinai. *Journal of Bacteriology* 150: 851–860.

Kessel, M., Cohen, Y., and **Walsby, A.E.** 1985. Structure and physiology of square-shaped and other halophilic bacteria from the Gavish Sabkha, pp. 267–287 in Friedman, G.M. and Krumbein, W.E. (editors), *Hypersaline Ecosystems. The Gavish Sabkha,* Ecological Studies 53, Springer-Verlag, New York.

Kocur, M. and **Bohacek, J.** 1972. DNA base composition of extremely halophilic cocci. *Archiv für Mikrobiologie* 82: 280–282.

Kocur, M. and **Hodgkiss, W.** 1973. Taxonomic status of the genus *Halococcus* Schoop. *International Journal of Systematic Bacteriology* 23: 151–156.

Kushner, D.J. 1978. Life in high salt and solute concentrations: Halophilic bacteria, pp. 317–368 in Kushner, D.J. (editor), *Microbial Life in Extreme Environments*, Academic Press, London.

Kushner, D.J. 1985. The Halobacteriaceae, pp. 171–215 in Woese, C.R. and Wolfe, R.S. (editors), *The Bacteria. A Treatise on Structure and Function*, Vol. VIII, *Archaebacteria*. Academic Press, New York.

Kushwaha, S.C. and **Kates, M.** 1979. Effect of glycerol on carotenogenesis in the extreme halophile, *Halobacterium cutirubrum*. *Canadian Journal of Microbiology* 25: 1288–1291.

Kushwaha, S.C., Juez-Perez, G., Rodriguez-Valera, F., Kates, M., and **Kushner, D.J.** 1982. Survey of lipids of a new group of extremely halophilic bacteria from salt ponds in Spain. *Canadian Journal of Microbiology* 28: 1365–1372.

Langworthy, T.A., Tornabene, T.G., and **Holzer, G.** 1982. Lipids of archaebacteria, pp. 228–244 in Kandler, O. (editor), *Archaebacteria*, Gustav Fischer, New York.

Langworthy, T.A., Holzer, G., Zeikus, J.G., and **Tornabene, T.G.** 1983. Iso- and anteiso-branched glycerol diethers of the thermophilic anaerobe *Thermodesulfotobacterium commune*. *Systematic and Applied Microbiology* 4: 1–17.

Lanyi, J.K. 1976. Membrane structure and salt dependence in extremely halophilic bacteria, pp. 295–303 in Heinrich, M.R. (editor), *Extreme Environments. Mechanisms of Microbial Adaptations*. Academic Press, New York.

Lanyi, J.K. 1980. Physical chemistry and evolution of salt tolerance in halobacteria. *Origins of Life* 10: 161–167.

Larsen, H. 1981. The family Halobacteriaceae, pp. 985–994 in Starr, M.P., Stolp, H., Trüper, H.G., Balows, A. and Schlegel, H.G. (editors), *The Prokaryotes: A Handbook on Habitats, Isolation, and Identification of Bacteria*, vol. 1, Springer-Verlag, New York.

Larsen, H. 1984. Family V. Halobacteriaceae Gibbons 269[AL], pp. 261–267 in Krieg, N.R. and Holt, J.G. (editors), *Bergey's Manual of Systematic Bacteriology*, vol. 1, 9th Edition, Williams and Wilkins, Baltimore.

Luehrsen, K., Nicholson, D.E., and **Fox, G.E.** 1985. Widespread distribution of a 7S RNA in archaebacteria. *Current Microbiology* 12: 69–72.

Mancinelli, R.L. and **Hochstein, L.I.** 1986. The occurrence of denitrification in extremely halophilic bacteria. *FEMS Microbiology Letters* 35: 55–58.

Mathrani, I.M. and **D.R. Boone.** 1985. Isolation and characterization of a moderately halophilic methanogen from a solar saltern. *Applied and Environmental Microbiology* 50: 140–143.

Mathrani, I.M., Boone, D.R., Mah, R.A., Fox, G.E., and **Lau, P.P.** 1988. *Methanohalophilus zhilinae* sp. nov., an alkaliphilic, halophilic, methylotrophic methanogen. *International Journal of Systematic Bacteriology* 38: 139–142.

Matsuoka, H., Suzuki, S., Aizawa, M., Kimura, Y., and **Ikegami, A.** 1981. Role of magnesium in the membrane systems of *Halobacterium halobium*. *Journal of Applied Biochemistry* 3: 425–436.

Meseguer, I. and **Rodriguez-Valera, F.** 1985. Production and purification of halocin H4. *FEMS Microbiology Letters* 28: 177–182.

Moore, R.L. and **McCarthy, B.J.** 1969a. Characterization of the deoxyribonucleic acid of various strains of halophilic bacteria. *Journal of Bacteriology* 99: 248–254.

Moore, R.L. and **McCarthy, B.J.** 1969b. Base sequence homology and renaturation studies of the deoxyribonucleic acid of extremely halophilic bacteria. *Journal of Bacteriology* 99: 255–262.

Morth, S. and **Tindall, B.J.** 1985a. Evidence that changes in the growth conditions affect the relative distribution of diether lipids in haloalkaliphilic archaebacteria. *FEMS Microbiology Letters* 29: 285–288.

Morth, S. and **Tindall, B.J.** 1985b. Variation of polar lipid composition within haloalkaliphilic archaebacteria. *Systematic and Applied Microbiology* 6: 247–250.

Mullakhanbhai, M.F. and **Larsen, H.** 1975. *Halobacterium volcanii* spec. nov., a Dead Sea halobacterium with a moderate salt requirement. *Archives of Microbiology* 104: 207–214.

Newton, G.l. and **Javor, B.J.** 1985. γ-Glutamylcysteine and thiosulfate are the major low molecular weight thiols in halobacteria. *Journal of Bacteriology* 161: 438–441.

Nicholson, D.E. and **Fox, G.E.** 1983. Molecular evidence for a close phylogenetic relationship among box-shaped halophilic bacteria, *Halobacterium vallismortis* and *Halobacterium marismortui. Canadian Journal of Microbiology* 29: 52–59.

Nissenbaum, A., Baedecker, M.J., and **Kaplan, I.R.** 1972. Organic geochemistry of Dead Sea sediments. *Geochimica et Cosmochimica Acta* 36: 709–727.

Norberg, P. and **B.v. Hofsten.** 1969. Proteolytic enzymes from extremely halophilic bacteria. *Journal of General Microbiology* 55: 251–256.

Oesterhelt, D. 1982. Anaerobic growth of halobacteria. *Methods in Enzymology* 88: 417–420.

Oesterhelt, D. and **Krippahl, G.** 1973. Light inhibition of respiration in *Halobacterium halobium. FEBS Letters* 36: 72–76.

Oesterhelt, D. and **Krippahl, G.** 1983. Phototrophic growth of halobacteria and its use for isolation of photosynthetically-deficient mutants. *Annales de Microbiologie (Institute Pasteur)* 134 B: 137–150.

Onishi, H., McCance, M.E., and **Gibbons, N.E.** 1965. A synthetic medium for extremely halophilic bacteria. *Canadian Journal of Microbiology* 11: 365–373.

Oremland, R.S., Marsh, L., and **Des Marais, D.J.** 1982. Methanogenesis in Big Soda Lake, Nevada: an alkaline, moderately hypersaline desert lake. *Applied and Environmental Microbiology* 43: 462–468.

Oren, A. 1983a. A thermophilic amyloglucosidase from *Halobacterium sodomense,* a halophilic bacterium from the Dead Sea. *Current Microbiology.* 8: 225–230.

Oren, A. 1983b. Bacteriorhodopsin-mediated CO_2 photoassimilation in the Dead Sea. *Limnology and Oceanography* 28: 33–41.

Oren, A. 1983c. *Halobacterium sodomense* sp. nov., a Dead Sea halobacterium with an extremely high magnesium requirement. *International Journal of Systematic Bacteriology* 33: 381–386.

Oren, A. and **Shilo, M.** 1981. Bacteriorhodopsin in a bloom of halobacteria in the Dead Sea. *Archives of Microbiology* 130: 185–187.

Parkes, K. and **Walsby, A.E.** 1981. Ultrastructure of a gas-vacuolate square bacterium. *Journal of General Microbiology* 126: 503–506.

Paterek, J.R. and **Smith, P.H.** 1985. Isolation and characterization of a halophilic methanogen from Great Salt Lake. *Applied and Environmental Microbiology* 50: 877–881.

Paterek, J.R. and **Smith, P.H.** 1988. *Methanohalophilus mahii* gen. nov., sp.nov., a methylotrophic methanogen. *International Journal of Systematic Bacteriology* 38: 122–123.

Pauling, C. 1982. Bacteriophages of *Halobacterium halobium*: isolation from fermented fish sauce and primary characterization. *Canadian Journal of Microbiology* 28: 916–921.

Pecher, T. and **Böck, A.** 1981. *In vivo* susceptibility of halophilic and methanogenic organisms to protein synthesis inhibitors. *FEMS Microbiology Letters* 10: 295–297.

Petter, H.F.M. 1931. On bacteria of salted fish. *Verhandelingen der Koninklijke Akademie van Wetenschappen te Amsterdam. Afdeeling Natuurkunde* 34: 1417–1423.

Pfeifer, F. 1988. Genetics of halobacteria, p. 105–133 in Rodriguez-Valera, F. (editor), *Halophilic Bacteria,* Vol. II. CRC Press, Boca Raton.

Pfeifer, F., Weidinger, G., and **Goebel, W.** 1981. Characterization of plasmids in halobacteria. *Journal of Bacteriology* 145: 369–374.

Plotkin, B., Boyd, M.S., Palmer, P.L., and **Sherman, W.V.** 1985. Temperature dependence of the photoresponse of *Halobacterium halobium. Current Microbiology* 12: 97–100.

Rayman, M.K., Gordon, R.C., and **MacLeod, R.A.** 1967. Isolation of a Mg_{2+} phospholipid from *Halobacterium cutirubrum. Journal of Bacteriology* 93: 1465–1470.

Reistad, R. 1975. Amino sugar and amino acid constituents of the cell walls of extremely halophilic cocci. *Archives of Microbiology* 102: 71–73.

Rodriguez-Valera, F. 1988. Characteristics and microbial ecology of hypersaline environments in Rodriguez-Valera, F. (editor), *Halophilic Bacteria,* Vol. I. CRC Press, Boca Raton.

Rodriguez-Valera, F., Ruiz-Berraquero, F., and **Ramos-Cormenzana, A.** 1979. Isolation of extreme halophiles from seawater. *Applied and Environmental Microbiology* 38: 164–165.

Rodriguez-Valera, F., Ruiz-Berraquero, F., and **Ramos-Cormenzana, A.** 1980. Isolation of extremely halophilic bacteria able to grow on defined inorganic media with single carbon sources. *Journal of General Microbiology* 119: 535–538.

Rodriguez-Valera, F., Ruiz-Berraquero, F., and **Ramos-Cormenzana, A.** 1981. Characteristics of the heterotrophic populations in hypersaline environments of different salt concentrations. *Microbial Ecology* 7: 235– 243.

Rodriguez-Valera, F., Juez, G., and **Kushner, D.J.** 1982. Halocins: salt-dependent bacteriocins produced by extremely halophilic rods. *Canadian Journal of Microbiology* 28: 151–154.

Rodriguez-Valera, F., Ventosa, A., Quesada, E., and **Ruiz-Berraquero, F.** 1982b. Some physiological features of a *Halococcus* sp. at low salt concentrations. *FEMS Microbiology Letters* 15: 249–252.

Rodriguez-Valera, F., Juez, G. and Kushner, D.J., 1983a. *Halobacterium mediterranei,* spec. nov., a new carbohydrate-utilizing extreme halophile. *Systematic and Applied Microbiology* 4: 369–381.

Rodriguez-Valera, F., Nieto, J.J., and **Ruiz-Berraquero, F.** 1983b. Light as an energy source in continuous cultures of bacteriorhodopsin-containing halobacteria. *Applied and Environmental Microbiology* 45: 868–871.

Rodriguez-Valera, F, Ventosa, A., Juez, G., and **Imhoff, J.F.** 1985. Variation of environmental features and microbial populations with salt concentration in a multipond saltern. *Microbial Ecology* 11: 107–115.

Rohrmann, G.F. and **Cheney, R.** 1983. Bacteriophages of *Halobacterium halobium*: virion DNAs and proteins. *Canadian Journal of Microbiology* 29: 627–629.

Romanenko, V.I. 1981. Square microcolonies in the surface saline water film of the Saxkoye Lake. *Mikrobiologiya* 50: 571–574.

Ross, H.N.M. and **Grant, W.D.** 1985. Nucleic acid studies on halophilic archaebacteria. *Journal of General Microbiology* 131: 165–173.

Sapienza, C. and **Doolittle, W.F.** 1982. Unusual physical organization of the *Halobacterium* genome. *Nature (London)* 295: 384–389.

Schinzel, R. and **Burger, K.J.** 1984. Sensitivity of halobacteria to aphidicolin, an inhibitor of eukaryotic a-type DNA polymerase. *FEMS Microbiology Letters* 25: 187–190.

Schnabel, H. 1984a. An immune strain of *Halobacterium halobium* carries the invertible L segment of phage ϕH as a plasmid. *Proceedings of the National Academy of Science U.S.A.* 81: 1017–1020.

Schnabel, H. 1984b. Integration of plasmid pϕHL into phage genomes during infection of *Halobacterium halobium* R1-L with phage ϕHL1. *Molecular and General Genetics* 197: 19–23.

Schnabel, H. and **Zillig, W.** 1982. *Halobacterium halobium* phage ϕH, p. 352 in Kandler, O. (editor), *Archaebacteria*, Gustav Fischer, New York.

Schnabel, H. and **Zillig, W.** 1984. Circular structure of the genome of phage ϕH in a lysogenic *Halobacterium halobium. Molecular and General Genetics* 193: 422–426.

Schnabel, H., Zillig, W., Pfaffle, M., Schnabel, R., Michel, H., and **Delius, H.** 1982a. *Halobacterium halobium* phage ϕH. *EMBO Journal* 1: 87–92.

Schnabel, H., Schramm, E., Schnabel, R., and **Zillig, W.** 1982b. Structural variability in the genome of phage ϕH of *Halobacterium halobium. Molecular and General Genetics* 188: 370–377.

Schoop, G. 1934a. Halophile microbes in foodstuffs. *Deutschen Tierartzlichen Wochenschrift* 42: 816–819.

Schoop, G. 1934b. Salt bacteria. *Deutschen Tierartzlichen Wochenschrift* 13: 205–207.

Schoop, G. 1935. Obligat Halophile Mikroben. *Zentralblatt für Bakteriologie* Abteilung I 134: 14–26.

Soliman, G.S.H. and **Trüper, H.G.** 1982. *Halobacterium pharaonis* sp. nov., a new, extremely haloalkalophilic archaebacterium with low magnesium requirement, pp. 318–329 in Kandler, O. (editor), *Archaebacteria*, Gustav Fischer, New York.

Spudich, J. 1985. Sensory transduction by phototactic halobacteria. *Einstein Quarterly Journal of Biology and Medicine* 3: 97–103.

Steber, J. and **Schleifer, K.H.** 1975. *Halococcus morrhuae*: a sulfated heteropolysaccharide as the structural component of the bacterial cell wall. *Archives of Microbiology* 105: 173–177.

Stetter, K.O., Lauerer, G., Thomm, M., and **Neuner, A.** 1987. Isolation of extremely thermophilic sulfate reducers: evidence for a novel branch of archaebacteria. *Science* 236: 822–824.

Stoeckenius, W. 1981. Walsby's square bacterium: fine structure of an orthogonal procaryote. *Journal of Bacteriology* 148: 352–360.

Stoeckenius, W. and **Bogomolni, R.A.** 1982. Bacteriorhodopsin and related pigments of halobacteria. *Annual Review of Biochemistry* 52: 587–616.

Stoeckenius, W., Lozier, R.H., and **Bogomolni, R.A.** 1979. Bacteriorhodopsin and the purple membrane of halobacteria. *Biochimica et Biophysica Acta* 505: 215–278.

Stoeckenius, W., Bivin, D., and **McGinnis, K.** 1985. Photoactive pigments in halobacteria from Gavish Sabkha, pp. 288–295 in Friedman, G.M. and Krumbein, W.E. (editors), *Hypersaline Ecosystems. The Gavish Sabkha.* Ecological Studies 53, Springer-Verlag, New York.

Stuart, L.S. 1935. The morphology of bacteria causing reddening of salted hides. *Journal of the American Leather Chemists Association* 30: 226–235.

Stuart, L.S. 1936. A note of halophilic chitinovorous bacteria. *Journal of the American Leather Chemists Association* 31: 119–120.

Stuart, L.S. 1941a. Effect of protein concentration and cysteine on growth of halophilic bacteria. *Journal of Agricultural Research* 61: 267–276.

Stuart, L.S. 1941b. The growth of halophilic bacteria in concentrations of sodium chloride above three molar. *Journal of Agricultural Research* 61: 259–266.

Stuart, L.S. and **James, L.H.** 1938a. The effect of Eh and sodium chloride concentration on the physiology of halophilic bacteria. *Journal of Bacteriology* 35: 381–396.

Stuart, L.S. and **James, L.H.** 1938b. The effect of sodium chloride on the Eh of protogenous media. *Journal of Bacteriology* 35: 369–380.

Takahashi, T., Tomioka, H., Kamo, N., and **Kobatake, Y.** 1985a. A photosystem other than PS370 also mediates the negative phototaxis of *Halobacterium halobium. FEMS Microbiology Letters* 28: 161–164.

Takahashi, T., Watanabe, M., Kamo, N., and **Kobatake, Y.** 1985b. Negative phototaxis from blue light and the role of third rhodopsinlike pigment in *Halobacterium cutirubrum*. *Biophysical Journal* 48: 235–240.

Tew, R.W. 1980. Halotolerant *Ectothiorhodospira* survival in mirabilite: experiments with a model of chemical stratification by hydrate deposition in saline lakes. *Geomicrobiology Journal* 2: 13–20.

Tindall, B.J. 1985. Qualitative and quantitative distribution of diether lipids in haloalkliphilic archaebacteria. *Systematic and Applied Microbiology* 6: 243–246.

Tindall, B.J. 1988. Prokaryotic life in the alkaline, saline, athalassic environment, p. 31–67 in Rodriguez-Valera, F. (editor), *Halophilic Bacteria*, Vol. I. CRC Press, Boca Raton.

Tindall, B.J. and **Trüper, H.G.** 1986. Ecophysiology of the aerobic halophilic archaebacteria. *Systematic and Applied Microbiology* 7: 202–212.

Tindall, B.J., Mills, A.A., and **Grant, W.D.** 1980. An alkalophilic red halophilic bacterium with a low magnesium requirement from a Kenyan soda lake. *Journal of General Microbiology* 116: 257–260.

Tindall, B.J., Ross, H.N.M., and **Grant, W.D.** 1984. *Natronobacterium* gen. nov. and *Natronococcus*, gen. nov., two new genera of haloalkalophilic archaebacteria. *Systematic and Applied Microbiology* 5: 41–57.

Tomlinson, G.A. and **Hochstein, L.I.** 1972. Isolation of carbohydrate-metabolizing, extremely halophilic bacteria. *Canadian Journal of Microbiology* 18: 698–701.

Tomlinson, G.A. and **Hochstein, L.I.** 1976. *Halobacterium saccharovorum* sp. nov., a carbohydrate-metabolizing, extremely halophilic bacterium. *Canadian Journal of Microbiology* 22: 587–591.

Tomlinson, G.A., Koch, T.K., and **Hochstein, L.I.** 1974. The metabolism of carbohydrates by extremely halophilic bacteria: Glucose metabolism via a modified Entner-Doudoroff pathway. *Canadian Journal of Microbiology* 20: 1085–1091.

Tomlinson, G.A., Jahnke, L.L., and **Hochstein, L.I.** 1986. *Halobacterium denitrificans* sp. nov., an extremely halophilic denitrifying bacterium. *International Journal of Systematic Bacteriology* 36: 66–70.

Tornabene, T.G. 1978. Non-aerated cultivation of *Halobacterium cutirubrum* and its effects on cellular squalenes. *Journal of Molecular Evolution* 11: 253–257.

Torreblanca, M., Rodriguez-Valera, F., Juez, G., Ventosa, A., Kamekura, M., and **Kates, M.** 1986. Classification of non-haloalkaliphilic halobacteria based on numerical taxonomy and polar lipid composition, and description of *Haloarcula* gen. nov. and *Haloferax* gen. nov. *Systematic and Applied Microbiology* 8: 89–99.

Torsvik, T. and **Dundas, I.A.** 1974. Bacteriophage of *Halobacterium salinarium*. *Nature (London)* 248: 680-681.

Torsvik, T. and **Dundas, I.A.** 1980. Persisting phage infection in *Halobacterium salinarium* str. 1. *Journal of General Virology* 47: 29–36.

Venkataraman, R. and **Sreenivasan, A.** 1954. Studies on the red halophilic bacteria from salted fish and salt. *Proceedings of the Indian Academy of Science* 39, Section B: 17–23.

Venkataraman, R. and **Sreenivasan, A.** 1956. Further studies on the red halophilic bacteria from solar salts and salted fish. *Proceedings of the Indian Academy of Science* 43: 197–206.

Wagner, G. 1984. Blue light effects in halobacteria, pp. 48–54 in Senger, H. (editor), *Blue Light Effects in Biological Systems*, Springer-Verlag, Berlin.

Wagner, G. and **Linhardt, R.** 1988. The retinal proteins of halobacteria, p. 85–104 in Rodriguez-Valera, F. (editor), *Halophilic Bacteria*, Vol. II. CRC Press, Boca Raton.

Wais, A.C. 1985. Cellular morphogenesis in a halophilic archaebacterium. *Current Microbiology* 12: 191–196.

Wais, A.C. and **Daniels, L.L.** 1985. Populations of bacteriophage infecting *Halobacterium* in a transient brine pool. *FEMS Microbiology Ecology* 31: 323–326.

Wais, A.C., Kon, M., MacDonald, R.E., and **Stollar, R.D.** 1975. Salt-dependent bacteriophage infecting *Halobacterium cutirubrum* and *Halobacterium halobium. Nature (London)* 256: 314–315.

Walsby, A.E. 1980. A square bacterium. *Nature (London)* 283: 69–71.

Werber, M.M. and **Mevarech, M.** 1978. Induction of a dissimilatory reduction pathway of nitrate in *Halobacterium* of the Dead Sea. A possible role for the 2 Fe-ferredoxin isolated from this organism. *Archives of Biochemistry and Biophysics* 186: 60–65.

Wieland, F. 1988. The cell surfaces of halobacteria, p. 55–65 in Rodriguez-Valera, F. (editor), *Halophilic Bacteria,* Vol. II. CRC Press, Boca Raton.

Wieland, F., Paul, G., and **Sumper, M.** 1985. Halobacterial flagellins are sulfated glycoproteins. *Journal of Biological Chemistry* 260: 15180–15185.

Woese, C. 1987. Bacterial evolution. *Microbiological Reviews* 51: 221–271.

Yu, I.K., Kawamura, F., and **Doi, R.H.** 1985. Isolation and characterization of an obligately halophilic methanogenic bacterium. *American Society of Microbiology Annual Meeting Abstract* I 18: 149.

Zhilina, T.N. 1983. New obligate halophilic methane-producing bacterium. *Microbiology* (English translation) 52: 290–297.

Zhilina, T.N. 1986. Methanogenic bacteria from hypersaline environments. *Systematic and Applied Microbiology* 7: 216–222.

Zhilina, T.N. and **Kevbrin, V.V.** 1985. Cultivation of *Methanococcus halophilus* on monomethylamine. *Mikrobiologiya* 54: 93–99.

Zillig, W., Holz, I., Klenk, H.-P., Trent, J., Wunderl, S., Janekovic, D., Imsel, E., and **Haas, B.** 1987. *Pyrococcus woesei,* sp. nov., an ultra-thermophilic marine archaebacterium, representing a novel order, *Thermococcales. Systematic and Applied Microbiology* 9: 62–70.

7

Phototrophic Bacteria

Phototrophic bacteria include five families of purple and green eubacteria that derive energy from light. Many phototrophic bacteria are autotrophs. In contrast to the cyanobacteria, they never produce O_2 during photosynthesis and they synthesize one or more bacteriochlorophylls instead of chlorophyll *a*. Phototrophic bacteria are typically associated with lighted, anaerobic environments although not all of these bacteria are obligate anaerobes. Many can alternatively grow as aerobic heterotrophs in the dark.

Hypersaline habitats may be particularly suited for the proliferation of phototrophic bacteria due to several factors: 1) microbial mats (consisting of a layer of cyanobacteria overlying a layer of phototrophic bacteria) often dominate the benthic surfaces in the euphotic zone; 2) brines hold less oxygen than normal seawater; and 3) density stratification due to salinity differences may trap shallow, anaerobic brines in the euphotic zone. Relatively little is known about the ecology and productivity of phototrophic bacteria in natural hypersaline environments although their distribution appears to be nearly ubiquitous. The following discussion of phototrophic bacteria emphasizes studies on laboratory cultures and outlines their taxonomic and metabolic differences and similarities.

Although representatives of all five families of phototrophic bacteria have been observed in hypersaline environments (Chromatiaceae and Ectothiorhodaceae [purple sulfur bacteria]; Rhodospirillaceae [purple non-sulfur bacteria]; Chlorobiaceae [non-filamentous green bacteria]; and Chloroflexaceae [filamentous green bacteria]), only members of the purple sulfur and purple non-sulfur bacteria have been well documented. A detailed review of both marine and halophilic phototrophic bacteria was presented by Imhoff (1988).

7.1 Green bacteria

Green bacteria of the genus *Prosthecochloris* were noted in the hypolimnion of Solar Lake at 4 m depth (Cohen et al., 1977) and in estuarine waters in the Soviet Union (Puchkova, 1984). For the isolates that were tested, salinity optima ranged from 2% to 5% NaCl, but 10% NaCl was tolerated. Strains of *Chlorobium* and *Pelodictyon* had similar salt optima but they tolerated up to 15% NaCl (Puchkova, 1984). Green bacteria have also been observed in Hot Lake (Washington, U.S.A.), a $MgSO_4$ lake, in nearly 40% salinity (Anderson, 1958). Filaments that resemble *Chloroflexus* in ultrastructure have also been identified in microbial mats of a hypersaline, coastal depression in Mexico (Stolz, 1984), as well as in salterns (Jørgensen et al., 1987). The phototrophic bacterial layer of *Aphanothece halophytica*-dominated microbial mats in an Israeli saltern (15% salinity) was composed primarily of green *Chloroflexus*-like filaments that accumulated elemental sulfur internally (Javor, unpublished data). *In vivo* absorption spectra of that layer showed a strong peak of bacteriochlorophyll *c* (748–752 nm) and fluorescence microscopy demonstrated that the filaments were not cyanobacteria.

7.2 Purple sulfur bacteria

Members of the genus *Ectothiorhodospira* are never found in freshwater. They prefer either marine or hypersaline habitats (Table 7.1). Like all members of the genus, the bacteria from hypersaline environments deposit elemental sulfur outside their cells when growing on sulfide. The NaCl optima of three halotolerant species (*E. mobilis, E. shaposhnikovii,* and *E. vacuolata*) are close to that of seawater while more extreme halophilism is found in *E. halophila, E. halochloris,* and *E. abdelmalekii. E. mobilis* is rather broadly halotolerant, with at least one strain capable of growth in 22% salt (Imhoff et al., 1978).

Ectothiorhodospira can grow by either photoautotrophy or by photoheterotrophy using several different low molecular weight compounds such as acetate and succinate. Typical sources of sulfur than can be utilized are sulfide, elemental sulfur, and thiosulfate. Evidence for the role of cytochrome C_{551} as a sulfide:acceptor oxidoreductase in *E. abdelmalekii* was presented by Then and Trüper (1983). Aerobic, dark heterotrophic growth was observed in *E. halophila, E. halochloris, E. vacuolata,* and *E. abdelmalekii* by Krasil'nikov and Kondrat'eva (1984), although all of these species except *E. halochloris* were previously reported to be strict anaerobes (Raymond and Sistrom, 1967; Imhoff and Trüper, 1981; Imhoff et al, 1981).

The three more halophilic species of *Ectothiorhodospira* have been isolated from the hypersaline, alkaline Wadi Natrun lakes of Egypt (Imhoff and Trüper, 1977; Imhoff et al., 1978; Imhoff, 1988), although *E. halophila* was first described from an alkaline lake in North America (Raymond and Sistrom,

Table 7.1 Halophilic and halotolerant phototrophic bacteria

Species	Optimum salinity, %	NaCl range, %	Optimum pH	Optimum °C	Bchl	Obligate anaerobe	Mol % G+C	Sulfur electron donor	Reference[a]
Ectothiorhodospira									
E. halophila	11–22	9–30	7.4–7.8	47	*a*	+	68.4	S^{2-}, SO_3^{2-}	6, 7, 8
E. halochloris	14–27	5–40	7.5–10	48	*b*	−	52.9	S^{2}	3, 8
E. abdelmalekii	12–18	≤30	>8.5	30–34	*b*	+	63.3–63.8	S^{2-}	4
E. shaposhnikovii[b]	nd	0.1–8	9.0–9.5	30	*a*	+	(61.2–62.8)	S^{2-}	2, 8
		0.5–15[c]						$S_2O_3^{2-}$	
E. vacuolata	3	<10	6.5–10	39	*a*	+	61.4–63.6	S^{2-}	4
								$S_2O_3^{2-}$	
E. mobilis[b]	2–8	5–22	(7.6–8.0)	(25–30)	*a*		(67.3–69.9)	S^{2-}	4, 8
Rhodospirillum									
R. salexigens	6–8	5–20	ca. 7	40	*a*	−	64	none	1
R. salinarum	12–18	2≥24	7.5–8	42	*a*	−	67.4–68.1	none	5
R. mediosalinum	4–7	0.15–15		30–35°C	(*a*)				10
Rhodobacter									
R. sulfidophilus	1–5	≥10							9
R. adriaticum	2.5–7.5	≥10							9

[a](1) Drews (1981); (2) Grant et al. (1979); (3) Imhoff and Trüper (1977); (4) Imhoff and Trüper (1981); (5) Nissen and Dundas (1984); (6) Raymond and Sistrom (1967); (7) Raymond and Sistrom (1969); (8) Trüper and Imhoff (1981); (9) Imhoff (1988); (10) Kompantseva and Gorlenko (1984).
[b]Tentative assignment of a halophilic or halotolerant strain. Where data are not reported, attributes of non-halophilic strains are given in parentheses.
[c]$NaCO_3 \cdot 10\ H_2O$.

1967, 1969). Some of the attributes of the three species are compared in Table 7.1.

Based on 16S rRNA oligonucleotide catalogs, two sub-groups among five of the species of *Ectothiorhodospira* have been recognized (Stackebrandt et al., 1984). *E. halophila, E. halochloris,* and *E. abdelmalekii* form one similar group, and *E. mobilis* and *E. shaposhnikovii* belong to a second group. *E. vacuolata* was not included in the comparison. The evolutionary implications for the first group in the genus are interesting because of the wide differences in the molar % G+C in the DNA (Table 7.1) and the heterogeneity of the major pigments.

Several distinct biochemical or physiological characteristrics are found among the different strains of halotolerant and halophilic species of *Ectothiorhodospira.* Under certain growth conditions, *E. vacuolata* produces gas vacuoles that cause the cells to float. The vacuoles disappear under conditions of low light and low sulfide (Imhoff et al., 1981).

An obligately anaerobic, phototrophic, alkalophilic (pH 9.0–9.5) bacterium resembling *E. shaposhnikovii* was isolated from Lake Bogoria (also called Lake Hannington), Kenya (Grant et al., 1979). It has a broader salt tolerance for sodium carbonate (0.5–15% $Na_2CO_3 \cdot 10\ H_2O$) than for sodium chloride (0.1–8% NaCl).

E. halochloris and *E. abdelmalekii* produce bacteriochlorophyll *b* while the other salt-tolerant species produce bacteriochlorophyll *a.* The esterifying alcohol in bacteriochlorophyll *b* of *E. halochloris* is Δ2,10 phytodienol, which is different from phytol (Δ2-phytaenol), the esterifying alcohol of this pigment in purple non-sulfur bacteria (Steiner et al., 1981).

E. halophila has soluble purple and yellow proteins that have not been found in other phototrophic bacteria (Meyer, 1985). Meyer noted that the proteins are very acidic, much like the proteins of other halophiles. Imhoff et al. (1983) also found that the membrane proteins from several halophilic species of *Ectothiorhodospira* contained a relatively high proportion of polar:non-polar amino acids.

The main compatible solute of *E. halochloris* is glycine betaine (Galinski and Trüper, 1982). In cells grown in 20–24% NaCl, the intracellular concentration of glycine betaine was 1.6 molal, or about 10% of the dry weight. In cells grown in 16% NaCl, the cells accumulated 1.4 molal glycine betaine. In cells cultured in 12% salt, only about 0.8 molal intracellular glycine betaine accumulated. Concentrations of intracellular amino acids and polyols were small, and the cells largely excluded monovalent cations. Because the measured concentrations of glycine betaine were insufficient to maintain the cells in osmotic balance with the medium, the authors concluded that other undetected compatible solutes must also be present. Subsequent work showed that a unique, cyclic amino acid, ectoine (1,4,5,6-tetrahydro-2-methyl-4-pyrimidinecarboxylic acid), plays a role in osmoregulation in the genus *Ectothiorhodospira* (Galinski et al., 1985; Schuh et al., 1985).

The lipid composition of the six species of *Ectothiorhodospira* listed in Table 7.1 was determined by Asselineau and Trüper (1982). Among the phospholipids, all synthesize cardiolipin, phosphatidylglycerol, and phosphatidylcholine. The latter phospholipid is rare in bacteria although it has been found in the purple non-sulfur bacteria. Phosphatidylethanolamine was found in all species except *E. halochloris* and *E. abdelmalekii*. An unidentified glycolipid was observed in all species except *E. vacuolata* and *E. mobilis*.

The fatty acids are dominated by $C_{18:1}$ followed by $C_{16:0}$ acids. Smaller amounts of $C_{16:1}$ and $C_{18:0}$ fatty acids are present. A C_{19} fatty acid with a cyclopropane ring constituted 31% of the total fatty acids in *E. halochloris, E. abdelmalekii*, and *E. mobilis*. This fatty acid is missing in *E. shaposhnikovii* and *E. vacuolata*. A $C_{17:2}$ or $C_{17:3}$ fatty acid was identified in *E. halochloris* and *E. abdelmalekii*.

An unidentified strain of *Ectothiorhodospira* from the alkaline, hypersaline Owens Lake was found to survive in a synthetic brine formulated to lake water composition (26% salts, pH 9.6, $a_w = 0.93$) and in mirabilite ($Na_2SO_4 \cdot 10\ H_2O$), but not in thenardite (Na_2SO_4) (Tew, 1980). The optimal water activity for the organism was 0.95. Apparently the loss of chemically-bound water in the sodium sulfate minerals had a detrimental effect on the ability of the cells to retain water.

Purple sulfur bacteria that deposit elemental sulfur internally include *Chromatium* spp. (Chromatiaceae), which are commonly found in environments of normal marine salinity. They may be poor competitors under hypersaline conditions. A marine isolate that grew optimally in 5% NaCl was found to tolerate >10% NaCl, although most marine strains of *Chromatium* require lower salt concentrations (Imhoff, 1988). A large *Chromatium* was observed in hypersaline enrichments from Wadi Natrun (30% total dissolved solids), but it was not isolated (Imhoff et al., 1978).

7.3 Purple non-sulfur bacteria

Among the moderately halophilic purple non-sulfur bacteria, the best known species is *Rhodospirillum salexigens* (see Table 7.1). The type strain, WS68, was originally isolated from partially evaporated pools of seawater with decaying seaweed (Golecki and Drews, 1980; Drews, 1981). It is an obligate halophile requiring 5–20% salt at neutral pH although it grows optimally in 6–8% NaCl. It can grow as an anaerobe in the light in mineral medium with acetate as the carbon source, or it can grow aerobically in the dark. It is unable to grow as a photoautotroph on sulfide or thiosulfate.

The thin, Gram-negative cell wall of *R. salexigens* lacks lipopolysaccharides or other carbohydrate-linked constituents (Golecki and Drews, 1980; Drews, 1981; Tadros et al., 1982; Evers et al., 1984). Like the membrane proteins of halophilic species of *Ectothiorhodospira*, the envelope protein of

R. salexigens has an excess of acidic over basic amino acids (18.3 mol %). The major fatty acids are C_{14}, C_{16}, and C_{18} acids.

In acetate mineral medium, Drews (1981) reported that nitrogen could be assimilated from NH_4^+ or glutamate. However, Rubin and Madigan (1986) showed that NH_4^+ did not support growth on acetate medium because the cells secreted a very basic substance that raised the pH above that which would support growth. In acetate medium, only certain amino acids could be assimilated as nitrogen sources unless the medium was supplemented with pyruvate. Glutamate was the preferred source of nitrogen. Glycine and asparagine were poor sources of nitrogen, and neither urea, nitrate, nor alanine were utilized. The authors noted that *R. salexigens* can fix N_2.

A second halophilic species of purple non-sulfur bacteria, *Rhodospirillum salinarum,* was described from a solar saltern (Nissen and Dundas, 1984). It was found in the NaCl crystallizer ponds co-existing with halobacteria. It grows as an anaerobic phototroph and as an aerobic heterotroph in the dark. Neither sulfide or elemental sulfur can serve as a reductant for phototrophic growth.

The salt optima and tolerances are different for phototrophic and heterotrophic growth in *R. salinarum*. For anaerobic growth, the optimal salt concentration is 12–18%, but good growth is still observed in 24% NaCl. For aerobic growth, the optimal salinity is only 6–12%, and little or no growth is observed in >20% salt. The authors noted that the osmoregulatory response of *R. salinarum* to increasing salinity did not involve an increase in internal polyalcohols or amino acids, but a slight increase in internal K^+ was noted. The authors suggested that glycine betaine is the major compatible solute.

A third halophilic species of *Rhodospirillum, R. mediosalinum,* isolated from a spring, grows optimally in 5–7% NaCl, but tolerates as much as 15% NaCl (Kompantseva and Gorlenko, 1984). Imhoff (1988) also reported the isolation of strains of *Rhodobacter* from a variety of hypersaline environments. Optimal growth of all strains of *R. sulfidophilus* was only 1–5% NaCl although some strains tolerated up to 10% salt. *R. adriaticus* had a slightly higher salt optimum (2.5–7.5% NaCl). *Rhodopseudomonas marina,* a common marine isolate, had a similar salinity optimum as *R. sulfidophilus.*

7.4 Microbial ecology

Phototrophic bacteria have been noted in both the plankton and benthos of a wide variety of hypersaline environments, including alkaline and neutral pH lakes, sabkha flats and ponds, and solar salterns. Imhoff (1988) gave a detailed account of interrelationships between physical and chemical conditions, and phototrophic bacterial growth and success in marine and some hypersaline habitats. Phototrophic bacteria may be responsible for significant contributions to the primary productivity of those habitats as well as the degradation of low molecular weight substances, especially near the oxic-anoxic interface in the dark. In Big Soda Lake, Nevada, purple sulfur bacteria (identified as *Thiocapsa*) were responsible for 25% of the primary productivity (Priscu et al., 1982). During stratification in Solar Lake, Sinai, the photosynthetic uptake of $^{14}CO_2$ by the bacterial plate at 2.0–2.5 m depth was calculated by Cohen et al. (1977) to be as high as 2.80–3.10 g $C \cdot m^{-3} \cdot d^{-1}$. High concentrations of *Chromatium violescens* (up to 10^6 $cells \cdot ml^{-1}$) and *Prosthecochloris* (up to 2×10^6 $cells \cdot ml^{-1}$) were found in the water column. These bacteria were not characterized in culture. Near the benthos in the anaerobic hypolimnion, the populations of phototrophic bacteria and cyanobacteria were mixed. Photosynthetic uptake in the 4.0–4.5 m depth zone during stratification was similar to that in the metalimnion (2.50–3.20 g $C \cdot m^{-3} \cdot d^{-1}$). Mixed populations of oxygenic photoautotrophs (including *Dunaliella*) and phototrophic bacteria were noted in shallow solar saltern ponds of 10–18% salinity. In a plankton sample of about 13% salinity taken in the summer, 46 $\mu g \cdot l^{-1}$ bacteriochlorophyll *a* was found mixed with nearly 200 $\mu g \cdot l^{-1}$ chlorophyll *a* (Javor, 1983).

The morphologic, metabolic, and taxonomic diversity of halotolerant and halophilic phototrophic bacteria has only been partially explored. Purple sulfur bacteria have often been noted in marine hypersaline sediments and microbial mats beneath a layer of cyanobacteria, but they have rarely been cultured. The competition for sulfide by *Chromatium* and the colorless sulfur bacterium *Beggiatoa* in the benthic microbial mats of a saltern was described (Jørgensen and Des Marais, 1986). Novel, filamentous purple bacteria have also been found in those mats (D'Amelio et al., 1987). The possession of distinct pigments and possibly unusual compatible solutes make phototrophic bacteria good potential candidates as biogenic markers in studies of the organogeochemistry of evaporite habitats. Hypersaline enrichments (15% salinity) from microbial mats of an Israeli saltern produced mixed populations of phototrophic bacteria (Javor, unpublished data). Some peaks in the *in vivo* absorption spectra of the enrichments corresponded to bacteriochlorophylls *a* and *c*, as well as *e* (712 nm) and an unknown pigment (776 nm). These novel phototrophs, as well as those described by D'Amelio et al. (1987), await isolation in pure culture. Enrichments and isolation of salt-tolerant and halophilic phototrophic bacteria from a variety of evaporite habitats should demonstrate that halophilism can be found among all the major groups, and perhaps even among novel groups, of these microorganisms.

References

Anderson, G.C. 1958. Some limnological features of a shallow saline meromictic lake. *Limnology and Oceanography* 3: 259–270.
Asselineau, J. and **Trüper, H.G.** 1982. Lipid composition of six species of the pho-

totrophic bacterial genus *Ectothiorhodospira. Biochimica et Biophysica Acta* 712: 111–116.

Cohen, Y., Krumbein, W.E., and **Shilo, M.** 1977. Solar Lake (Sinai). 3. Bacterial distribution and production. *Limnology and Oceanography* 22: 621–634.

D'Amelio, E.D., Cohen, Y., and **Des Marais, D.J.** 1987. Association of a new type of gliding, filamentous, purple phototrophic bacterium inside bundles of *Microcoleus chthonoplastes* in hypersaline cyanobacterial mats. *Archives of Microbiology* 147: 213–220.

Drews, G. 1981. *Rhodospirillum salexigens,* spec. nov., an obligately halophilic phototrophic bacterium. *Archives of Microbiology* 130: 325–327.

Evers, D., Weckesser, J., and **Drews, G.** 1984. Protein on the cell surface of the moderately halophilic phototrophic bacterium *Rhodospirillum salexigens. Journal of Bacteriology* 160: 107–111.

Galinski, E.A. and **Trüper, H.G.** 1982. Betaine, a compatible solute in the extremely halophilic phototrophic bacterium *Ectothiorhodospira halochloris. FEMS Microbiology Letters* 13: 357–360.

Galinski, E.A., Pfeiffer, H.-P., and **Trüper, H.G.** 1985. 1, 4, 5, 6-Tetrahydro-2-methyl-4-pyrimidinecarboxylic acid: a novel cyclic amino acid from halophilic phototrophic bacteria of the genus *Ectothiorhodospira. European Journal of Biochemistry* 149: 135–140.

Golecki, J.R. and **Drews, G.** 1980. Cellular organization of the halophilic phototrophic bacterium strain WS68. *European Journal of Cell Biology* 22: 654–660.

Grant, W.D., Mills, A.A., and **Schofield, A.K.** 1979. An alkalophilic species of *Ectothiorhodospira* from a Kenyan soda lake. *Journal of General Microbiology* 110: 137–142.

Imhoff, J.F. 1986. Osmoregulation and compatible solutes in eubacteria. *FEMS Microbiology Reviews* 39: 57–66.

Imhoff, J.F. 1988. Halophilic phototrophic bacteria, p. 85–108 in Rodriguez-Valera, F. (editor), *Halophilic Bacteria,* Vol. I. CRC Press, Boca Raton.

Imhoff, J.F. and **Trüper, H.G.** 1977. *Ectothiorhodospira halochloris* sp. nov., a new extremely halophilic phototrophic bacterium containing bacteriochlorophyll *b. Archives of Microbiology* 114: 115–121.

Imhoff, J.F. and **Trüper, H.G.** 1981. *Ectothiorhodospira abdelmalekii* sp. nov., a new halophilic and alkalophilic phototrophic bacterium. *Zeitblatt fur Bakteriologie, Mikrobiologie und Hygiene* I. Abteil Originale C 2: 228–234.

Imhoff, J.F., Hashwa, F., and **Trüper, H.G.** 1978. Isolation of extremely halophilic phototrophic bacteria from the alkaline Wadi Natrun, Egypt. *Archiv fur Hydrobiologie* 84: 381–388.

Imhoff, J.F., Tindall, B.J., Grant, W.D., and **Trüper, H.G.** 1981. *Ectothiorhodospira vacuolata* sp. nov., a new phototrophic bacterium from soda lakes. *Archives of Microbiology* 130: 238–242.

Imhoff, J.F., Kushner, D.J., and **Anderson, P.J.** 1983. Amino acid composition of proteins of halophilic phototrophic bacteria of the genus *Ectothiorhodospira. Canadian Journal of Microbiology* 29: 1675–1679.

Javor, B.J. 1983. Nutrients and ecology of the Western Salt and Exportadora de Sal saltern brines, pp. 195–205 in Schreiber, B.C. and Harner, H.L. (editors), *Sixth International Symposium on Salt,* vol. 1, The Salt Institute, Toronto.

Jørgensen, B.B. and **Des Marais, D.J.** 1986. Competition for sulfide among colorless and purple sulfur bacteria in cyanobacterial mats. *FEMS Microbiology Ecology* 38: 179–186.

Jørgensen, B.B., Cohen, Y., and **Des Marais, D.J.** 1987. Photosynthetic action spectra and adaptation to spectral light distribution in a benthic cyanobacterial mat. *Applied and Environmental Microbiology* 53: 879–886.

Kompantseva, E.I. and **Gorlenko, V.M.** 1984. A new species of moderately halophilic purple bacterium *Rhodospirillum mediosalinum* sp. nov. *Mikrobiologiya* 53: 954–961.

Krasil'nikova, E.N. and **Kondrat'eva, E.N.** 1984. Growth of different purple sulfur bacterial species belonging to the genus *Ectothiorhodospira* in the dark. *Mikrobiologiya* 53: 526–528.

Meyer, T.E. 1985. Isolation and characterization of soluble cytochromes, ferredoxins and other chromophoric proteins from the halophilic phototrophic bacterium, *Ectothiorhodospira halophila. Biochimica et Biophysica Acta* 806: 175–183.

Nissen, H. and **Dundas, I.D.** 1984. *Rhodospirillum salinarium* sp. nov., a halophilic photosynthetic bacterium isolated from a Portuguese saltern. *Archives of Microbiology* 138: 251–256.

Priscu, J.C., Axler, R.P., Carlton, R.G., Reuter, J.E., Arneson, P.A., and **Goldman, C.R.** 1982. Vertical profiles of primary productivity, biomass and physico-chemical properties in meromictic Big Soda Lake, Nevada, U.S.A. *Hydrobiologia* 96: 113–120.

Puchkova, N.N. 1984. Green sulfur bacteria inhabiting shallow saline water bodies. *Mikrobiologiya* 53: 324.

Raymond, J.C. and **Sistrom, W.R.** 1967. The isolation and preliminary characterization of a halophilic photosynthetic bacterium. *Archiv für Mikrobiologie* 59: 255–268.

Raymond, J.C. and **Sistrom, W.R.** 1969. *Ectothiorhodospira halophila*: a new species of the genus *Ectothiorhodospira. Archiv für Mikrobiologie* 69: 121–126.

Rubin, C.A. and **Madigan, M.R.** 1986. Amino acid and ammonia metabolism by the halophilic nonsulfur purple bacterium *Rhodospirillum salexigens. FEMS Microbiology Letters* 34: 73–77.

Schuh, W., Puff, H., Galinski, E.A., and **Trüper, H.G.** 1985. Die Kristallstruktur des Ectoin, einer neuen osmoregulatorisch werksamen Aminosauer. *Zeitschrift für Naturforschung* 40c: 780–784.

Stackebrandt, E., Fowler, V.J., Schubert, W., and **Imhoff, J.F.** 1984. Toward a phylogeny of purple sulfur bacteria: the genus *Ectothiorhodospira. Archives of Microbiology* 137: 366–370.

Steiner, R., Schafer, W., Blos, I., Wieschhoff, H., and **Scheer, H.** 1981. Δ2,10-Phytadienol as esterifying alcohol of bacteriochlorophyll *b* from *Ectothiorhodospira halochloris. Zeitschrift für Naturforschung* 36c: 417– 420.

Stolz, J.F. 1984. Fine structure of the stratified microbial community at Laguna Figueroa, Baja California, Mexico: II. Transmission electron microscopy as a diagnostic tool in studying microbial communities *in situ*, pp. 23–38 in Cohen, Y., Castenholz, R.W. and Halvorson, H.O. (editors), *Microbial Mats: Stromatolites*, MBL Lectures in Biology, vol. 3, Alan R. Liss, New York.

Tadros, M.H., Drews, G., and **Evers, D.** 1982. Peptidoglycan and protein, the major cell wall constituents of the obligate halophilic bacterium *Rhodospirillum salexigens. Zeitschrift für Naturforschung* 37c: 210–212.

Tew, R.W. 1980. Halotolerant *Ectothiorhodospira* survival in mirabilite: experiments with a model of chemical stratification by hydrate deposition in saline lakes. *Geomicrobiology Journal* 2: 13–20.

Then, J. and **Trüper, H.G.** 1983. Sulfide oxidation in *Ectothiorhodospira abdelmalekii*: evidence for the catalytic role of cytochrome C-551. *Archives of Microbiology* 135: 254–258.

Then, J. and **Trüper, H.G.** 1984. Utilization of sulfide and elemental sulfur by *Ectothiorhodospira halochloris. Archives of Microbiology* 139: 295–298.

Trüper, H.G. and **Imhoff, J.F.** 1981. The genus *Ectothiorhodospira*, pp. 274– 278 in Starr, M.P., Stolp, H., Trüper, H.G., Balows, A. and Schlegel, H.G. (editors), *The Prokaryotes. A Handbook on Habitats, Isolation, and Identification of Bacteria*, vol. 1, Springer-Verlag, New York.

8

Cyanobacteria

Cyanobacteria (blue-green algae) constitute a separate group of phototrophic eubacteria. They are distinct from the other groups of phototrophic procaryotes by certain aspects of their cell organization as well as by their synthesis of chlorophyll *a* and phycobilin pigments which are used for oxygenic photosynthesis. In addition, some cyanobacteria can perform anoxygenic photosynthesis. There is greater morphological diversity among the cyanobacteria than among many other groups of bacteria. Several taxonomic surveys of cyanobacteria from hypersaline environments have been published and are compiled in this volume. The physiology and biochemistry of several isolated species of halophilic cyanobacteria are discussed below. Studies of the primary productivity of halophilic cyanobacteria in their natural habitats are presented in Chapters 12–20.

8.1 Distribution

Cyanobacteria are common inhabitants of extremely hypersaline habitats, both those of permanently high salinity and those in which evaporation-dilution cycles cause broad fluctuations in salinity. In a survey of microbial mats of hypersaline pools of the Sinai, Potts (1980) noted that coccoid and other non-filamentous cyanobacteria dominated the pool peripheries. The coccoid genera *Entophysalis* and *Pleurocapsa* were found in association with gypsum crusts. In the central regions of the pools, filamentous forms dominated, including *Oscillatoria* and *Microcoleus*. In total, 41 species of cyanobacteria were recorded from the hypersaline coastal pools, including 24 species in Solar Lake and 13 species from Ras Mohammed pool.

Both the *Pleurocapsa* mats and the *Microcoleus* mats in a sabkha environment had nearly equivalent chlorophyll *a* contents (46 and 55 $\mu g \cdot cm^{-3}$, respectively) (Potts, 1980; Gerdes et al., 1985a). The pheophytin content was only about 1% of the total chlorophyll *a* (Potts, 1980). In contrast, a Solar Lake mat contained 260 $\mu g \cdot cm^{-3}$ chlorophyll *a*, of which 50% of the pigment was pheophytin.

The taxonomy and distribution of cyanobacteria in Gavish Sabkha were detailed by Erlich and Dor (1985) and Gerdes et al. (1985). Three different mat types with distinct assemblages were recognized: compact, laminated mats; soft, floccose mats; and thin, slimy mats. Cyanobacteria dominated the biomass of the environment in salinities greater than 10%. Eleven species were commonly found in ≥9–13.5% salinity by Erlich and Dor. Only two species were found in salinities of 25–33% (*Aphanothece halophytica* and *Schizothrix arenaria*) (Table 8.1). Their catalog of figures of the cyanobacteria from the Gavish Sabkha should be useful for recognizing and comparing cyanobacteria from other hypersaline habitats.

The cyanobacterial mats of Solar Lake are of several different types, reflecting their depth distribution in the pond and their species composition (Krumbein et al., 1977; Jørgensen et al., 1983). These include flat mats in

Table 8.1 Salinity distribution of cyanobacteria in some hypersaline habitats[a]

Species	Salinity, %				Reference[b]
	9–13.5	15–17.5	18–20.5	25–33	
Aphanothece halophytica	+	++	+++	++	1, 2
A. stagnina	+	+	+++		2
Aphanocapsa marina	++	+++	++		2
Chroococcus minor	+	+	++		2
Chroococcus sp.		+	+		2
Dactylococcopsis salina		+			6
Entophysalis granulosa		++	+		2
Gloeocapsa polydermica			+		2
Gloeothece confluens			+		2
Gomphosphaeria aponina	+	+	+		2
Lyngbya, 3 spp.	+				2
Microcoleus chthonoplastes	+	+	+		2, 3, 4
Oscillatoria limnetica		+	+		5
O. salina		+	+		5
O. nigro-viridis	+				2
O. tenuis	+	+			2
Phormidium sp.				+	1
Pleurocapsa fuliginosa	++	++	+++		2
Schizothrix arenaria	+	++	+	++	2
S. calcicola	+++	++	+++		2
S. nasri	+	+	+		2
Spirulina subsalsa	+++	++	++		2
S. labyrinthiformis	+++	+			2

[a]Indicated by relative abundance; +++ = most abundant.
[b]References: (1) Brock (1976); (2) Erlich and Dor (1985); (3) Javor (1983); (4) Jørgensen et al. (1983); (5) Krumbein et al. (1977); (6) Walsby et al. (1983).

shallow water, pinnacle or blister mats on the upper slope, gelatinous mats at the thermocline, films on the lower slope, and floccose mats at the bottom. The cyanobacteria of the shallow mat consist of many of the species listed in Table 8.1, including representatives of both coccoid (*Aphanothece, Aphanocapsa,* and *Entophysalis*) and filamentous (*Microcoleus, Oscillatoria, Phormidium,* and *Spirulina*) forms. The cyanobacteria of the pinnacle mat consist of coccoid strains while the cyanobacterial community of the bottom floccose zone consists of *Oscillatoria limnetica* and *O. salina*. Seasonal variation in the mat composition was noted. The salinity of Solar Lake varies from 6.8% to about 18% throughout the year. The photic zone of each mat reflects its composition and compactness, ranging from 0.8 mm in the shallow mat to 10 mm in the gelatinous mat. The composition of each mat on a millimeter-scale was given by Jørgensen et al. (1983). The productivity of these mats is described in Chapter 14.

A halophilic cyanobacterium from Solar Lake, *Dactylococcopsis salina,* was also described (Walsby et al., 1983, van Rijn and Cohen, 1983). It grows in the plankton at 1–4 m depth during lake stratification. In culture it can grow in salinities of 5–20% and at temperatures up to 45 °C. Its optimal salinity is 7.5–15% and it survives 3% salinity but without growth. Experiments with brine shrimp from the Solar Lake showed that *D. salina* populations were effectively grazed by adult brine shrimp. This species is unique to Solar Lake (Potts, 1980).

The cyanobacterial population of the Great Salt Lake, Utah, is largely confined to the more dilute sections of the South Arm with the exception of *Aphanothece halophytica* and a species of *Phormidium* or *Oscillatoria* (Brock, 1976; Post, 1977, 1981; Felix and Rushforth, 1979). *A. halophytica* is always found in biostromes (microbial mats) or benthic communities, and never in the plankton.

The cyanobacterial species composition of solar salterns has not been well documented. Davis (1978) identified coccoid (*Anacystis* spp., *Aphanothece halophytica* [*Coccochloris elabens*], *Entophysalis*) and filamentous (*Oscillatoria, Schizothrix, Spirulina,* and *Porphyrosiphon*) forms, but their distributions with respect to salinity were not given in detail. Other identified cyanobacteria include coccoid species similar to *Aphanocapsa* or *Coelospherium, Dactylococcopsis* or *Synechococcus,* and *Xenococcus* (Golubic, 1980) as well as filamentous species *Microcoleus chthonoplastes* and *Phormidium* sp. (Javor, 1983; Jørgensen and Des Marais, 1986).

Other references to cyanobacteria actually growing in high salinities are not well detailed. Bauld (1981) gave a list of cyanobacteria in microbial mats in saline lakes and the maximum salinity of those lakes. This documentation does not take into consideration whether the microorganisms actually are thriving in the more dilute epilimnion or only during periods of lower salinity of the lakes. These considerations are particularly notable in the Antarctic lake environments. Imhoff et al. (1978, 1979) also noted cyanobacteria in hypersaline (up to nearly 40% salinity), alkaline lakes. No details of their taxonomy or physiology were given besides the notation of a *Synechococcus* from an enrichment of Lake Hamra (23.8% TDS). It grew optimally in 10–20% salt and tolerated 5–30% salt. It could grow anoxigenically.

Spirulina platensis forms exceptionally dense blooms in some saline, alkaline lakes such as the African soda lakes. However, the total salinity of such lakes may not exceed by much that of marine seawater. For example, Lake Bogoria, with a conductivity of 72 $mmho \cdot cm^{-1}$ (Melack, 1981), probably has a salinity corresponding to approximately 0.9–1.0 M NaCl or Na_2CO_3 (see Figure 1.2). Other *Spirulina*-rich lakes of Kenya had even lower conductivities.

8.2 Taxonomy

The proper taxonomy of cyanobacteria is more than just a trivial pursuit. Unlike other bacteria, cyanobacterial identification cannot be confirmed by a simple battery of biochemical and growth tests. Taxonomic keys by different authors sometimes give different names for morphologically similar cyanobacteria. The morphology of cyanobacteria can be altered by environmental conditions. For example, Gerdes et al. (1985) isolated numerous halotolerant cyanobacteria from the Gavish Sabkha. Although details with respect to most of the isolates were not given, the following general features were reported. Some strains grew in artificial seawater medium in 23% salinity. Increased salinity resulted in the increase of cell dimensions and increase in slime production. One strain of *Synechococcus* (possibly *Aphanothece* or *Aphanocapsa*) increased in diameter from 6 μm to 15 μm when transferred from low-salt to high-salt medium. It was noted that it even changed to filamentous growth forms in high salt, suggesting that cell division is inhibited. Although many strains were halotolerant, they tended to grow better in more dilute salinities.

Some of the problems of taxonomy of halophilic and halotolerant cyanobacteria are described by Brock (1976) and Golubic (1980). *Aphanothece halophytica* has also been called *Aphanocapsa, Coccochloris elabens, Stichococcus,* and possibly *Synechococcus*. Morphologically distinct forms co-exist in hypersaline habitats. Only a detailed description of their physiological, biochemical, and ecological attributes could provide a basis for determining their species or strain assignments. The following brief description summarizes some of the easily detectable differences between common coccoid cyanobacteria of hypersaline habitats (Golubic, 1980): *Aphanothece, Synechococcus,* and *Gloeothece* are usually rod-shaped single cells that divide in a single plane. *Aphanothece* grows embedded in an amorphous mucilage while *Synechococcus* does not. Jørgensen et al. (1983) noted that an isolate of *Aphanothece* from Solar Lake only produced slime capsules under conditions of low phosphate but not in the presence of high phosphate concentrations. *Xenococcus* also produces a mucilaginous sheath. *Aphanocapsa, Synechocystis,* and *Gloeocapsa* are usually spherical cells that divide in two or three planes. *Dactylococcopsis* is a long, spindle-shaped cell that divides transversely.

Taxonomy of filamentous forms of the Oscillatoriaceae (possibly the only family of filamentous cyanobacteria found in extremely hypersaline habitats) also can be problematic. The filamentous forms are typically unsheathed (*Oscillatoria, Schizothrix,* and *Spirulina*) or sheathed (LPP forms: *Lyngbya-Phormidium-Plectonema*). *Microcoleus chthonoplastes* is encased in a multiple-filament sheath. However, individual trichomes can glide out of the sheath and be mistaken for *Oscillatoria* or *Schizothrix* under the microscope. The most distinctive member of the group is *Spirulina,* which grows as a corkscrew-shaped filament.

Members of the heterocyst-producing Nostocales are largely absent from environments that remain continuously, extremely hypersaline. *Calothrix* mats cover large areas of somewhat hypersaline but periodically inundated salt marshes (Javor and Castenholz, 1984). Hammer et al. (1983) noted the presence of *Nodularia spumigena* and *Anabaena wisconsinense* in hypersaline lakes of Saskatchewan.

The heterocyst is a specialized cell involved in N_2 fixation. N_2 fixation (nitrogenase activity) has been detected in hypersaline habitats, but the salinities were not reported (Potts, 1980). It is unknown whether the lack of nostocalean cyanobacteria in these environments reflects the lack of halophilic species or whether such habitats are not typically N-limited.

8.3 *Aphanothece halophytica*

More is known about *Aphanothece halophytica* than about any other halophilic cyanobacterium. Because it grows in a mucilaginous coat that also encourages bacterial growth (Figure 8.1), it is difficult to isolate axenic cultures. Yopp et al. (1978a) used a variety of techniques to obtain pure cultures: temperature elevation and antibiotics to inhibit eucaryotes, and osmotic shock, antibiotics, density gradients, and ultraviolet radiation to rid the cells of other bacteria.

A. halophytica has been isolated from a variety of environments, including the Great Salt Lake, solar salterns, and Solar Lake (Figure 8.2). Its responses to salt depend on its source. Brock (1976) found that an isolate from the Great Salt Lake could grow slowly in medium saturated with NaCl (about 30% salinity), but grew optimally in about 16–23% salinity. Isolates from solar salt ponds grew optimally in 5.8–11.6% salinity (Yopp et al., 1978a) or in 5–15% (Kao et al., 1973). They grew slowly in 25–30% salinity. Yopp et al. (1978a) observed that cells grown in 1–2 M NaCl were cocccoid, bicellular, and small. In 4 M NaCl, cells were elongated, cylindrical, often bicellular, and larger. In saturated NaCl, cells were elongated, in short filaments, and large. The isolate from Solar Lake showed fastest growth in medium prepared from

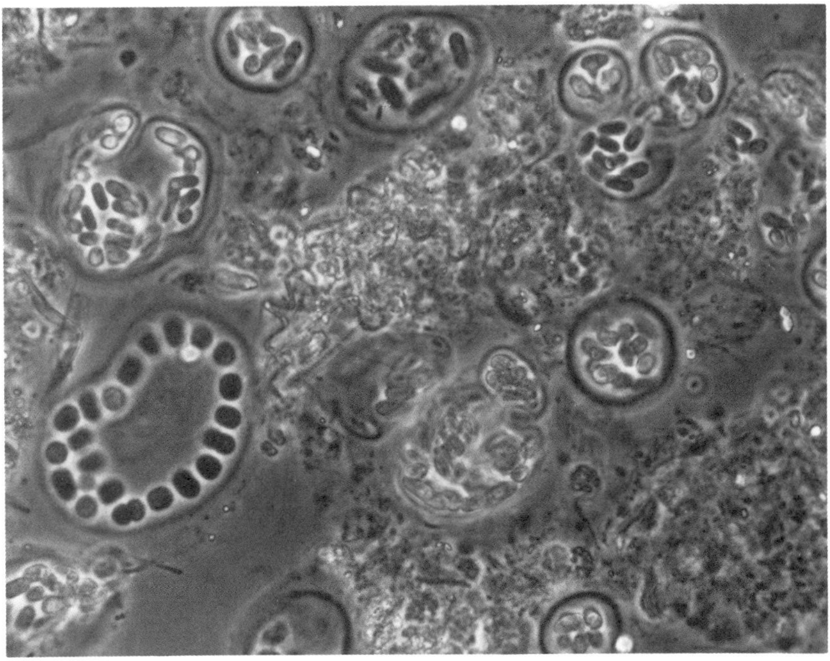

Figure 8.1 *Aphanothece halophytica* in mucilaginous packets. Phase contrast. From a microbial mat in the Exportadora de Sal saltern (Mexico). Cells are about 2–4 μm wide.

ocean seawater or seawater plus 5% NaCl (Cohen, 1975). It grew moderately well in Solar Lake salinities (15–18% salinity) at 15°C but not at 35°C. It grew well in seawater plus 5% NaCl at 35°C but not at 50°C. The temperature of the bottom of Solar Lake reaches 60.5°C during stratification. This strain was also described as motile and gas-vacuolated (Simon, 1981), features that have not been noted in the more halophilic isolates.

A. halophytica isolated from the Leslie salt ponds (San Francisco Bay, California, U.S.A.) has been well characterized. It grows optimally at 43°C, with increasing generation times as salinity is raised from 2 M to 4 M NaCl (Tindall et al., 1978). The cells do not grow below pH 6, and pH values above 8 are inhibitory. The cells can use NO_3^-, NO_2^-, NH_4^+, or urea as sole nitrogen sources, but not glycine. They cannot grow as photoheterotrophs using glycerol, glucose, or sucrose. Photoinhibition was detected at light intensities >5000 lux. Sensitivity to antibiotics is similar to that of other Chroococcales (Yopp et al., 1979).

The protein composition of the cells varies with age and salinity (Tindall et al., 1977). In 1 M NaCl, protein content increased from 64% to 76% of the dry weight during the growth cycle. This is the highest value ever recorded

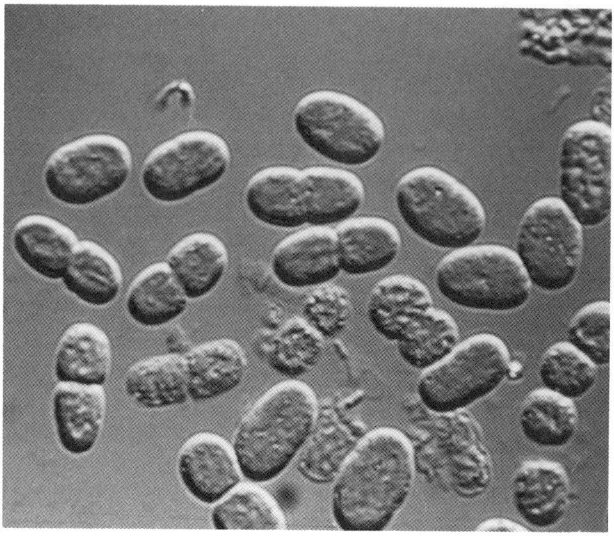

Figure 8.2 Photomicrograph of a pure culture of *Aphanothece halophytica* isolated from Great Salt Lake. Nomarski interference microscopy. (From Brock, 1976.)

for a cyanobacterium. In 2 M NaCl, protein content increased from 57% to 72% during growth, but in 3 M NaCl it decreased from 60% to 43% during the growth cycle. The composition of the bulk protein showed a dominance of acidic over basic amino acids (ratio = 2.3–2.6). The pigment c-phycocyanin had an acidic:basic amino acid ratio of 3.06, which was significantly higher than that of *Coccochloris elabens* (1.74). and a freshwater species (1.68). Kao et al. (1973) also found an excess of acidic amino acids in c-phycocyanin in a halophilic isolate they described as *C. elabens*.

A. halophytica isolated by Yopp et al. (1978a) is a true halophile. It lyses in distilled water. Neither KCl, LiCl, CsCl, nor glycerol can replace its requirement for NaCl. When the cells are transferred from low to high salt they deplasmolyze. Yopp et al. (1978b) found that intracellular water potential was adjusted primarily by the accumulation of K^+, although carbohydrates, polyols, and amino acids increased two- to four-fold during osmotic adjustment to higher salt. However, enzymes were inhibited by high salts but not by glycerol. Miller et al. (1976) determined that intracellular concentrations of K^+ were as high as 1 M and that they varied directly with salinity. Intracellular Na^+ content was low (about 0.38 mmoles·g^{-1} dry weight).

Other data on osmotic relations of cyanobacteria contrast with those of Miller et al. (1976) and Yopp et al. (1978b). Halophilic or halotolerant cyanobacteria have been found to produce quaternary ammonium compounds (glycine betaine or glutamate betaine) in combination with one or more sugars (sucrose, trehalose, glucose, or fructose) as primary osmolytes. Marine strains (those incapable of growth in $\geq$11% salinity) produce glucosylglycerol as their primary osmolyte (Mackay et al., 1983, 1984). In a study of four cyanobacteria from hypersaline habitats (*A. halophytica, C. elabens, Dactylococcopsis salina,* and *Synechococcus* DUN 52), Reed et al. (1984) found the internal K^+ concentration was not high (80–320 mmol·dm^{-3} cell volume) and only showed minor changes with external salinity. Intracellular carbohydrate concentrations were also low. Glycine betaine showed the greatest change with salinity and the greatest concentration (up to 1.64 mol·dm^{-3} cell volume). The accumulation of betaine glycine as the major osmolyte was also confirmed in *A. halophytica* from two Australian hypersaline habitats and in halophilic strains of *Synechococcus* (Mohammad et al., 1983; Mackay et al., 1983, 1984). Glycine betaine reduced the inhibitory effects of NaCl in an assay for glutamine synthase activity in halophilic *Synechococcus* DUN 52 (Warr et al., 1984).

At least one protein in *A. halophytica* requires salt or a compatible solute to maintain its structure. The multi-unit enzyme, D-ribulose 1,5-bisphosphate carboxylase, requires 250 mM NaCl, 300 mM KCl, or polyols to retain its 16 subunits intact (Codd et al., 1979; Asami et al., 1983, Incharden Sakdi et al., 1985).

Reports of anoxygenic photosynthesis by *Oscillatoria limnetica* from Solar Lake (see below and Cohen et al., 1975a, 1975b) were later followed by reports of similar activity by other Solar Lake isolates, including *A. halophytica* (Garlick et al., 1977; Oren et al., 1979; Cohen, 1984; Cohen et al., 1986; Jørgensen et al., 1986). Other halophilic species capable of this mode of photosynthesis, which uses S^{2-} instead of H_2O as a source of electrons, include *O. salina, Microcoleus chthonoplastes, Oscillatoria* from Bardawil Lagoon (Sinai), and *Phormidium* from Wadi Natrun and Mexico. Both *A. halophytica* and *O. limnetica* also can use H_2 as an electron donor for photosynthesis (Belkin and Padan, 1978). There have been no published reports of anoxygenic photosynthesis in *A. halophytica* from hypersaline environments besides Solar Lake. *Dactylococcopsis salina,* which stays in the oxygenated region of Solar Lake, is incapable of anoxygenic photosynthesis (Walsby et al., 1983; Cohen et al., 1986).

The fatty acid composition of *A. halophytica* has been investigated in both the saltern strain of Yopp et al. (1978a) and the Solar Lake strain. When the strain of Yopp et al. was grown in 2 M NaCl, the fatty acids were predominantly $C_{16:0}$ and $C_{18:1}$. (Colclasure et al., 1974). An increase in salinity influenced the ratios of saturated-monounsaturated-polyunsaturated acids, but details were not given. The major fatty acids of the polar lipids of the Solar Lake isolates grown in 2× seawater salinity were $C_{16:0}$ (41%), $C_{16:1}$ (29%), $C_{18:1}$ (15%), $C_{16:2}$ (11%), and minor amounts of C_{14} and other C_{18} fatty acids (Oren et al., 1985). The presence of significant $C_{16:2}$ fatty acids in *A. halophytica* is unique among halophilic cyanobacteria. Other cyanobacteria surveyed by

Oren et al. (1985), including *O. limnetica, Microcoleus, Phormidium* from Wadi Natrun, and *Spirulina platensis* (not a true halophile), synthesized very little $C_{16:2}$ fatty acids (4% of the total fatty acids).

8.4 *Oscillatoria limnetica*

Another halophilic cyanobacterium that has been investigated in detail is *Oscillatoria limnetica* of Solar Lake. It grows well in medium made with Solar Lake water or in seawater plus 5% NaCl, but a precise salinity optimum has not been described (Cohen, 1975). It grows well at 35°C and moderately well at 50°C. Fair growth was measured at 26°C and no growth was detected at 15°C. The high temperature tolerance reflects the temperatures it encounters in the hypolimnion of Solar Lake.

The ability of *O. limnetica* to grow by facultative anoxygenic photosynthesis and to switch between oxygenic and anoxygenic modes has been well documented (Cohen et al., 1975a, 1975b, 1986; Garlick et al., 1977; Oren et al., 1977; Belkin and Padan, 1978; Cohen, 1984). This ability allows *O. limnetica* to take advantage of S^{2-} or H_2 that emanates from decomposition processes below the mat surface, and that is trapped in the stratified bottom water and mats of habitats such as Solar Lake. The photosynthetic oxidation of sulfide leads to the accumulation of extracellular elemental sulfur. *O. limnetica* has also been found in microbial mats of a solar saltern in Mexico (Cohen et al., 1986).

In the dark under anaerobic conditions, *O. limnetica* can ferment polyglucose to lactate in the absence of sulfur (Oren and Shilo, 1979). In the presence of elemental sulfur, anaerobic respiration of reserve polyglucose occurs, which results in the reduction of S^0 to sulfide. Externally added glucose does not stimulate the reaction nor does thiosulfate or sulfate. Neither the fermentation nor sulfur respiration supports growth in the dark. These modes of anaerobic heterotrophy involving the breakdown of endogenous substrates are apparently associated with cell maintenance. Similar dark metabolism has not been reported in *A. halophytica* or other salt-tolerant cyanobacteria.

The fatty acid profile of *O. limnetica* grown in 2× seawater medium under both aerobic and anaerobic conditions was described by Oren et al. (1985). Under aerobic conditions the fatty acids in the polar lipid fraction were dominated by $C_{16:0}$ (43% of total) and $C_{18:1}$ (36%) fatty acids with lesser amounts of $C_{16:1}$ (16%) and $C_{18:0}$ (3%) fatty acids. C_{14} acids and polyunsaturated C_{16} and C_{18} acids each constituted <1% of the total in the polar lipids. Growth under anaerobic conditions produced a slight shift to lower amounts of $C_{16:1}$ fatty acid, and a slightly increased content of $C_{16:0}$ and $C_{18:1}$ acids. Unsaturated fatty acids constituted about 53% of the fatty acids in the polar lipids under aerobic conditions and 51% of the fatty acids in the polar lipids under anaerobic conditions.

Oscillatoria salina, another halophilic isolate from Solar Lake, grew better in Solar Lake water medium (salinity not given) at 35 °C than in seawater-based medium at that temperature (Cohen, 1975). In the presence of 0.45–0.90 mM sulfide, the isolate showed even higher rates of anoxygenic photosynthesis than *O. limnetica* (Garlick, et al., 1977).

8.5 Planktonic vs. benthic habitats

The relationships between cyanobacteria and the hypersaline habitats in which they thrive have only been partially described. In particular, the biological and chemical differences between planktonic and benthic cyanobacterial communities deserves further scrutiny. The author has noted morphologically distinct strains of *A. halophytica* in the plankton than in the benthos of solar salterns, both in fresh material and in enrichments in similar media. The strain from mats produced much more muciliage. Fattom et al. (1984) described the hydrophobic nature of mat-forming cyanobacteria (including *A. halophytica* and *O. limnetica*). This attribute allows the cells to adhere to surfaces or microbial mats. A planktonic cyanobacterium was shown to be hydrophilic.

The noted correlations between growth in low phosphate and mucilage excretion by *A. halophytica* and the dominance of microbial mats rather than plankton in phosphate-limited solar salterns (see Chapter 13) are evidence that phosphate availability is a determinant of plankton vs. benthic community development. Nitrogenase activity has also been found in hypersaline microbial mats, suggesting that this community might be important for fixing N_2 in hypersaline ponds and lakes. Microbial mats have sometimes fossilized as distinct stromatolitic sediments which remain as tell-tale signs of cyanobacterial mat communities. Based on such evidence, it can be concluded that nutrient availability has probably played an important role in determining the type of organic matter that has been generated and preserved in both recent and ancient hypersaline sediments.

References

Asami, S., Takabe, T., Akazawa, T., and **Codd, G.A.** 1983. Ribulose 1,5-bisphosphate carboxylase from the halophilic cyanobacterium *Aphanothece halophytica*. *Archives of Biochemistry and Biophysics* 225: 713–721.

Bauld, J. 1981. Occurrence of microbial mats in saline lakes. *Hydrobiologia* 81: 87–111.

Belkin, S. and **Padan, E.** 1978. Hydrogen metabolism in the facultative anoxygenic cyanobacteria (blue-green algae) *Oscillatoria limnetica* and *Aphanothece halophytica*. *Archives of Microbiology* 116: 109–111.

Brock, T.D. 1976. Halophilic blue-green algae. *Archives of Microbiology* 107: 109–111.

Codd, G.A., Cook, C.M., and **Stewart, W.D.P.** 1979. Purification and subunit structure of D-ribulose 1,5-bisphosphate carboxylase from the cyanobacterium *Aphanothece halophytica*. *FEMS Microbiology Letters* 6: 81–86.

Cohen, Y. 1975. *Dynamics of Procaryotic Photosynthetic Communities of the Solar Lake*, pp. 99–104. Ph.D. Thesis. Hebrew University of Jerusalem.

Cohen, Y. 1984. The Solar Lake cyanobacterial mats: strategies of photosynthetic life under sulfide, pp. 133–148 in Cohen, Y., Castenholz, R.W. and Halvorson, H.O. (editors), *Microbial Mats: Stromatolites*, MBL Lectures in Biology, vol. 3, Alan R. Liss, New York.

Cohen, Y., Padan, E., and **Shilo, M.** 1975a. Facultative anoxygenic photosynthesis in the cyanobacterium *Oscillatoria limnetica*. *Journal of Bacteriology* 123: 855–861.

Cohen, Y., Jørgensen, B.B., Padan, E., and **Shilo, M.** 1975b. Sulfide dependent anoxygenic photosynthesis in the cyanobacterium *Oscillatoria limnetica*. *Nature (London)* 257: 489–492.

Cohen, Y., Jørgensen, B.B., Revsbech, N.P., and **Poplawski, R.** 1986. Adaptation to hydrogen sulfide of oxygenic and anoxygenic photosynthesis among cyanobacteria. *Applied and Environmental Microbiology* 51: 398–407.

Colclasure, G.C., Yopp, J.H., and **Tindall, D.R.** 1974. Characterization of the fatty acids of the obligately halophilic blue-green alga, *Aphanothece halophytica*. *American Journal of Botany (Supplement)* 61: 21–22.

Davis, J.S. 1978. Biological communities of a nutrient enriched salina. *Aquatic Botany* 4: 23–42.

Erlich, A. and **Dor, I.** 1985. Photosynthetic microorganisms of the Gavish Sabkha, pp. 296–321 in Friedman, G.M. and Krumbein, W.E. (editors), *Hypersaline Ecosystems. The Gavish Sabkha*. Ecological Studies 53, Springer-Verlag, New York.

Fattom, A. and **Shilo, M.** 1984. Hydrophobicity as an adhesion mechanism of benthic cyanobacteria. *Applied and Environmental Microbiology* 47: 135–143.

Felix, E.A. and **Rushforth, S.R.** 1979. The algal flora of the Great Salt Lake, Utah, U.S.A. *Nova Hedwigia* 31: 163–194.

Garlick, S., Oren, A., and **Padan, E.** 1977. Occurrence of facultative anoxygenic photosynthesis among filamentous and unicellular cyanobacteria. *Journal of Bacteriology* 129: 623–629.

Gerdes, G., Krumbein, W.E., and **Holtkamp, E.** 1985. Salinity and water activity related zonation of microbial communities and potential stromatolites of the Gavish Sabkha, pp. 238–266 in Friedman, G.M. and Krumbein, W.E. (editors), *Hypersaline Ecosystems. The Gavish Sabkha*, Ecological Studies 53, Springer-Verlag, New York.

Golubic, S. 1980. Halophily and halotolerance in cyanophytes. *Origins of Life* 10: 169–183.

Hammer, U.T., Shamess, J. and **Haynes, R.C.** 1983. The distribution and abundance of algae in saline lakes of Saskatchewan, Canada. *Hydrobiologia* 105: 1–26.

Imhoff, J.F., Hashwa, F., and **Trüper, H.G.** 1978. Isolation of extremely halophilic phototrophic bacteria from the alkaline Wadi Natrun, Egypt. *Archiv für Hydrobiologie* 84: 381–388.

Imhoff, J.F., Sahl, H.G., Soliman, G.S.H., and **Trüper, H.G.** 1979. The Wadi Natrun: chemical composition and microbial mass developments in alkaline brines of eutrophic desert lakes. *Geomicrobiology Journal* 1: 219–234.

Incharden Sakdi, A., Takabe, T., and **Akazawa, T.** 1985. Factors affecting the dissociation and reconstitution of ribulose-1,5-bisphosphate carboxylase/oxygenase from *Aphanothece halophytica*. *Archives of Biochemistry and Biophysics* 237: 445–453.

Javor, B.J. 1983. Planktonic standing crop and nutrients in a saltern ecosystem. *Limnology and Oceanography* 28: 153–159.

Javor, B.J. and **Castenholz, R.W.** 1984. Productivity studies of microbial mats, Laguna Guerrero Negro, Mexico, pp. 149–170 in Cohen, Y., Castenholz, R.W. and Halvorson, H.O. (editors), *Microbial Mats: Stromatolites*, MBL Lectures in Biology, vol. 3, Alan R. Liss, New York.

Jørgensen, B.B., Revsbech, N.P., and **Cohen, Y.** 1983. Photosynthesis and structure of benthic microbial mats: microelectrode and SEM studies of four cyanobacterial communities. *Limnology and Oceanography* 28: 1075–1093.

Jørgensen, B.B. and **Des Marais, D.J.** 1986. Competition for sulfide among colorless and purple sulfur bacteria in cyanobacterial mats. *FEMS Microbiology Ecology* 38: 179–186.

Jørgensen, B.B., Cohen, Y., and **Revsbech, N.P.** 1986. Transition from anoxygenic to oxygenic photosynthesis in a *Microcoleus chthonoplastes* cyanobacterial mat. *Applied and Environmental Microbiology* 51: 408–417.

Kao, O.H.W., Berns, D.S., and **Town, W.R.** 1973. The characterization of c-phycocyanin from an extremely halo-tolerant blue-green alga, *Cocchloris elabens*. *Biochemical Journal* 131: 39–50.

Krumbein, W.E., Cohen, Y., and **Shilo, M.** 1977. Solar Lake (Sinai). 4. Stromatolitic cyanobacterial mats. *Limnology and Oceanography* 22: 635–656.

Mackay, M.A., Norton, R.S., and **Borowitzka, L.J.** 1983. Marine blue-green algae have a unique osmoregulatory system. *Marine Biology* 73: 301–307.

Mackay, M.A., Norton, R.S., and **Borowitzka, L.J.** 1984. Organic osmoregulatory solutes in cyanobacteria. *Journal of General Microbiology* 130: 2177–2191.

Melack, J.M. 1981. Photosynthetic activity of phytoplankton in tropical African soda lakes. *Hydrobiologia* 81:71–85.

Miller, D.M., Jones, J.H., Yopp, J.H., Tindall, D.R., and **Schmid, W.E.** 1976. Ion metabolism in a halophilic blue-green alga, *Aphanothece halophytica*. *Archives of Microbiology* 111: 145–149.

Mohammad, F.A.A., Reed, R.H., and **Stewart, W.D.P.** 1983. The halophilic cyanobacterium *Synechocystis* DUN 52 and its osmotic responses. *FEMS Microbiology Letters* 16: 287–290.

Oren, A., Padan, E., and **Avron, M.** 1977. Quantum yields for oxygenic and anoxygenic photosynthesis in the cyanobacterium *Oscillatoria limnetica*. *Proceedings of the National Academy of Science U.S.A.* 74: 2152–2156.

Oren, A. and **Shilo, M.** 1979. Anaerobic heterotrophic dark metabolism in the cyanobacterium *Oscillatoria limnetica*: sulfur respiration and lactate fermentation. *Archives of Microbiology* 122: 77–84.

Oren, A., Padan, E., and **Malkin, S.** 1979. Sulfide inhibition of photosystem II in cyanobacteria (blue-green algae) and tobacco chloroplasts. *Biochimica et Biophysica Acta* 546: 270–279.

Oren, A., Fattom, A., Padan, E., and **Tietz, A.** 1985. Unsaturated fatty acid composition and biosynthesis in *Oscillatoria limnetica* and other cyanobacteria. *Archives of Microbiology* 141: 138–142.

Post, F.J. 1977. The microbial ecology of the Great Salt Lake. *Microbial Ecology* 3: 143–165.

Post, F.J. 1981. Microbiology of the Great Salt Lake North Arm. *Hydrobiologia* 81: 59–69.

Potts, M. 1980. Blue-green algae (Cyanophyta) in marine coastal environments of the Sinai peninsula; distribution, zonation, stratification and taxonomic diversity. *Phycologia* 19: 60–73.

Reed, R.H., Richardson, D.L., Warr, S.R.C., and **Stewart, W.D.P.** 1984. Carbohydrate accumulation and osmotic stress in cyanobacteria. *Journal of General Microbiology* 130: 1–4.

Simon, R.D. 1981. Gliding motility in *Aphanothece halophytica*: analysis of wall protein in *mot* mutants. *Journal of Bacteriology* 148: 315–321.

Tindall, D.R., Yopp, J.H., Schmid, W.E., and **Miller, D.M.** 1977. Protein and amino acid composition of the obligate halophile *Aphanothece halophytica* (Cyanophyta). *Journal of Phycology* 13: 127–133.

Tindall, J.H., Miller, D.M., and **Schmid, W.E.** 1978. Physico-chemical parameters governing the growth of *Aphanothece halophytica* (Chroococcales) in hypersaline media. *Phycologia* 17: 179–185.

van Rijn, J. and **Cohen, Y.** 1983. Ecophysiology of the cyanobacterium *Dactylococcopsis salina*: effects of light intensity, sulphide and temperature. *Journal of General Microbiology* 129: 1849–1856.

Walsby, A.E., van Rijn, J., and **Cohen, Y.** 1983. The biology of a new gas-vacuolate cyanobacterium, *Dactylococcopsis salina* sp. nov., in Solar Lake. *Proceedings of the Royal Society of London* B 217: 417–447.

Warr, S.R.C., Reed, R.H., and **Stewart, W.D.P.** 1984. Osmotic adjustment of cyanobacteria: the effects of NaCl, KCl, sucrose and glycine betaine on glutamine synthetase activity in a marine and a halotolerant strain. *Journal of General Microbiology* 130: 2169–2175.

Yopp, J.H., Tindall, D.R., Miller, D.M., and **Schmid W.E.** 1978a. Isolation, purification and evidence for a halophilic nature of the blue-green alga *Aphanothece halophytica* Fremy (Chroococcales). *Phycologia* 17: 172–178.

Yopp, J.H., Miller, D.M., and **Tindall, D.R.** 1978. Regulation of intracellular water potential in the halophilic blue-green alga *Aphanothece halophytica* (Chroococcales), pp. 619–624 in Caplan, S.R. and Ginzburg, M. (editors), *Energy and Structure of Halophilic Microorganisms*, Development in Halophilic Microorganisms, vol. 1, Elsevier-North-Holland Biomedical Press, New York.

Yopp, J.H., Albright, G., and **Miller, D.M.** 1979. Effects of antibiotics and ultraviolet radiation on the halophilic blue-green alga, *Aphanothece halophytica*. *Botanica Marina* 22: 267–272.

9

Dunaliella and Other Halophilic, Eucaryotic Algae

Dunaliella is the best known and possibly the most ubiquitous eucaryotic microorganism in hypersaline environments. It is a green alga that often assumes an orange coloration due to the synthesis of β-carotene in strong brines. It is an obligately phototrophic, oxygenic, aerobic, unicellular organism that does not tolerate the stagnating, anaerobic conditions that are often associated with hypersaline brines. *Dunaliella* is primarily found in the plankton, not in benthic communities and never in microbial mats. As primary producers, *Dunaliella* may introduce new organic matter into brines up to NaCl saturation.

The following discussion highlights the growth conditions tolerated by *Dunaliella* and lesser known halotolerant algae, their mechanisms of osmoregulation, and their lipid compositions. In spite of the detailed knowledge of the physiology and biochemistry of *Dunaliella*, there remain questions concerning the inability of other salt-tolerant eucaryotic algae to dominate the primary productivity of hypersaline habitats, and the exact environmental conditions that promote *Dunaliella* vs. cyanobacterial development in salinities up to gypsum saturation.

9.1 *Dunaliella*

The best known halotolerant or halophilic eucaryotic algae are species of the genus *Dunaliella* (a Chlorophyta or green alga), including *D. salina*, *D. viridis*, *D. parva*, *D. tertiolecta*, and *D. minuta*. *D. salina* (a large red cell, 11 $\times$ 14 μm) and *D. viridis* (a smaller green cell, 8 $\times$ 12 μm) have been reported from numerous hypersaline marine and athalassic environments (Figure 9.1). *D.*

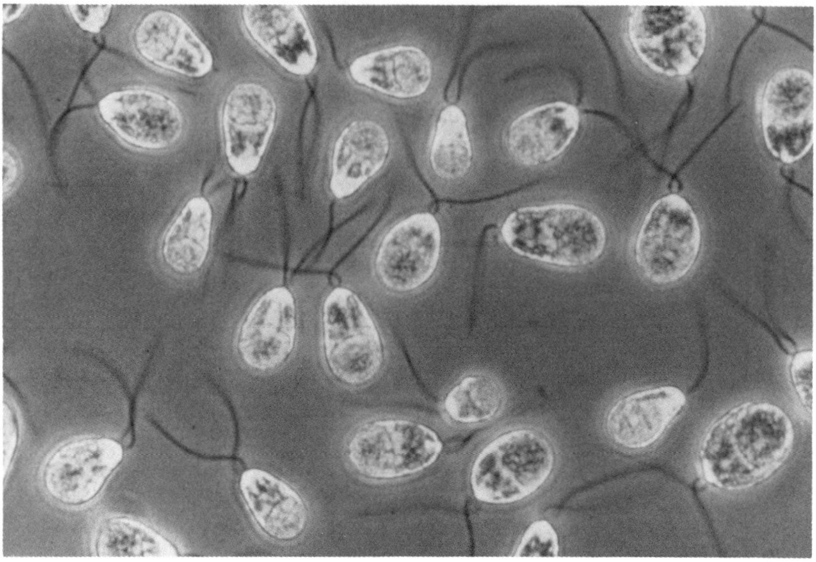

Figure 9.1 Photomicrograph of a pure culture of *Dunaliella salina*. Courtesy of Microbio Resources, Inc.

parva is the only species of *Dunaliella* found in the Dead Sea. Vegetative cells of *Dunaliella* are ellipsoid, flexible, and lack a cell wall. They have two equal flagella (Figure 9.1), a cup-shaped chloroplast, an anterior nucleus, an eyespot, and a pyrenoid surrounded by starch grains. Cultures are reported to have the smell of violets.

There is a very large body of literature on *Dunaliella* of hypersaline environments. The literature can be broadly organized into three areas: 1) general physiology, ecology, and culture; 2) osmoregulation; and 3) lipid composition, with particular references to β-carotene. The following discussion outlines the general characteristics of *Dunaliella* and other eucaryotic algae of hypersaline environments stressing the environmental factors that affect their success. More extensive reviews are given by Brown and Borowitzka (1979), Borowitzka (1981), Ben-Amotz and Avron (1983a), Munns (1983), and Borowitzka and Borowitzka (1988).

Dunaliella has several growth forms: vegetative cells, palmelloid cells, aplanospores, and encysted zygotes. Borowitzka (1981) outlined the different life stages. Vegetative cells of *D. salina* can be found in salinities of 4–35%. Palmelloid forms (non-motile vegetative cells) can occur when the salinity decreases to <1% and aplanospores (haploid, asexual resting cysts) can occur when the salinity decreases below 4%. Sexual reproduction can also occur when the salinity decreases.

Salts and nutrients The major species of most hypersaline environments, *D. salina* and *D. viridis*, have different salt optima. *D. viridis* grows optimally in 5.8–8.9% salinity and tolerates up to 23.2% salinity (Borowitzka et al., 1977; Brown and Borowitzka, 1979). *D. salina* grows best in 12% salinity and tolerates up to 35% (Loeblich, 1972). The higher salt tolerance of *D. salina* accounts for its dominance in the North Arm of the Great Salt Lake, Utah (33% TDS) (Post, 1977, 1981). *D. viridis* dominates the phytoplankton of the more dilute South Arm (12% TDS). Brock (1975) found that it was difficult to distinguish between the two species in Great Salt Lake enrichment cultures. Cultures of presumably both species grew optimally at lower salinities (8–17%) than those of their environments in the lake.

The strain of *D. tertiolecta* studied by Wegmann (1981) tolerated 0.5–34% NaCl while the isolate used by McLachlan (1960) only grew between 0.4 and 12% salinity. There is not a specific requirement for NaCl by *Dunaliella* since glucose or glycine can replace NaCl in the culture medium (Ben-Amotz and Avron, 1972).

Both the optimal pH and temperatures depend on the species and the source of strains. Halophilic *Dunaliella* species typically tolerate a pH range of 6 to 9 (Baas-Becking, 1931; Gibor, 1956; Loeblich, 1972; Brown and Borowitzka, 1979), although tolerance of pH 11 by *D. salina* was noted (Borowitzka and Borowitzka, 1988). Gibor (1956) found that both *D. salina* and *D. viridis* tolerated a temperature range of 8°–35°C and grew optimally at 14°–30°C. The strain of *D. salina* used by Loeblich (1972) grew optimally at 27°C (20 hr generation time) and incubation of the cells at 18°C caused the generation time to double. The strain used by Van Auken and McNulty (1973) tolerated 5°–40°C. Brown and Borowitzka (1979) reported that *D. viridis* grew optimally at 37°C, *D. salina* at 30°C, and *D. tertiolecta* at 20°C.

Gibor (1956) found that both *D. salina* and *D. viridis* (solar saltern isolates) grew much better with NO_3^--N than NH_4^+-N. Post (1977) noted that *D. salina* from the Great Salt Lake preferred NH_4^+-N over NO_3^--N. In *D. salina*, nitrate deficiency caused the cells to increase their carotenoid content and turn red (Loeblich, 1972). In *D. tertiolecta*, a change from NO_3^- to NH_4^+ caused a shift in carbon metabolism to produce more amino acids (Wegmann, 1981).

D. salina and *D. viridis* grew optimally in medium with about 0.3 mM phosphate and tolerated up to 3.7 mM (Gibor, 1956). Phosphate deficiency (0–4 μM) caused *D. salina* cells to turn red while 0.02–2 mM phosphate caused them to be green (Loeblich, 1972). In the Dead Sea, negligible concentrations of phosphate (in combination with very high salinities) caused *Dunaliella* to disappear from the lake (Oren and Shilo, 1982, 1985).

D. salina prefers CO_2 over HCO_3^- (Loeblich, 1972; Ginzburg and Ginzburg, 1981) while a smaller, green *Dunaliella* (probably *D. viridis*) prefers HCO_3^- (Ginzburg and Ginzburg, 1981). *D. salina* synthesizes carbonic anhydrase, the enzyme which converts bicarbonate to CO_2 (Loeblich, 1972). It was noted that the addition of HCO_3^- to cultures at high salinity resulted in the growth of cells in the green phase by increasing the content of chlorophyll *a* and decreasing the content of carotenoids.

Inhibition of carbonic anhydrase with acetazolamide (Diamox) in the presence of Tris-HCl buffer reduced the growth rate of *D. salina* but did not affect it in the presence of glycine buffer (Loeblich, 1972). Loeblich suggested that the algae may have been able to utilize the carboxyl group from glycine for growth. No dark heterotrophic growth was detected on glycerol. Loeblich cited an earlier report by Mil'ko (1963) that showed *Dunaliella* was unable to sustain heterotrophic growth on glucose, ethyl acetate, or acetic acid. Gibor (1956) found no dark heterotrophic growth by *Dunaliella,* but showed that glucose and other organic substrates stimulated growth in the light of both *D. salina* and *D. viridis.*

Both Ca^{2+} and Mg^{2+} retard the development of *D. viridis,* with Ca^{2+} being more toxic (Baas-Becking, 1931). Enrichments for *D. viridis* were successful in medium containing 0–10 mM Ca^{2+} and 0–100 mM Mg^{2+}. Mg^{2+} was found to detoxify the effects of Ca^{2+}, and higher concentrations of Mg^{2+} were necessary when the concentration of NaCl was higher. For. *D. tertiolecta,* the inhibitory effects of Mg^{2+} and Ca^{2+} were prevented when the Mg^{2+}:Ca^{2+} molar ratio was 4.0 (McLachan, 1960). McLachan suggested that the dominance of Mg^{2+} over Ca^{2+} in alkaline environments may inhibit the growth of *D. tertiolecta* in those habitats.

Light Optimal light conditions for both growth and photosynthesis have been established for several strains of *Dunaliella.* Growth of a strain of *D. salina* studied by Loeblich (1972) was saturated at 850 foot-candles of light. Maximal photosynthetic rates at this salinity were observed when the light was $\geq$2500 foot candles. Photosynthesis was not inhibited by light up to 42,000 foot candles. Higher light intensities were also correlated with carotenogenesis. Under the growth conditions employed by Van Auken and McNulty (1973), *D. salina* grew well under ca. 500–3700 foot candles, and optimally under 3200 foot candles of light.

Ginzburg and Ginzburg (1981) showed that it was difficult to establish optimal growth conditions for different strains of halotolerant and halophilic *Dunaliella* because of the integrated effects of temperature, salinities, CO_2 levels, and light. For example, higher NaCl concentrations could be tolerated under conditions of high light and high pCO_2. Higher light also permitted cells to grow at higher temperatures. If these factors are combined with the effects of light quality and day length, as well as ionic and nutrient composition of the medium or natural environment, it is indeed difficult to establish the optimal set of parameters for growth of halotolerant and halophilic species of *Dunaliella.*

Osmoregulation *Dunaliella* can thrive in hypersaline habitats because of its ability to adjust its cytoplasm to osmotic equilibrium with the environment.

Enzymes assayed from halophilic species of *Dunaliella* are sensitive to NaCl (Johnson et al., 1968; Heimer, 1973; Borowitzka and Brown, 1974). Gimmler et al. (1984) showed that enzymes with non-ionic substrates were less salt-sensitive than those that interacted with electrolytes in *D. parva*. Salt inactivation was caused by Cl^-, not by Na^+ and K^+.

Numerous studies have established that glycerol is the compatible solute produced by *Dunaliella* growing in high salt, and that the processes regulating glycerol synthesis and breakdown are unique to these and at least one other halophilic green alga, *Asteromonas gracilis* (Wegmann, 1971; Ben-Amotz and Avron, 1973, 1982, 1983a; Borowitzka and Brown, 1974; Ben-Amotz, 1975; Borowitzka et al., 1977; Enhuber and Gimmler, 1980; Kaplan et al., 1980; Brown et al., 1982; Norton et al., 1982; Degani et al., 1985). Related studies have demonstrated the permeability of the membranes of *Dunaliella*, the flux of water and ions across the membrane, and the intracellular volume of cells as a response to changing ionic environments (Ginzburg, 1969, 1981a, 1981b; Gimmler et al., 1977; Gimmler and Schirling, 1978; Degani and Avron, 1982; Brown et al., 1982; Curtain et al., 1983; Balnokin et al., 1983; Balnokin and Mendevev, 1984; Balnokin and Mazel', 1985; Katz and Avron, 1985).

Glycerol is synthesized in response to the high external osmotic environment. Two unique enzymes are involved in the osmoregulatory process: dihydroxyacetone reductase and dihyroxyacetone kinase. Glycerol may constitute >50% of the dry weight of *Dunaliella* (Ben-Amotz and Avron, 1982). In cells cultured in 1.5 M NaCl, 2.1 M glycerol was found in the cytoplasm of the cells. In cells grown in 4 M salt, 110 picograms of glycerol per cell was measured (Ben-Amotz and Avron, 1973, 1982). Intracellular Na^+ was found to be <100 mM (Katz and Avron, 1985) although earlier reports gave higher values (Gimmler and Schirling, 1978; Ginzburg, 1981a, 1981b). Very little glycerol appears to leak from the cells upon transfer to hypotonic medium at normal growth temperatures (Degani et al., 1985), although Enhuber and Gimmler (1980) found significant leakage. At elevated temperatures (47°C), cells lose significant glycerol (Ben-Amotz and Avron, 1983a). The cells shrink or swell in response to a change in osmotic environment, but they are completely re-equilibrated within 90 min (Ben-Amotz and Avron, 1973; Ben-Amotz, 1975).

Carotenoids and other lipids As a response to a variety of environmental factors, *D. salina* and *D. bardawil* can synthesize enough β-carotene to render the cells orange or red. *D. salina* is redder when grown in high light, high salinity, acidic pH, or in N- or P-deficient medium (Loeblich, 1972, 1982). Cells in 5–10% NaCl were green, in 15% NaCl were yellow-green, and in 20–25% NaCl were red-orange. The chlorophyll *a* content of the cells increased with respect to salinity up to 15% NaCl, and then leveled off. In contrast, carotenoid content continued to increase in cells with respect to salinity in medium with >10% NaCl. In 5–10% NaCl, the carotenoid:chlorophyll *a* ratio (w/w) was about 3, but in 25% NaCl it was 12. *D. bardawil* (called *D. salina* by Borowitzka and Borowitzka [1988]) can produce even greater amounts of β-carotene (Ben-Amotz and Avron, 1982; 1983b). Cultured cells contained 30% glycerol, 30% protein, 18% lipid, 11% carbohydrate, 9% β-carotene, and 1% chlorophyll based on dry weight. For *D. salina* cells grown under similar conditions, β-carotene constituted only about 0.3% of the dry weight. β-carotene is stored in lipid globules in *D. bardawil.* Its synthesis is induced by high light, high NaCl, nitrate deficiency, or extreme temperatures. The ability of *D. bardawil* to synthesize large quantities of β-carotene has led to its commercial cultivation to harvest the pigment (Ben-Amotz and Avron, 1980).

The polar and non-polar lipids of *Dunaliella* have been investigated to discern the algal response to different environmental conditions and to catalog the variety of lipids that could be organic geochemical markers in hypersaline sediments (Wright, 1979; Tornabene et al., 1980; Evans et al., 1982; Fried et al., 1982; Evans and Kates, 1984; Ben-Amotz et al., 1985). About 30–50% of the cellular material can be lipids. In *D. bardawil,* the polar:non-polar lipid ratio was about 50:50 while in *D. salina* the ratio was about 70:30. The major components among the non-polar lipids are glycolipids. Saturated and unsaturated C_{16} and C_{18} fatty acids are prevalent, with $C_{16:0}$ dominant. *D. salina* synthesizes trace amounts of C_{20} and C_{22} fatty acids but *D. parva, D. bardawil,* and *D. tertiolecta* do not. *D. salina* also produces C_{17} and highly branched ($C_{20:6}$) fatty acids. It produces at least six sterol derivatives. *D. tertiolecta* synthesizes ergosterol (Wright, 1979), which is generally believed to be a major sterol of yeast and fungi. Bulk lipid synthesis in *D. salina* was induced by lowering the salinity from 2 M to 0.5 M NaCl, and by giving the cells adequate concentrations of nitrate (Ben-Amotz et al., 1985).

Distribution *Dunaliella* is nearly ubiquitous in hypersaline environments although it was notably absent in the Dead Sea when the salinity increased and phosphate was limiting (Oren and Shilo, 1982) and in a solar saltern that was also phosphate-limited (Javor, 1983). *Dunaliella* is also absent in Solar Lake (Cohen et al., 1977), presumably due to phosphate limitation. The maximal density of *Dunaliella* cells in a Spanish solar saltern, 10^5 cells·ml^{-1}, was found in 30% TDS. In ca. 21% and 35% TDS, 10^4 cells·ml^{-1} were found, and in 15% and 40% TDS, only 10^3 cells·ml^{-1} were counted (Rodriguez-Valera et al., 1985). In the Great Salt Lake, up to 2×10^5 cells·ml^{-1} of *D. viridis* were found in the South Arm (12% TDS) in the late spring while up to 10^5 cells·ml^{-1} of *D. salina* were counted in the North Arm (33% TDS) (Post, 1977).

While survival of *Dunaliella* in extremely high salinity is dependent upon osmoregulation (the synthesis of glycerol as an osmolyte), survival in dilute brines is often accompanied by alternate cell forms and reproduction. Dilution also causes *D. parva* cells to lose motility and sink, which is possibly a survival mechanism to escape a dilute layer that would float on top of the brine (Dor, 1985).

The high lipid content of *Dunaliella* and the somewhat distinct composition of the lipids make it possible to estimate its contribution to lipid accumulation and diagenesis in both modern and ancient hypersaline sediments. Lipids constituted 45% of the total organic matter (0.75% by weight) sedimented from a salt plus brine sample of Pink Lake (Victoria, Australia) and β-carotene constituted about one-third of the lipids (Aasen et al., 1969). Such lipid-rich organic matter could be a primary source of hydrocarbons in some ancient hypersaline sediments.

9.2 Other unicellular algae

Diatoms Diatoms, common marine algae, are also nearly ubiquitous inhabitants of hypersaline environments, but they never appear to dominate. Little is known of their physiology and ecology in these habitats. They were noted in several environments in the Sinai: up to 10% salinity in Bardawil Lagoon (Erlich, 1975), up to ca. 18% salinity in Solar Lake (Cohen et al., 1977; Krumbein et al., 1977; Fischer, 1979; Jørgensen et al., 1983), and up to ca. 20.5% salinity in Gavish Sabkha (Erlich and Dor, 1985). They are also common in solar salterns (up to ca. 21.3% salinity) (Davis, 1978) and in up to about 12.9% salinity in the Great Salt Lake (Felix and Rushforth, 1979; see also Chapter 18). Species of *Navicula* and *Nitzschia* are represented in all of these environments and *Amphora coffeaeformis* has been identified in most.

The process of osmoregulation in diatoms is not well known. Fischer (1979) observed plasmolysis in diatoms upon exposure to lower salt concentrations. Solar Lake diatoms responded slower to increased salinity than diatoms from intertidal environments. Plasmolyzed cells survived exposure to higher salinities. Osmoregulation in the euryhaline, intertidal species *Cyclotella cryptica* and *C. menghiniana* involves the synthesis of proline and uptake of K^+ (Ben-Amotz and Avron, 1983a). A strain of *Navicula* was found to accumulate proline as well as an oligosaccharide. *Cylindrotheca fusiformis* accumulates mannose.

The lipid composition of diatoms of hypersaline habitats is not well known. A species of *Nitzschia* cultured from the hypersaline, alkaline Mono Lake (see Chapter 20) was reported by Ben-Amotz et al. (1985). When grown in 1.4 M NaCl, 22% of the ash-free dry weight was lipid. About half of the ash-free dry weight was of unknown composition, being neither protein, carbohydrate, nor lipid. No glycerol was detected. The dominant fatty acids produced by this strain of *Nitzschia* were $C_{16:1}$ and $C_{20:6}$ (also a minor fatty acid of *D. salina*). A $C_{21:0}$ fatty acid was produced by this strain of *Nitzschia* but not by *D. salina*.

Erlich and Dor (1985) noted that in Gavish Sabkha the most salt-tolerant diatoms were always found in association with cyanobacteria. The authors stated that they probably profited from the slimy sheath environment, but the actual demonstration of the uptake of carbohydrates, nitrogen and phosphorus nutrients, or other substances by halotolerant diatoms has not been shown. True halophily, or the requirement for high salt, has not been found among diatoms. The authors stated that *Nitzschia lembiformis* from Gavish Sabkha was found almost exclusively at salinities $\geq$15%. This species was not found in either Solar Lake or Bardawil Lagoon, but it did grow in artificial brine ponds near the Dead Sea. Fossils of *N. lembiformis* were found among Pleistocene evaporites of the Jordan Rift Valley. Whether this diatom is a true halophile or whether it is competitively excluded from lower salinities remains to be shown.

Other algae Osmoregulation in other halotolerant algae was reviewed by Ben-Amotz and Avron (1983a). *Asteromonas gracilis* (also called *Stephanoptera gracilis*) is a green alga (Prasinophyceae). It shares many characteristics with *Dunaliella* (Ben-Amotz and Grunwald, 1981; Ben-Amotz and Avron, 1982, 1983a). Lacking a wall, its cells are flexible and thus accomodate changes in cell volume during transitions and adaptation to higher or lower salinities. It can grow in 0.5–4.5 M NaCl (saturation). Maximum growth rate occurs in 0.5–2.5 M NaCl. Like *Dunaliella,* it accumulates glycerol as an osmolyte and its synthesis requires the same unique enzymes. Glycerol may constitute >50% of the dry weight of the cells. Up to 5.50 M glycerol has been measured in *A. gracilis* cells. The cells retain glycerol in salinities below 3.5 M NaCl. At higher salinities, significant amounts of glycerol leak from the cells. Exposure to high temperatures also causes glycerol to be released. Exposure to 47°C for several minutes caused the cells to lose half their glycerol while exposure to 60°C caused them to lose all their glycerol. *A. gracilis* has been found in salt marshes, brine pools, ocean seawater, and brackish water. It is unknown why it is never a dominant form in moderately hypersaline habitats. Other halotolerant eucaryotic algae, mostly members of the Chlorophyceae, are listed by Davis (1978), Felix and Rushforth (1979), and Erlich and Dor (1985).

The factors that restrict eucaryotic algae from many hypersaline environments probably include the inability to osmoregulate under given conditions, the inability to assimilate nutrients that may be scarce, and the exclusion from habitats in which the temperature rises above 40°–45°C. Periodic desiccation may also restrict some eucaryotic algae, although *Dunaliella* is well adapted to such habitats. Other limiting factors could include limitations to dark metabolism, the production of antibiotics by other microorganisms, deleterious or inadequate concentrations of specific major or minor ions and vitamins, and the inability to tolerate sulfide. The factors affecting the success, distribution, and chemical composition of diatoms of hypersaline habitats can be used to interpret some evaporites since diatom frustules are often preserved in Cretaceous and younger sediments.

References

Aasen, A.J., Eimhjellen, K.E., and **Liaaen-Jensen, S.** 1969. An extreme source of β-carotene. *Acta Chemica Scandinavica* 23: 2544–2545.

Baas-Becking, L.G.M. 1931. Salt effects on swarmers of *Dunaliella viridis* Teod. *Journal of General Physiology* 14: 765–779.

Balnokin, Yu. V., Mededev, A.V., and **Bodnar, I.V.** 1983. Potassium transport systems in the halophilic algae *Dunaliella* cells. *Fiziologiya Rasteny (Moscow)* 30: 955–963.

Balnokin, Yu. V. and **Medvedev, A.V.** 1984. Transport of sodium, potassium and hydrogen ions through the plasmalemma of potassium-deficient cells of the halophilic alga *Dunaliella maritima. Fiziologiya Rasteny (Moscow)* 31: 805–809.

Balnokin, Yu. V. and **Mazel', Yu. Ya.** 1985. Permeability to sodium ions of the plasma membrane of halophilic *Dunaliella* algae. *Fiziologiya Rasteny (Moscow)* 32: 33–41.

Ben-Amotz, A. 1975. Adaptation of the unicellular alga *Dunaliella parva* to a saline environment. *Journal of Phycology* 11: 50–54.

Ben-Amotz, A. and **Avron, M.** 1972. Photosynthetic activities of the halophilic alga *Dunaliella parva. Plant Physiology* 49: 240–243.

Ben-Amotz, A. and **Avron, M.** 1973. The role of glycerol in the osmotic regulation of the halophilic alga *Dunaliella parva. Plant Physiology* 51: 875–878.

Ben-Amotz, A. and **Avron, M.** 1980. Glycerol, β-carotene and dry algal meal production by commercial cultivation of *Dunaliella,* pp. 603–610 in Shelef, G. and Soeder, C.J. (editors), *Algae Biomass: Production and Use.* Holland Biomedical Press. Amsterdam.

Ben-Amotz, A. and **Avron, M.** 1982. Glycerol production by *Dunaliella. Experientia* 38: 49–52.

Ben-Amotz, A. and **Avron, M.** 1983a. Accumulation of metabolites by halotolerant algae and its industrial potential. *Annual Review of Microbiology* 37: 95–119.

Ben-Amotz, A. and **Avron, M.** 1983b. On the factors which determine massive β-carotene accumulation in the halotolerant alga *Dunaliella bardawil. Plant Physiology* 72: 593–597.

Ben-Amotz, A. and **Grunwald, T.** 1981. Osmoregulation in the halotolerant alga *Asteromonas gracilis. Plant Physiology* 67: 613–616.

Ben-Amotz, A., Tornabene, T.G., and **Thomas, W.H.** 1985. Chemical profile of selected species of microalgae with emphasis on lipids. *Journal of Phycology* 21: 72–81.

Borowitzka, L. 1981. The microflora. Adaptations to life in extremely saline lakes. *Hydrobiologia* 81: 33–46.

Borowitzka, L.J. and **Brown, A.D.** 1974. The salt relations of marine and halophilic species of the unicellular green alga, *Dunaliella.* The role of glycerol as a compatible solute. *Archives of Microbiology* 96: 37–52.

Borowitzka, M.A. and **Borowitzka, L.J.** 1988. Dunaliella, p. 27–58 in Borowitzka, M.A. and Borowitzka, L.J. (editors), *Micro-algal Biotechnology.* Cambridge University Press, Cambridge.

Borowitzka, L.J., Kessley, D.S., and **Brown, A.D.** 1977. The salt relations of *Dunaliella.* Further observations on glycerol production and its regulation. *Archives of Microbiology* 113: 131–138.

Brock, T.D. 1975. Salinity and the ecology of *Dunaliella* from Great Salt Lake. *Journal of General Microbiology* 89: 285–292.

Brown, A.D. and **Borowitzka, L.J.** 1979. Halotolerance of *Dunaliella,* pp. 139–190 in Levandowsky, M. and Hutner, S.H. (editors), *Biochemistry and Physiology of Protozoa,* vol. 1, Academic Press, New York.

Brown, F.F., Sussman, I., Avron, M., and **Degani, H.** 1982. NMR studies of glycerol permeability in lipid vesicles, erythrocytes and the alga *Dunaliella. Biochimica et Biophysica Acta* 690: 165–173.

Cohen, Y., Krumbein, W.E., and **Shilo, M.** 1977. Solar Lake (Sinai). 2. Distribution of photosynthetic microorganisms and primary production. *Limnology and Oceanography* 22: 609–620.

Curtain, C.C., Looney, F.D., Regan, D.L., and **Ivancic, N.M.** 1983. Changes in the ordering in the membrane of *Dunaliella* in response to osmotic-pressure changes. *Biochemical Journal.* 213: 131–136.

Davis, J.S. 1978. Biological communities of a nutrient enriched salina. *Aquatic Botany* 4: 23–42.

Degani, H. and **Avron, M.** 1982. The diffusional water permeability in the halotolerant alga *Dunaliella* as measured by nuclear magnetic resonance. *Biochimica et Biophysica Acta* 690: 174–177.

Degani, H., Sussman, I., Peschek, G.A., and **Avron, M.** 1985. Carbon-13 NMR and proton NMR studies of osmoregulation in *Dunaliella. Biochimica et Biophysica Acta* 846: 313–323.

Dor, I. 1985. Long-lasting effect of dilution on the cell volume, motility, division rate and vertical distribution of *Dunaliella parva. Journal of Experimental Marine Biology and Ecology* 91: 183–197.

Enhuber, G. and **Gimmler, H.** 1980. The glycerol permeability of the plasmalemma of the halotolerant green alga *Dunaliella parva* (Volvocales). *Journal of Phycology* 16: 524–532.

Erlich, A. 1975. The diatoms from the surface sediments of the Bardawil Lagoon (northern Sinai)—Paleoecological significance. *Nova Hedwigia (Beiheft)* 53: 253–282.

Evans, R.W., Kates, M., Ginzburg, M., and **Ginzburg, B.-Z.** 1982. Lipid composition of halotolerant algae, *Dunaliella parva* Lerche and *Dunaliella tertiolecta. Biochimica et Biophysica Acta* 712: 186–195.

Evans, R.W. and **Kates, M.** 1984. Lipid composition of halophilic species of *Dunaliella* from the Dead Sea. *Archives of Microbiology* 140: 50–56.

Felix, E.A. and **Rushforth, S.R.** 1979. The algal flora of the Great Salt Lake, Utah, U.S.A. *Nova Hedwigia* 31: 163–194.

Fischer, H. 1979. Osmotic behaviour of diatoms of a hypersaline lake in comparison with tidal diatoms. *Nova Hedwigia (Beiheft)* 64: 251–264.

Fried, A., Tietz, A., Ben-Amotz, A. and **Eichenberger, W.** 1982. Lipid composition of the halotolerant alga, *Dunaliella bardawil. Biochimica et Biophysica Acta* 713: 419–426.

Gibor, A. 1956. The culture of brine algae. *Biological Bulletin* 111: 223–229.

Gimmler, H., Schirling, R., and **Tobler, U.** 1977. Cation permeability of the plasmalemma of the halotolerant alga *Dunaliella parva* I. Cation induced osmotic volume changes. *Zeitschrift für Pflanzenphysiologie* 83: 145–158.

Gimmler, H. and **Schirling, R.** 1978. Cation permeability of the plasmalemma of the halotolerant alga *Dunaliella parva* II. Cation content and glycerol concentration of the cells as dependent on external NaCl concentration. *Zeitschrift für Pflanzenphysiologie* 87: 435–444.

Gimmler, H., Kaaden, R., Kurchner, U., and **Weyand, A.** 1984. The chloride sensitivity of *Dunaliella parva* enzymes. *Zeitschrift für Pflanzenphysiologie* 114: 131–150.

Ginzburg, M. 1969. The unusual membrane permeability of two halophilic unicellular organisms. *Biochimica et Biophysica Acta* 173: 370–376.

Ginzburg, M. 1981a. Measurements of ion concentrations and fluxes in *Dunaliella parva. Journal of Experimental Botany* 32: 321–332.

Ginzburg, M. 1981b. Measurements of ion concentrations in *Dunaliella parva* subjected to hypertonic shock. *Journal of Experimental Botany* 32: 333–340.

Ginzburg, M. and **Ginzburg, B.-Z.** 1981. Interrelationships of light, temperature, sodium chloride and carbon source in growth of halotolerant and halophilic strains of *Dunaliella. British Phycological Journal* 16: 313–324.

Heimer, Y.M. 1973. The effects of sodium chloride, potassium chloride and glycerol on the activity of nitrate reductase of a salt-tolerant and two non-tolerant plants. *Planta* 113: 279–281.

Javor, B.J. 1983. Planktonic standing crop and nutrients in a saltern ecosystem. *Limnology and Oceanography* 28: 153–159.

Johnson, M.K., Johnson, E.J., MacElroy, R.D., Speer, H.L., and **Bruff, B.S.** 1968. Effects of salts on the halophilic alga *Dunaliella viridis. Journal of Bacteriology* 95: 1461–1468.

Jørgensen, B.B., Revsbech, N.P., and **Cohen, Y.** 1983. Photosynthesis and structure of benthic microbial mats: microelectrode and SEM studies of four cyanobacterial communities. *Limnology and Oceanography* 28: 1075–1093.

Kaplan, A., Schreiber, U., and **Avron, M.** 1980. Salt-induced metabolic changes in *Dunaliella salina. Plant Physiology* 65: 810–813.

Katz, A. and **Avron, M.** 1985. Determination of intracellular volume and sodium concentration in *Dunaliella. Plant Physiology* 78: 817–820.

Krumbein, W.E., Cohen, Y., and **Shilo, M.** 1977. Solar Lake (Sinai). 4. Stromatolitic cyanobacterial mats. *Limnology and Oceanography* 22: 635–656.

Loeblich, L.A. 1972. *Studies on the brine flagellate Dunaliella salina.* Ph.D. Thesis. University of California. San Diego. 142 pp.

Loeblich, L.A. 1982. Photosynthesis and pigments influenced by light intensity and salinity in the halophile, *Dunaliella salina* (Chlorophyta). *Journal of the Marine Biological Association of the United Kingdom* 62: 493–508.

McLachlan, J. 1960. The culture of *Dunaliella tertiolecta* Butcher—a euryhaline organism. *Canadian Journal of Microbiology* 6: 367–379.

Munns, R., Greenway, H., and **Kirst, G.O.** 1983. Halotolerant eukaryotes, pp. 59–135 in Lange, O.L., Nobel, P.S., Osmond, C.S. and Ziegler H. (editors), *Physiological Plant Ecology* III, Springer-Verlag, New York.

Norton, R.S., Mackay, M.A., and **Borowitzka, L.J.** 1982. The physical state of osmoregulatory solutes in unicellular algae. *Biochemical Journal* 202: 699–706.

Oren, A. and **Shilo, M.** 1982. Population dynamics of *Dunaliella parva* in the Dead Sea. *Limnology and Oceanography* 27: 201–211.

Oren, A. and **Shilo, M.** 1985. Factors determining the development of algal and bacterial blooms in the Dead Sea: a study of simulation experiments in outdoor ponds. *FEMS Microbiology Ecology* 31: 229–237.

Post, F.J. 1977. The microbial ecology of the Great Salt Lake. *Microbial Ecology* 3: 143–165.

Post, F.J. 1981. Microbiology of the Great Salt Lake North Arm. *Hydrobiologia* 81: 59–69.

Rodriguez-Valera, F., Ventosa, A., Juez, G., and **Imhoff, J.F.** 1985. Variation of environmental features and microbial populations with salt concentration in a multipond saltern. *Microbial Ecology* 11: 107–115.

Tornabene, T.G., Holzer, G., and **Peterson, S.L.** 1980. Lipid profile of the halophilic alga, *Dunaliella salina. Biochemical and Biophysical Research Communications* 96: 1349–1356.

Van Auken, O.W. and **McNulty, I.B.** 1973. The effect of environmental factors on the growth of a halophilic species of algae. *Biological Bulletin* 145: 210–222.

Wegmann, K. 1971. Osmotic regulation of photosynthetic glycerol production in *Dunaliella. Biochimica et Biophysica Acta* 234: 317–323.

Wegmann, K. 1981. Influence of ecological factors on the carbon metabolism in *Dunaliella tertiolecta,* pp. 263–272 in Akoyunoglou, G. (editor), *Photosynthesis VI. Pho-*

tosynthesis and Productivity, Photosynthesis and Environment. Balaban International Science Services. Philadelphia.

Wright, J.L.C. 1979. The occurrence of ergosterol and (22*E*, 24*R*)-24-ethylcholesta-5, 7, 22-trien-3β-ol in the unicellular chlorophyte *Dunaliella tertiolecta*. *Canadian Journal of Chemistry* 57: 2569–2571.

10

Protozoa

Protozoa are commonly found in hypersaline lakes and solar salt ponds, but little is known about their physiological adaptations to high salt, specific ions, temperature fluctuations, and other chemical and physical parameters associated with these extreme environments (Fenchel, 1987). Table 10.1 lists the taxa of protozoans identified in a variety of hypersaline environments. The most comprehensive survey of salt-tolerant protozoa is that of Post et al. (1983). The dominance of ciliates in the table may not reflect their true dominance in these environments, but rather the ease in identifying individual taxa due to distinct morphological characteristics. Figures of the identified forms are given in Pack (1919), Wilbert and Kahan (1981), and Post et al. (1983).

In freshwater, protozoans regulate their internal osmotic pressure with contractile vacuoles that expel water (Kitching, 1967; Laybourn-Parry, 1984). Marine species do not usually need to osmoregulate since they are in osmotic equilibrium with their environment. Marine amoebae and flagellates typically lack contractile vacuoles although these organelles are usually found in marine ciliates. It is not known whether marine ciliates use these contractile vacuoles to osmoregulate, or whether they are vestiges from their freshwater origin.

Mast and Hopkins (1941) described the changes in water content through vacuolar elimination in *Amoeba mira* in medium ranging from 0.1 to 1.0× seawater. They stated that these protozoans grew in up to 10× seawater, but they did not describe the process of water regulation under hypersaline conditions. Kitching (1967) cited work published in 1924 (in Russian) by Gayewskaya that demonstrated that protozoa from brine pools either had numerous non-contracting vacuoles, no vacuoles at all, or vacuoles that underwent a very low frequency of contraction.

Table 10.1 Taxonomy, distribution, and range of salinity of protozoans of hypersaline environments[a]

	W. Australia[b]	Great Salt Lake[c]	Solar Lake[d]	Dead Sea[e]	Solar Salterns[f]
CILIATES					
Blepharisma halophila	16.5–31.0				
Chiliophrya utahensis[g]	27.8				
Chondylostoma spp.	22.0–31.1		+		CA +
Cladotricha sigmoidea	≤20.0				
Enchelydon trepida			+		
Euplotes spp.	29.0		+		BA +
Fabrea salina	≤17.0				
Frontonia marina			+		
Halteria grandinella			+		
Holosticha diademata			+		
Litonotus spp.			+		BA ≤10.2
Metacystis truncata	16.5–19.9				
Nassula sp.	16.5				
Palmarella salina	27.8				
Parauronema virginianum			+		
Podophyra sp.	16.5–23.2				
Prorodon utahensis		23.0			
Pseudocohnilembus marinus			+		
Rhopalophyra salina	17.0–33.2				
Stephanopogon apogon			+		
Tachysoma sp.			+		
Trachelocerca conifer	16.5–22.3				
Trematosoma bocqueti	19.2				
Uroleptus packii		23.0			
Uronchia transfuga			+		
Uronema spp.	15.0–17.3		+		
ZOOFLAGELLATES					
Bodo spp.	20.2–saturated				
Monosigma spp.	17.0–23.0				
Phyllomitus sp.	27.8–saturated				
Rhinchomonas nasuta	15.0–20.0				
Tetramites spp.	up to saturated				PR +
Acanthoecopsis unguiculata[h]					
Unidentified spp.		33		15	
SARCODINES					
Amoeba limax		23			
Heteramoeba sp.	12.5–21.0				
Nagleria spp.	up to saturated				
Unidentified spp.	15–22	33		15	

[a]Salinities noted in %. + = Occurrence, but no salinity noted.
[b]Hutt Lagoon marine embayment, W. Australia (Post et al., 1983).
[c]Great Salt Lake, Utah, U.S.A. (Vorhies, 1917; Pack, 1919; Post, 1977).
[d]Solar Lake, Sinai, 15–18% (Wilbert and Kahan, 1981).
[e]Elazari-Volcani (1943).
[f]CA = San Francisco, California, U.S.A. (Carpelan, 1957); BA = Bahamas (Davis, 1978); PR = Puerto Rico (Golubic, 1980).
[g]*Chiliophrya utahensis* = *Prorodon utahensis*.
[h]Found in Organic Lake, Antarctica: 16.5–22.8% TDS, −14° to 9.5°C (Franzmann et al., 1987).

Kitching (1967) pointed out that marine protozoans are permeable to Na^+ and Cl^- and that they maintain a high K^+:Na^+ ratio internally through the use of a Na^+ pump. Neither the internal ion concentration nor the compatible solutes of protozoans in hypersaline media have been described. K^+ flux in a soil amoeba in dilute medium was measured by Klein (1959). The role of cations in membrane excitation and swimming behavior of *Fabrea salina* in medium with 1.1 M NaCl and 0.1 M $MgCl_2$ was described by Dryl et al. (1982) and Kubalski (1983a, 1983b). In these three studies normal swimming behavior was associated with regulation of Ca^{2+} and K^+ by the cells.

Pack (1919) described the behavioral response to high salinity of two ciliates from Great Salt Lake, Utah. When the salinity increased from 6% to 15% *Uroleptus packii* decreased in length from 0.11 mm to 0.07 mm. Its cysts decreased in diameter from 30 μm to 25 μm. *Prorodon utahensis* decreased in length from 0.08 mm to 0.06 mm. Its cysts decreased in size from 27 μm to 25 μm. The rate of movement of the feeding cirri decreased with increasing salinity. In 4.4% salinity the movement was too fast to measure. In 15% salinity, a single stroke lasted 0.2 sec. In 20% salt a single stroke lasted 1 sec while in saturated solution a single stroke of the cirri lasted 15 sec. Pack also noted the feeding cirri increased in length between 4.4% salinity (10–11 μm) and saturated medium (17–20 μm). Other physiological responses to higher salt included slower food ingestion, slower and less frequent vacuolar discharge, slower reproduction, and more rigid bodies.

Elazari-Volcani (1943) described amoebae isolated from Dead Sea muds that grew optimally in 15–18% salt and survived in saturated solution. In 6% salt the rate of reproduction decreased from the optimal rate in 15–18% salt.

The large variety of protozoans found thriving in hypersaline environments attests to the relative success of this group of microorganisms in these habitats. They thrive by ingesting bacteria and algae, and in some cases, by cannibalism. Further investigations of the physiology of these protozoans would demonstrate whether they are truly halophilic or merely halotolerant. Such studies would also elucidate the nature of osmotic regulation of protozoans and their role in carbon and nutrient cycling in hypersaline environments.

References

Carpelan, L.H. 1957. Hydrobiology of the Alviso salt ponds. *Ecology* 38: 375–390.

Davis, J.S. 1978. Biological communities of a nutrient enriched salina. *Aquatic Botany* 4: 23–42.

Dryl, S., Demar-Gervais, C., and **Kubalski, A.** 1982. Role of external cations in excitability of marine ciliate *Fabrea salina*. *Acta Protozoologica* 21: 55–60.

Elazari-Volcani, B. 1943. A dimastigamoeba in the bed of the Dead Sea. *Nature (London)* 152: 275–277.

Fenchel, Tom. 1987. *Ecology of Protozoa. The Biology of Free-living Phagotropic Protists.* Science Tech Publishers, Madison. 208 pp.

Franzmann, P.D., Burton, H.R., and **McMeekin, T.A.** 1987. *Halomonas subglaciescola,* a new species of halotolerant bacteria isolated from Antarctica. *International Journal of Systematic Bacteriology* 37: 27–34.

Golubic, S. 1980. Halophily and halotolerance in cyanophytes. *Origins of Life* 10: 169–183.

Kitching, J.A. 1967. Contractile vacuoles, ionic regulation, and excretion, pp. 307–336 in Chen, T.-T. (editor), *Research in Protozoology*, vol. 1, Pergamon Press, New York.

Klein, R.L. 1959. Transmembrane flux of K^{42} in *Acanthamoeba* sp. *Journal of Cellular and Comparative Physiology* 53: 241–258.

Kubalski, A. 1983a. Electrical properties of the cell membrane of the marine ciliate *Fabrea salina. Acta Protozoologica* 22: 219–228.

Kubalski, A. 1983b. The inhibiting effect of barium on potassium induced ciliary reversal in a marine ciliate *Fabrea salina. Acta Protozoologica* 22: 229–236.

Laybourn-Parry, J. 1984. *A Functional Biology of Free-Living Protozoa.* University of California Press. Berkeley, 218 pp.

Mast, S.O. and **Hopkins, D.L.** 1941. Regulation of the water content of *Amoeba mira* and adaptation to changes in the osmotic concentrations of the surrounding medium. *Journal of Cellular and Comparative Physiology* 41: 31–48.

Pack, D.A. 1919. Two ciliata of Great Salt Lake. *Biological Bulletin* 36: 273–282.

Post, F.J. 1977. The microbial ecology of the Great Salt Lake. *Microbial Ecology* 3: 143–165.

Post, F.J., Borowitzka, L.J., Borowitzka, M.A., Mackay, B., and **Moulton, T.** 1983. The protozoa of a Western Australian hypersaline lagoon. *Hydrobiologia* 105: 95–113.

Vorhies, C.T. 1917. Notes on the fauna of Great Salt Lake. *American Naturalist* 51: 494–499.

Wilbert, N. and **Kahan, D.** 1981. Ciliates of Solar Lake on the Red Sea shore. *Archiv für Protistenkunde* 124: 70–95.

11

Yeasts and Fungi

Yeasts and fungi are suited for osmophilic life *par excellence*. There is a very large body of literature on the osmophilic nature of these organisms in the food industry with respect to their metabolic abilities in manufacturing high solute food products such as soy sauce, as well as in food spoilage. In spite of the high salt tolerance of many taxa (Tables 11.1 and 11.2), there have been very few reports of yeasts and fungi from extremely hypersaline environments. In the present chapter, the halotolerance of these microorganisms is outlined, some of the documented occurrences of yeasts and fungi in natural and man-made brines are discussed, and physiological and biochemical attributes for osmophilic life in this group are described. Some of the factors that permit the coexistence and competition between yeasts, fungi, and bacteria are also discussed. It is striking that despite ability to grow in environments of very high osmotic pressure, fungi and yeasts are relatively unknown from hypersaline environments. It remains to be proven whether the relative lack of representation by yeasts and fungi in natural hypersaline environments reflects the bias of investigators, or whether these organisms are poor competitors against bacteria in evaporite habitats.

11.1 General environmental factors influencing growth

The majority of osmophilic yeasts can grow facultatively as anaerobes, although they grow most rapidly under aerobic conditions (Tilbury, 1980a). *Debaryomyces hansenii*, one of the most osmotolerant yeasts known, is exceptional in its inability to grow well or at all by anaerobic fermentation. Osmophilic fungi, which have a better ability of penetrating solid substrates

Table 11.1 Salt tolerances of yeasts

Yeast	Maximum salt tolerance	Source	Reference
Candida parapsilosis	20%	Marine	Norkrans, 1966
C. tropicalis	18%	Food brines	Noda et al., 1980
C. mogii	≥18%	Salted fish	Zvyagintseva and Gorodnyanskaya, 1978
Debaryomyces hansenii	24%	Marine	Norkrans, 1966
D. membranaefaciens var. *hollandicus*	24%	Food brines	Mrak and Phaff, 1948
D. guilliermondii var. *nova zeelandicus*	24%	Food brines	Mrak and Phaff, 1948
D. tyrocola	15%	Food brines	Mrak and Phaff, 1948
D. subglobosus	22%	Marine	Ross and Morris, 1962
D. kloeckeri	24%	Marine	Ross and Morris, 1962
Hansenula sp.	≥15%	Food brines	Kroemer and Krumbholz, 1932
Metschnikowia bicuspidata var. *australis*	≥12%	Marine	Lachance et al., 1976 Phaff and Starmer, 1980
Mycoderma vini	21%	—	Kroemer and Krumbholz, 1932
M. decolorans	15%	Food brines	Mrak and Phaff, 1948
Pichia etchellsii	20%	Marine	Norkrans, 1966
P. membranaefaciens	15%	Food brines	Mrak and Phaff, 1948
	24%	Marine	Ross and Morris, 1962
Rhodotorula rubra	≥16%	Marine	Norkrans, 1966
R. glutinis	23%	Marine	Ross and Morris, 1962
Saccharomyces rouxii	20–22%	Food brines	Onishi, 1963
	≥18%	Food brines	Noda et al., 1980; Horisberger et al., 1985
S. cerevisiae	12%	Marine	Norkrans, 1966
S. mellis	23%	—	Koppensteiner and Windisch, 1971
S. heterogenicus	23%	—	Koppensteiner and Windisch, 1971
Torulopsis candida	18%	Marine	Pal et al., 1979
T. famata	≥12%	Marine	Norkrans, 1966
	19%	Marine	Ross & Morris, 1962
T. versatilis	18%	Food brines	Noda et al., 1980
T. colliculosa	13%	—	Tilbury, 1980b
T. dattila	15%	—	Tilbury, 1980b
T. etchellsii	21%	—	Tilbury, 1980b
T. spp.	≥18%	Food brines	Onishi, 1963
Willia spp.	21%	—	Kroemer and Krumbholz, 1932
Zygosaccharomyces sp.	≥15%	Food brines	Kroemer and Krumbholz, 1932

than yeasts, are obligate aerobes. Yeasts and fungi thrive best on carbohydrates or hydrocarbons. They typically can obtain their nitrogen from amino acids or ammonia, but they generally have a limited capability of breaking down nitrogenous compounds. Environments with high C:N ratios may select for yeasts and fungi along with bacteria, whereas environments with low C:N ratios may select for bacterial development alone.

Yeasts typically tolerate temperatures from 0° to 40°C, with an average optimum of about 27°C (Tilbury, 1980b). Bacteria as a group also tolerate

Table 11.2 Salt tolerances of osmophilic fungi[a]

Fungus	Maximum salt tolerance	Source	Reference
Aspergillus spp.	≥5%	Desert soil	Abdel-Hafez, 1981
Aspergillus. spp.	20–25%	Marine	Siepmann, 1959
A. repens	≥25%	Marine	Siepmann, 1959
A. ochraceus	satd	–	Pitt and Hocking, 1977
A. wentii	satd	Salted fish	Andrews and Pitt, 1987
A. penicilloides	satd	Salted fish	Andrews and Pitt, 1987
Basipetospora halophila	satd	Salted fish	Andrews and Pitt, 1987
Cladosporium sp.	36%	Great Salt Lake	Cronin and Post, 1977
Drechslera spp.	≥5%	Desert soil	Abdel-Hafez, 1981
Eurotium repens	25%	Salted fish	Andrews and Pitt, 1987
Exophiala werneckii	satd	Salted fish	Andrews and Pitt, 1987
Fusarium spp.	20–25%	Marine	Siepmann, 1959
Glomus fasciculatum	≥13%[b]	Mycorrhizae	Pond et al., 1984
G. mosseae	≥11%[b]	Mycorrhizae	Pond et al., 1984
G. etunicatus	≥10%[b]	Mycorrhizae	Pond et al., 1984
Penicillium spp.	20–25%	Marine	Siepmann, 1959
P. notatum	25–30%	Desert soil	Radwan et al., 1984
Polypaecilium pisce	25%	Salted fish	Andrews and Pitt, 1987
Ulocladium spp.	≥5%	Desert soil	Abdel-Hafez, 1981
Wallemia sebi	satd	Bread	Pitt and Hocking, 1977

satd, saturated.
[a] Values reported as a_w were recalculated to % NaCl according to Robinson and Stokes (1955).
[b] Fungi isolated in medium with 0.42% salinity from soils of the given salinities.

this temperature range although many bacteria thrive at much higher temperatures. Moderately halophilic bacteria have an optimal temperature of about 30°C (Rodriguez-Valera et al., 1980). Extremely halophilic bacteria are mesophilic, with optimal temperatures about 40°–45°C, and maximum temperatures near 55°C (Larsen, 1981). Based on temperatures and oxygen availability, density-stratified hypersaline brines which experience extreme solar heating would be expected to have few or no fungi, and yeast populations would only develop in the cooler sediments.

Yeasts and fungi generally prefer acidic to neutral pH conditions (range = pH 2–7, optimum = pH 4.0–4.5; Tilbury, 1980b), which often result from the excretion of organic acids during both bacterial and fungal decomposition processes. In high salt habitats, yeasts and fungi have a more narrow range of pH tolerance and lower ability to grow under acidic conditions (Tilbury, 1980b). The inability to culture osmophilic yeasts and fungi under alkaline conditions does not always reflect the environmental pH in which they are found in nature. Pal et al. (1979) noted that a seawater isolate of *Torulopsis candida* grew optimally at pH 3–7, and not at all at pH 8, although the pH of the environment from which it was isolated was greater than 8. In Mono Lake, a hypersaline body of pH 9.7, numerous fungal hyphae, gemmae, and arthrospores were observed, and active parasitism of diatoms by *Chytridomycetes* was noted (Winkler, 1977).

It is possible that yeasts, fungi, and bacteria produce antibiotic substances that limit the development of certain microbes in their environment. Antibiotic production in extremely hypersaline environments has not been well studied. Bacterial-yeast antagonism was reported in soy sauce fermentation (18% NaCl) in which the bacterial production of acetic and lactic acids greatly inhibited the growth of yeasts (*Saccharomyces rouxii* and *Torulopsis versatilis*) (Noda et al., 1980). Radwan et al. (1985) measured antibiotic activity against bacteria produced by halophilic strains of *Penicillium notatum* and *P. purpurogenum* isolated from saline soils. In a standard medium with 0% or 10% salt, antibiotic production only occurred in the dilute medium, although better growth was obtained in high salt. Different carbon sources and growth at pH values between 3 and 8 gave the same results. Only when potassium was included in the medium did the *Penicillium* strains produce antibiotics in medium with 10% salt. Other substances with antibiotic activity against yeasts and fungi include other weak organic acids and their salts, such as benzoic acid, esters of polyhydroxybutyrate, and propionic acid, as well as some sulfur compounds, including sulfur dioxide, sulfite, and bisulfite (Tilbury, 1980b). A protein-like antibiotic (halocin) produced by halobacteria has been described (Rodriguez-Valera, 1982) but its natural occurrence and mode of action against yeast and fungi is unknown.

11.2 Halotolerance and halophily

It has been widely accepted that yeasts and fungi may be halotolerant but not obligately halophilic. In some cases, they show better growth under conditions of moderate to high salt. *Torulopsis candida,* a yeast isolated from Arabian Gulf seawater, grew best in 2 M NaCl at 37°C, although it also grew moderately well in dilute (1.45% salts) medium (Pal et al., 1979). Yamagata and Fujita (1974) found that *T. etchellsii* and *T. glabrata* had enhanced ability to ferment sugars in the presence of 2 or 10% NaCl. Radwan et al. (1984) found several strains of *Penicillium* that grew optimally in 9–10% salt and one strain of *P. notatum* that grew optimally in 10–15% NaCl. Andrews and Pitt (1987) described several xerophilic fungi that grew better on NaCl than on sugars, including *Polypaecilium pisce, Exophiala wenekii,* and *Aspergillus wentii* (see Table 11.2). However, there was no obligate requirement for NaCl.

Some degree of halotolerance may be associated with temperature dependence. Onishi (1963) noted that *T. halonitratophila* only grew on medium with >6% NaCl at 30°C, but at 20°C it grew in dilute medium. Salt tolerance can also be raised by the inclusion of certain sugars (Koppensteiner and Windisch, 1971) or fish extract (a complex form of nitrogen) (Ross and Morris, 1962) in the medium.

True salt dependence or halophily has been recorded in the yeast *Metschnikowia bicuspidata* var. *australis,* a parasite of the brine shrimp *Artemia salina* that lives in salt ponds with 10–12% NaCl (Phaff and Starmer, 1980). These authors reported that it only grew on medium supplemented with 10 or 12% NaCl, although the temperature and organic content of the medium were not given. It is the only yeast reported with an obligate salt requirement. Spencer et al. (1964) reported the isolation of a similar (or same) yeast (*M. kamienskii*) from brine shrimp using dilute medium at 30° and 37°C.

Onishi (1963) demonstrated that salt tolerance or salt requirement is "trainable" in *Saccharomyces rouxii* by growing cells in 0 or 18% NaCl, and transferring each culture to both 0 and 18% salt media. Upon the first transfer from low to high salt or from high to low salt, there was a significant loss in viability, but surviving cells regained ability to grow well upon repeated transfer to the same salinity. Further experimentation with the salt-dependent *M. bicuspidata* under different medium and culture conditions would demonstrate the nature of its obligate osmophily or halophily.

Osmotolerance and halotolerance in yeasts and fungi have been studied by comparing metabolism and growth in ionic and non-ionic solutions of comparable water activity or osmotic strength. Kroemer and Krumbholz (1932) tested the osmotolerance of a variety of osmophilic yeasts by growing them in media with various concentrations of NaCl, $NaNO_3$, KCl, KNO_3, or glycerol. By recalculating osmolarity or a_w from the molar values given in that report, the yeasts showed a greater osmotolerance in the presence of glycerol than in the salt media, and a slightly better tolerance of chloride salts over nitrate salts. Na^+ and K^+ were equally tolerated when their salts were compared by their osmolar rather than molar strength. A greater tolerance for lower water activity adjusted with sugars or polyols rather than salts has subsequently been demonstrated in numerous yeasts and fungi (Norkrans, 1968; Koppensteiner and Windisch, 1971; Pitt and Hocking, 1977; Tilbury 1980a, 1980b).

Onishi (1963) tested solute tolerance of an osmophilic strain of *S. rouxii* isolated from soy sauce mash. When the yeast was transferred from low salt to 18% salt, there was considerable loss of viability, but when it was transferred from the dilute medium to 50% glucose (which has the same osmotic pressure as 18% NaCl), there was no loss of viability. Like the studies listed above, this experiment demonstrated that water activity alone does not limit osmophilic yeasts in hypersaline media. This strain grew in $\leq$ 3 M NaCl (a_w = 0.89) and in at least 4 M KCl (a_w = 0.87). Mg^{2+} was less toxic than Ca^{2+} Onishi further demonstrated that the cells were highly permeable to K^+ when cultured in 18% NaCl but not when they were cultured in 50% glucose.

Osmoregulation Onishi (1963) showed that *S. rouxii* cells suspended in an 18% NaCl solution at 30°C showed a large loss of viability unless glucose was added to the solution. At 0°–5°C, cells remained viable without the added glucose. It was concluded that viability in NaCl solutions at physiological temperatures was related to cell permeability to K^+ ions, which could be at least partly regulated by active metabolism. In this strain some metabolic capabilities (maltose fermentation or possibly maltose uptake) were greatly inhibited by high salt, but others (glucose fermentation) were not. Norkrans (1968) suggested that the inability of *D. hansenii* to survive at very high salt concentrations may be a result of an inability to take up exogenous substrates.

When grown in medium with high solute concentrations, osmophilic yeasts and fungi maintain their internal osmotic pressure by the synthesis of simple polyhedric alcohols: glycerol, erythritol, arabitol, and/or mannitol (Gustafsson and Norkrans, 1976; Edgley and Brown, 1978; Adler and Gustafsson, 1980; Adler et al., 1981; Luard, 1982; Hocking and Norton, 1983; Adler et al., 1985; Nobre and da Costa, 1985; Andre et al., 1988). One or more polyols may be synthesized during growth, and certain polyols may be selectively synthesized during a particular phase of growth (e.g., logarithmic phase vs. stationary phase). The amount of polyol produced is proportional to the salinity of the culture medium (Gustafsson and Norkrans, 1976; Adler et al., 1985). In addition to polyols, negatively charged amino acids (aspartate and glutamate) possibly contribute to the osmotic balance in fungi (Luard, 1982).

The ability to regulate K^+ and Na^+ also reflects halotolerance. Onishi (1963) demonstrated that *S. rouxii* was very leaky to K^+ ions when suspended in NaCl solutions. *D. hansenii*, which demonstrates greater osmotolerance and halotolerance than *S. cerevisiae*, has a better ability to extrude Na^+ and take up K^+ (Norkrans and Kylin, 1969). Similarly, the fungus *Chrysosporium fastidium* is more salt tolerant than *Penicillium chrysogenum*, and it demonstrated a better ability to extrude Na^+ and maintain a higher K^+:Na^+ ratio than the less tolerant species (Luard, 1982).

11.3 Yeasts and fungi from saline environments

Among the reports of yeasts and fungi actually thriving in natural hypersaline environments are the occurrence of a dematiaceous hyphomycete (*Cladosporium* sp.) growing and producing conidia on a submerged piece of pine wood in the Great Salt Lake (29–36% salinity; see Chapter 18) (Cronin and Post, 1977); identified and unidentified fungi in Mono Lake (9% salinity; see Chapter 20 and Winkler, 1977); the infection of brine shrimp (*Artemia salina*) with the yeast *Metschnikowia kamienskii* or *M. bicuspidata* var. *australis* in solar saltern brines (10–12% salinity) (Spencer et al., 1964; Lachance et al., 1976; Phaff and Starmer, 1980); and the isolation of high salt-tolerant strains from the bottom sediments of Antarctic Lake Bonney (*Dendryphiella salina*) (Waguri, 1976) and the deep waters of Antarctic Lake Vanda (including two strains of *Penicillium*, one phycomycete, and one other fungus). The bottom sediments of Lake Vanda yielded a large variety of fungi, primarily strains of *Aspergillus* and *Penicillium* (Sugiyama et al., 1967). The bottom water of Lake Vanda has 10% salt and is 25°C, and the bottom water of Lake Bonney has 11–20% salt and is −3° to −5°C (see Chapter 16). Although these reports suggest that these yeasts and fungi appear to be physiologically active under the prevailing conditions, there have been no demonstrations of their activity *in situ*. There remains the possibility that the microorganisms were imported from more dilute sources and remained inactive but viable under hypersaline (and in some cases anaerobic) conditions.

Yeasts and fungi have also been documented from a variety of saline desert soils (Abdel-Hafez, 1981; Pond et al., 1984; Radwan et al., 1984, 1985; Hunter-Cervera and Sotos, 1986) and salt marshes of elevated salinity (Moustafa, 1975a, 1975b; Abdel-Hafez et al., 1977). Radwan et al. (1984) isolated four strains of *Penicillium* that grew optimally in about 10% NaCl with maximum salt tolerances ranging up to 15% to 30% salt (Table 11.2). The salinity tolerances of the identified strains were not tested in the other cited studies nor were there reports of *in situ* activity. Most of the samples collected were associated with the root zones of higher plants. The most numerically common taxa cited were *Aspergillus* and *Penicillium*, although several other genera were also reported.

Using sucrose-enriched medium to estimate the populations of osmophilic fungi in coastal environments, Moustafa (1975a) compared colony counts and diversity in hypersaline (8.0–24.0% soluble salt per g dry soil) and less saline (0.8–6.4% soluble salts) sediment samples. The hypersaline samples typically had lower colony counts (475–998 total counts per g dry soil) than non-hypersaline soils (up to 1525 counts per g dry soil), but diversity was about the same for the hypersaline (9–16 species, 7–12 genera per sample) and less saline sediments (6–15 species, 3–12 genera per sample). While the hypersaline sediments were collected from a salt depression, less saline sediments were collected from a salt marsh and from coastal sandy soil. Interestingly, the salt marsh sediments yielded fewer osmotolerant (sugar-tolerant) fungi than the salt pan sediments. Moustafa (1975b) found that some of the heterogeneity in osmophilic fungal density in salt marshes not only correlated with soil salinity, but also with proximity to certain species of plants. This was interpreted as a reflection of the organic content of localized soils.

Osmotolerant yeasts and fungi have also been directly isolated from seawater (Siepmann, 1959; Ross and Morris, 1962; Norkrans, 1966; Pal et al., 1979). Siepmann (1959) isolated 56 strains with varying degrees of salt tolerance. One-third of the isolates had a maximum tolerance of 10–15% NaCl, one-fourth of the isolates had a maximum tolerance of 15–20% NaCl, one-fifth of the strains tolerated 20–25% NaCl, and one strain (*Aspergillus repens*) tolerated ≥25% NaCl (Table 12.2). *A. repens* grew optimally in 6% salt. Ross and Morris (1962) isolated 10 species of yeasts from seawater including *Debaryomyces kloeckeri* and *D. subglobosus* (but not *D. hansenii*) with NaCl tolerances of 22–24%. The other isolated strains tolerated 9–22% NaCl (Table 11.1).

11.4 *Debaryomyces hansenii*, a halotolerant yeast

Norkrans (1966) isolated several seawater yeasts with varying degrees of halotolerance. The most halotolerant was *Debaryomyces hansenii*, which tolerated up to 24% salt. Following Norkrans' investigation, *D. hansenii* has been the focus of numerous studies of mechanisms of halotolerance in yeasts. Although this species has not yet been documented to occur in natural hypersaline habitats, it is worthwhile summarizing the findings concerning the physiological and biochemical adaptation in this yeast as a reference to what characteristics can be expected from halotolerant (or possibly obligately halophilic) yeasts and fungi in evaporite environments.

Table 11.3 summarizes some of the physiological and biochemical characteristics of *D. hansenii*. Most strains can grow on a variety of carbohydrates and hydrocarbons, but they are restricted in their sources of nitrogen. *D. hansenii* grows poorly or not at all by anaerobic fermentation. The temperature

Table 11.3 Some physiological and biochemical characteristics of *Debaryomyces hansenii*

Characteristic	Reference
General growth characteristics	
Carbon sources metabolized:	
sugars, starch, ethanol, glycerol, sorbitol	Tilbury, 1980b
glucosamine supports slow growth	Lindman, 1981
aromatic compounds	Cerniglia and Crow, 1981
hydrocarbons	Bos and de Bruyn, 1973; Zvyagintseva and Gorodnyanskaya, 1978
Vitamin-free growth: some strains	Tilbury, 1980b
Nitrogen sources metabolized:	
ammonia and urea utilized; nitrate, nitrite, L-asparagine, hydroxylamine, and hydrazine not assimilated	Choudary and Rao, 1984
Anaerobic growth: little or none	Tilbury, 1980b
Anaerobic growth decreases above 12% salt	Norkrans, 1968
Upper temperature limit: ≥37°C	Tilbury, 1980b
Lower temperature limit: 6°C	Norkrans, 1966
pH tolerance: 3–8.5	Norkrans, 1966
Salt tolerance and osmoregulation	
NaCl tolerance for growth: 0–24%	Norkrans, 1966
Survival in 28% NaCl for 36 days: negative	Norkrans, 1966
Osmoregulation:	
Glycerol dominant in log phase	Gustafsson and Norkrans, 1976; Adler and Gustafsson, 1980
Arabitol dominant in stationary phase	Adler and Gustafsson, 1980
Amino acid pool not affected by salinity	Adler and Gustafsson, 1980
Ion regulation:	
Maintains high K^+:Na^+ internally	Norkrans, 1968; Norkrans and Kylin, 1969

range of growth is typical for yeasts. The strain used by Norkrans (1966) proved capable of growth at moderately alkaline pH.

D. hansenii produces glycerol as a compatible solute during logarithmic growth and produces arabitol when it enters the stationary phase. In 16% NaCl, exponentially growing cells had 1.3 M glycerol (Gustafsson, 1979) and as much as about 4 M glycerol (Adler et al., 1985). By late logarithmic phase, cells become leaky to glycerol and much of it was lost to the medium (Adler et al., 1985). The glycerol in the medium was then metabolized by the yeast so that by late stationary phase there was no glycerol in the cells nor in the medium. Adler and Lijenberg (1981) found that cells grown in high salt also had increased permeability to various low-molecular-weight glycols.

Growth appears to be slower and less efficient at high salt concentrations. Norkrans (1966) noted that growth in high salt was characterized by long lag periods and low yield. In comparing the ATP pools and molar growth yields of cells grown in 4 mM and 2.7 M NaCl, Gustafsson (1979) found that the ATP pool during late logarithmic phase was slightly higher in the cells in saline medium (3.5 mM vs. 3.1 mM), but the molar growth yield was dramatically lower (59 vs. 91). Cells in high salt had a decreased capacity to maintain high internal K^+:Na^+ ratios (Norkrans, 1968). In cells grown without NaCl, the internal K^+:Na^+ ratio (measured in μg·mg dry wt) was 28.4:0.87. In cells grown in 16% NaCl, the K^+:Na^+ ratio was 8.8:8.6. In the less halotolerant *S. cerevisiae*, cells grown in 12% NaCl had a K^+:Na^+ ratio of 6.5:46.5.

Although the intracellular enzymes of *D. hansenii* are presumably sensitive to salt, alkaline phosphatase (an extracellular enzyme) showed maximum activity in 1–3 M NaCl, very high activity in 4 M NaCl, and reduced activity in $\leq$0.5 M NaCl (Adler, 1978). Glycerol could not substitute for NaCl and in fact inhibited the phosphatase more than dilute buffer alone. Mg^{2+} (up to at least 100 mM) increased enzyme activity. There was no difference in alkaline phosphatase actvity between cells grown in high and low salt. Adler concluded that this enzyme behaves much like the alkaline phosphatase of halobacteria.

The ultrastructure and membrane composition of *D. hansenii* may provide some clues to mechanisms of halotolerance. Electron microscopic studies of cells grown in 12 or 20% salt show unique storage vacuoles and channels that are not found in other yeasts (Gezelius and Norkrans, 1970; Lindman and Norkrans, 1982). Adler and Lijenberg (1981) compared the lipid content of cells grown in high and low salt. In high salt the amount of total and free sterols and total fatty acids increased, and the mole percent of mono- and di-unsaturated C_{18} acids (oleic and linoleic acids) increased while the proportion of tri-unsaturated C_{18} acid (linolenic acid) decreased. Similar trends were observed in the less osmotolerant *S. cerevisiae*. Changes in lipid composition were also accompanied by increased membrane permeability in *D. hansenii*. Gorodnyanskaya and Zvyagintseva (1978) found that changes in lipid composition correlated with salt concentrations and C:N ratios in the medium.

Although the effect of NaCl concentration on the cell wall composition of *D. hansenii* has not been described, that of the halotolerant *S. rouxii* has been (Horisberger et al., 1985). When the yeast was grown both in 0% and 18% NaCl, the major component of the cell envelope was glucan with lesser amounts of mannan and chitin. In low-salt cells, the glucan had a more complex structure. In other respects there were no significant differences in the cell walls of the yeast grown in high and low salt.

D. hansenii forms ascospores with unique ornamentation (Banno and Mikata, 1985). Spore formation is an adaptation to facilitate dispersion and survival under adverse conditions, but the factors that promote and inhibit spore formation and germination in this halotolerant species have not been documented.

11.5 Conclusions

It appears that yeasts and fungi are well suited for life in natural habitats of high salt content. Their lack of representation in the literature may not reflect their inability to colonize these environments, but rather the relatively little effort that has been spent to find them. True halophily may occur among yeasts and fungi, but most studies show there is no absolute requirement for NaCl among salt-tolerant, osmophilic strains.

References

Abdel-Hafez, S.I.I. 1981. Halophilic fungi of desert soils of Saudi Arabia. *Mycopathologia* 75: 76–80.

Abdel-Hafez, S.I., Moubasher, A.H., and **Abdel-Fattah, H.M.** 1977. Studies on mycoflora of salt marshes in Egypt. IV. Osmophilic fungi. *Mycopathologia* 63: 143–151.

Adler, L. 1978. Properties of alkaline phosphatase of the halotolerant yeast *Debaryomyces hansenii. Biochimica et Biophysica Acta* 522: 113–121.

Adler, L. and **Gustafsson, L.** 1980. Polyhydric alcohol production and intracellular amino acid pool in relation to halotolerance of the yeast *Debaryomyces hansenii. Archives of Microbiology* 124: 123–130.

Adler, L. and **Lijenberg, C.** 1981. Sterol content, fatty acid composition of phospholipids, and permeability of labeled ethylene glycols in relation to salt-tolerance of yeasts. *Physiologica Plantarum* 53: 368–374.

Adler, L., Falk, K.-E., Norkrans, B., and **Ångstrom, J.** 1981. Polyol and water in the salt-tolerant yeast *Debaryomyces hansenii* as studied by ^{1}H nuclear magnetic resonance. *FEMS Microbiology Letters* 11: 269–271.

Adler, L., Blomberg, A., and **Nilsson, A.** 1985. Glycerol metabolism and osmoregulation in the salt-tolerant yeast *Debaryomyces hansenii. Journal of Bacteriology* 162: 300–306.

Andre, L., Nilsson, A., and **Adler, L.** 1988. The role of glycerol in osmotolerance of the yeast *Debaryomyces hansenii. Journal of General Microbiology* 134: 669–677.

Andrews, S. and **Pitt, J.I.** 1987. Further studies on the water relations of xerophilic fungi, including some halophiles. *Journal of General Microbiology* 133: 233–238.

Banno, I. and **Mikata, K.** 1985. Scanning electron microscopy of ascospores from various strains of *Debaryomyces hansenii*. *Institute for Fermentation Research Communications (Osaka)* 0: 63–69.

Bos, P. and **de Bruyn, J.C.** 1973. The significance of hydrocarbon assimilation in yeast identification. *Antonie van Leeuwenhoek Journal of Microbiology and Serology* 39: 99–107.

Cerniglia, C.E. and **Crow, S.A.** 1981. Metabolism of aromatic hydrocarbons by yeasts. *Archives of Microbiology* 129: 9–13.

Choudary, P.V. and **Rao, G.R.** 1984. Molecular analysis of inorganic nitrogen assimilation in yeasts. *Archives of Microbiology* 138: 183–186.

Cronin, E.A. and **Post, F.J.** 1977. Report of a dematiaceous hyphomycete from the Great Salt Lake, Utah. *Mycologia* 69: 846–847.

Edgley, M. and **Brown, A.D.** 1978. Response of xerotolerant and non-tolerant yeasts to water stress. *Journal of General Microbiology* 104: 343–345.

Gezelius, K. and **Norkrans, B.** 1970. Ultrastructure of *Debaryomyces hansenii*. *Archiv für Mikrobiologie* 70: 14–25.

Gorodnyanskaya, L.I. and **Zvyagintseva, I.S.** 1978. Fatty acid and fraction composition of lipids of osmotolerant yeasts. *Mikrobiologiya* 47: 456–459.

Gustafsson, L. 1979. The ATP pool in relation to the production of glycerol and heat during growth of the halotolerant yeast *Debaryomyces hansenii*. *Archives of Microbiology* 120: 15–24.

Gustafsson, L. and **Norkrans, B.** 1976. On the mechanism of salt tolerance. Production of glycerol and heat during growth of *Debaryomyces hansenii*. *Archives of Microbiology* 110: 177–183.

Hocking, A.D. and **Norton, R.S.** 1983. Natural abundance ^{13}C nuclear magnetic resonance studies on the internal solutes of xerophilic fungi. *Journal of General Microbiology* 129: 2915–2925.

Horisberger, M., **Rouvet-Vauthey, M.**, **Richli, U.**, and **Farr, D.-R.** 1985. Cell wall architecture of the halophilic yeast *Saccharomyces rouxii*: an immunological study. *European Journal of Cell Biology* 37: 70–77.

Hunter-Cervera, J. and **Sotos, L.** 1986. Screening for a "new" enzyme in nature: haloperoxidase production by Death Valley dematiaceous hyphomycetes. *Microbial Ecology* 12: 121–127.

Koppensteiner, G. and **Windisch, S.** 1971. Der osmotische Wert als begrenzender Faktor für das Wachstum und die Garung von Hefen. *Archiv für Mikrobiologie* 80: 300–314.

Kroemer, K. and **Krumbholz, G.** 1932. Untersuchen über osmophile Sprosspilze. *Archive für Mikrobiologie* 3: 384–396.

Lachance, M.-A., **Miranda, M.**, **Miller, M.W.**, and **Phaff, H.J.** 1976. Dehiscence and active spore release in pathogenic strains of the yeast *Metschnikowia bicuspidata* var. *australis*: possible predatory implication. *Canadian Journal of Microbiology* 22: 1756–1761.

Larsen, H. 1981. The family Halobacteriaceae, pp. 985–994 in Starr, M.P., Stolp, H., Trüper, H.G., Balows, A. and Schlegel, H.G. (editors), *The Prokaryotes. A Handbook of Habitats, Isolation, and Identification of Bacteria*, Springer-Verlag, New York.

Lindman, B. 1981. Effect of NaCl on kinetics of D-glucosamine uptake in yeasts differing in halotolerance. *Antonie van Leeuwenhoek Journal of Microbiology and Serology* 47: 297–306.

Lindman, B. and **Norkrans, B.** 1982. Membranous interrelationships: an ultrastructural study of the halotolerant yeast *Debaryomyces hansenii*. *Protoplasma* 110: 66–70.

Luard, E.J. 1982. Accumulation of intracellular solutes by two filamentous fungi in response to growth. A low steady state osmotic potential. *Journal of General Microbiology* 128: 2563–2574.

Moustafa, A.F. 1975a. Contribution to the fungal flora of Kuwait. *Transactions of the British Mycological Society* 65: 547–553.

Moustafa, A.F. 1975b. Osmophilous fungi in the salt marshes of Kuwait. *Canadian Journal of Microbiology* 21: 1573–1580.

Mrak, E.M. and **Phaff, H.J.** 1948. Yeasts. *Annual Review of Microbiology* 2: 1–46.

Nobre, M.F. and **da Costa, M.S.** 1985. The accumulation of polyols by the yeast *Debaryomyces hansenii* in response to water stress. *Canadian Journal of Microbiology* 31: 1061–1064.

Noda, F., Hayashi, K., and **Mizunuma, T.** 1980. Antagonism between osmophilic lactic acid bacteria and yeasts in brine fermentation of soy sauce. *Applied and Environmental Microbiology* 40: 452–457.

Norkrans, B. 1966. Studies on marine occurring yeasts: growth related to pH, NaCl concentration and temperature. *Archiv für Microbiologie* 54: 374– 392.

Norkrans, B. 1968. Studies on marine occurring yeasts: respiration, fermentation and salt tolerance. *Archiv für Mikrobiologie* 62: 358–372.

Norkrans, B. and **Kylin, A.** 1969. Regulation of potassium to sodium ratio and of the osmotic potential in relation to salt tolerance in yeasts. *Journal of Bacteriology* 100: 838–845.

Onishi, H. 1963. Osmophilic yeasts. *Advances in Food Research* 12: 53–94.

Pal, H.S., Khosrawi, L., and **Bond, C.** 1979. Halophilic n-alkane utilizing yeast from the Arabian Gulf. *Biotechnology Letters* 1: 445–450.

Phaff, H.J. and **Starmer, W.T.** 1980. Specificity of natural habitats for yeasts and yeast-like organisms, pp. 79–102 in Skinner, F.A., Passmore, S.M. and Davenport, R.R. (editors), *Biology and Activities of Yeasts*, Academic Press, New York.

Pitt, J.I. and **Hocking, A.D.** 1977. Influence of solute and hydrogen ion concentration on the water relations of some xerophilic fungi. *Journal of General Microbiology* 101: 35–40.

Pond, E.C., Menge, J.A., and **Jarrell, W.M.** 1984. Improved growth of tomato in salinized soil by vesicular-arbuscular mycorrhizal fungi collected from saline soils. *Mycologia* 76: 74–84.

Radwan, S.S., El-Essawy, A.A., and **Helal, G.A.** 1984. Salinity-loving fungi in Egyptian soils. 1. Numbers, identities, and halophilism. *Zentralblatt für Microbiologie* 139: 435–440.

Radwan, S.S., El-Essawy, A.A., and **Helal, G.A.** 1985. Salinity-loving fungi in Egyptian soils. 2. Preliminary notes on antibiotics from halophilic *Penicillium* spp. *Zentralblatt fur Mikrobiologie* 140: 149–154.

Robinson, R.A. and **Stokes, R.H.** 1968. *Electrolyte Solutions.* Buttersworth. London. 571 pp.

Rodriguez-Valera, F., Ruiz-Berraquero, F., and **Ramos-Cormenzana, A.** 1980. Behaviour of mixed populations of halophilic bacteria in continuous cultures. *Canadian Journal of Microbiology* 26: 1259–1263.

Rodriguez-Valera, F., Juez, G., and **Kushner, D.J.** 1982. Halocins: salt-dependent bacteriocins produced by extremely halophilic rods. *Canadian Journal of Microbiology* 28: 151–154.

Ross, S.S. and **Morris, E.O.** 1962. Effect of sodium chloride on the growth of certain yeasts of marine origin. *Journal of the Science of Food and Agriculture* 13: 467–475.

Siepmann, R. 1959. Ein Beitrag zur saprophytischen Pilzflora des Wattes der Wesermundung. *Veroffenlichungen des Institutes für Meeresforschung in Bremerhaven* 6: 283–301.

Spencer, J.F.T., Phaff, J.H., and **Gardner, N.R.** 1964. *Metschnikowia kamienskii* sp. n., a yeast associated with brine shrimp. *Journal of Bacteriology* 88: 758–762.

Sugiyama, J., Sugiyama, Y., Iizuka, H., and **Torii, T.** 1967. Report of the Japanese summer parties in Dry Valleys, Victoria Land, 1963–1965. IV. Mycological studies of the Antarctic fungi. Part 2. Mycoflora of Lake Vanda, an ice-free lake. *Antarctic Record* 28: 23–32.

Tilbury, R.H. 1980a. Xerotolerant (osmophilic) yeasts, pp. 153–179 in Skinner, F.A., Passmore, S.M. and Davenport, R.R. (editors), *Biology and Activities of Yeasts.* Academic Press. New York.

Tilbury, R.H. 1980b. Xerotolerant yeasts at high sugar concentrations, pp. 103–128 in Gould, G.W. and Corry, J.E.C. (editors), *Microbial Growth and Survival in Extremes of Environment,* Society of Applied Bacteriology Technical Service, no. 15, Academic Press, New York.

Waguri, O. 1976. Isolation of microorganisms from salt lakes in the Dry Valley, Antarctica, and their living environment. *Antarctic Record* 57: 80–96.

Winkler, D.W. (editor). 1977. *An Ecological Study of Mono Lake, California.* Institute of Ecology Publication 12, University of California, Davis, 190 pp.

Yamagata, K. and **Fujita, T.** 1974. Salt-tolerant yeasts isolated from the pickling of narazuke. *Hakko Kogaku Zasshi* 52: 217–224.

Zvyagintseva, I.S. and **Gorodnyanskaya, L.I.** 1978. Lipolytic activity of osmotically tolerant yeasts. *Mikrobiologiya* 47: 230–233.

12

Deep Sea Hypersaline Basins

Several deep sea hypersaline basins have been investigated to understand their geochemical processes as well as their potential and actual biological activity. These environments include the Orca Basin and East Flower Garden bank (Gulf of Mexico), and the Red Sea hot brines (see map, Figure 1.1, for locations). They are all believed to be dissolution products of older evaporite deposits, the brines being trapped on the ocean floor by density stratification. While they all share certain features (high Na^+ and Cl^- concentrations and anoxic conditions), they differ in other biologically important and telltale features such as concentrations of nutrients, sulfur species, metals, and hydrocarbons, as well as stable isotope signatures. Tables 12.1 and 12.2 summarize some of these differences. The following discussion outlines the major features of each of these environments. It emphasizes the criteria used to determine whether or not there is recent biological activity in the brines and associated sediments, or whether the biological by-products are imported into the brines.

12.1 Orca Basin

Both the Orca Basin and the East Flower Garden bank brines are examples of evaporite diagenesis and dissolution in continental slope environments. By examining the roles of microorganisms in these habitats (available substrates, products, rates of activity, and types of microorganisms), it should be possible to construct a model to evaluate the importance of halophilic and halotolerant bacteria in the processes of petroleum genesis, maturation, and degradation in the Gulf of Mexico and similar environments.

Table 12.1 The inorganic chemistry of deep sea brines[a]

Site	Depth, m	°C	Salinity, %	Cl^-	Na^{2+}	Mg^{2+}	Ca^{2+}	K^+	SO_4^{2-}	S^{2-}	Reference[b]
Orca Basin	2200–2400	5.65	25.8	149.5	91.5	1.05	1.09	0.63	3.66	very low	4, 5, 8
East Flower Garden bank (1977)	72	22	19.5	131	79.3	1.07	1.62	0.36	4.42	0.070	2
Red Sea, Discovery Deep	>2000	22–60	25.6	155	92.7	0.79	4.65	2.13	0.70	0	1, 6, 7
Tyro Basin	>3300	15	26.5	157	104	1.42	1.34	1.30	nd	high	3, 9
Average seawater	–	–	3.54	19.4	10.8	1.29	0.41	0.40	2.72	–	4

nd, no data.
[a]All chemical components listed in $g \cdot kg^{-1}$ except total salinity (%).
[b] (1) Brewer et al. (1969); (2) Brooks et al. (1979); (3) de Lange et al. (1983); (4) Sackett et al. (1979); (5) Shokes et al. (1977); (6) Trüper (1969); (7) Watson and Waterbury (1969); (8) Wiesenburg et al. (1985); (9) A. Yayanos, pers. comm.

Table 12.2 Nutrients and hydrocarbon chemistry of deep sea brines

Site	PO_4-P, μm	NO_3-N, μm	NH_4-N, μm	DOC, $mg \cdot l^{-1}$	CH_4, $\mu l \cdot l^{-1}$	C_2H_6, $\mu l \cdot l^{-1}$	C_3H_8, $\mu l \cdot l^{-1}$	C_1:C_2+C_3	$\delta^{13}C$-CH_4, ‰	Reference[a]
Orca Basin	22–63	0	500	3.3–3.7	16,800	29.1	<0.022	730	−73	4, 5, 6
East Flower Garden bank (1977)	2.36	0.8	nd	4.56[b]	2730	414	14.7	6.4	−42	2
									−51.3	1
Red Sea	nd	nd	nd	1.03–2.15	5–158	0.007–0.228[c]	0.01–0.02	40–1040	–	3, 8
Average seawater[d]	2.5	23	<0.1	0.5–0.6	50	0.5	0.5	50	–	5, 6

nd, no data.
[a] (1) Bernard et al. (1976); (2) Brooks et al. (1979); (3) Burke et al. (1981); (4) LaRock et al. (1979); (5) Sackett et al. (1979); (6) Shokes et al. (1977); (7) Watson and Waterbury (1969).
[b] 1976 analysis.
[c] Saturated and unsaturated hydrocarbons.
[d] Gulf of Mexico seawater overlying the Orca Basin.

The Orca Basin is an anoxic body of water in the northern Gulf of Mexico (27°N, 91°W) in a submarine basin 25 km long and encompassing 400 km^2. The basin slopes down from 1800 m to 2400 m depth with about 9 km^3 of brine filling the bottom 180 m (Shokes et al., 1977; Wiesenburg et al., 1985). The brine has a total salinity of 258 $g \cdot kg^{-1}$, with NaCl constituting 93% (w/w) of the salts. In contrast, NaCl constitutes only 85% of the total salts in average seawater. Sediment porewaters decrease in Cl^- content with depth, indicating that the source of dissolution brine is not from below the basin, but from some point surrounding the basin (Sackett and Bernard, 1977; Shokes et al., 1977). The Orca Basin has been hypersaline for at least 8000 years (Addy and Behrens, 1980).

Dissolved Mn (about 250 μM) and Fe (about 30 μM) are very high in the brine, while sediment concentrations of $CaCO_3$ (15–20%), Fe (3–4%), Zn (90–110 ppm), Cu (30–40 ppm), and Ni (50–55 ppm) are comparable to nearby, non-hypersaline slope sediments. Orca Basin sediments are depleted in Mn (700–900 ppm) relative to nearby slope sediments. Mn is apparently mobilized in the anoxic, reducing environment of the basin brines and sediments (Trefry and Presley, 1979). Although very little free sulfide has been detected, the high iron sulfide content of the sediments (up to 0.7%) suggests that sulfide is efficiently trapped by iron (Wiesenburg et al., 1985; Sheu and Presley, 1986a, 1986b).

Besides the presence of sulfide trapped by iron, other circumstantial evidence suggests that sulfate-reducing bacteria may be active in the Orca Basin sediments. An earlier report of sulfate depletion in a core sample may not represent the true Orca Basin profiles since it was taken from slumped slope sediments (Sackett and Bernard, 1977). In a sediment core analyzed by Wiesenburg et al. (1985), there was a slight sulfate depletion at 25 cm depth (41 mM) while the rest of the core had sulfate values of 45–47 mM. The sediment contained abundant organic carbon (2–3% by weight). Kennett and Penrose (1978) found fragments of Holocene seaweed with attached calcareous polychaetes at a core depth of 10 m in an Orca Basin sediment. Such a finding indicates that anaerobic degradation must be extremely slow in this environment. Wiesenburg et al. (1985) concluded that rates of *in situ* activity of sulfate-reducing bacteria could be calculated if the $\delta^{34}S$ values of pyrite were known.

Such measurements would have to take into consideration the laminated nature of the sediments. Addy and Behrens (1980) reported that the top 485 cm of sediments were black with three grey layers totalling 70 cm. Below 485 cm, the sediments were grey. They interpreted the environment of deposition to have been oxic during the early stages and anaerobic only during the later stages represented by the top 485 cm of sediment. The grey laminae in the black layers were interpreted as turbidite inflows. Sheu and Presley (1986b) found that Orca Basin sediments had alternating black and grey laminae, with the black layers relatively enriched in $CaCO_3$ (19% vs. 15%), organic carbon

(1.8% vs. 0.7%), and iron sulfides (0.7% vs. 0.4%). Black mud was also enriched in Mn, Co, and Ni (Addy and Behrens, 1980). Occasional red layers of oxidized iron (hematite) were encountered (Sheu and Presley, 1986a, 1986b), which suggested that there were occasional periods of oxygen introduction through the seawater-brine interface during an essentially anoxic period. Such events could have accompanied the influx of turbidites noted by Addy and Behrens (1980). A detailed analysis of $\delta^{34}S$ values of sulfides in the Orca Basin sediments through the laminae would not only demonstrate that sulfate reduction had occurred, but that it may be locally and temporally variable in this otherwise stable environment.

LaRock et al. (1979) measured microbial biomass and activity (by uptake of radioactive uridine) in and above the Orca Basin brines (but not in the sediments). The highest cell counts were found in the seawater above the pycnocline ($5.4–8.6 \times 10^5 \cdot ml^{-1}$), while the hypersaline brines had modest counts ($1.4–1.7 \times 10^5 \cdot ml^{-1}$) except near the bottom ($2.9 \times 10^5 \cdot ml^{-1}$). This sample was taken 62 m from the sediment interface which means that even higher cell densities may exist in the deeper brines. ATP levels in the aerobic seawater above the brines were 0.93–2.47 nanograms per liter and they were relatively higher in the transition zone above the dense brines (4.12–5.89 $ng \cdot l^{-1}$, salinity not given). The transition zone waters showed the highest uridine uptake rates of all the samples measured above and in the brines. The authors suggested that autotrophic bacteria (i.e., sulfur and ammonia oxidizers) may be the dominant group in the region above the brine-seawater interface.

Wiesenburg et al. (1977) reported that values of nitrate decreased just above the pycnocline (23 μM to 0) while nitrite peaked in this zone (14.7 μM). Such evidence of denitrification indicates that autotrophic processes are probably accompanied by heterotrophic processes. Wiesenburg et al. (1985) found the only measurable free sulfide (2.8 μM) only just below the pycnocline, although sulfide generated by sulfate-reducing bacteria may have reacted with dissolved Fe at other depths.

Activity of heterotrophic bacteria was estimated by LaRock et al. (1979) from the incorporation by ^{14}C-acetate in samples taken at different depths. Fairly constant rates both in and above the brines (1.7–2.5 $ng \cdot l^{-1} \cdot d^{-1}$) were found except in the deepest sample (3.3 $ng \cdot l^{-1} \cdot d^{-1}$). These rates are apparently uncorrected for the activity of the radiolabeled acetate in the seawater and brine. In addition, acetate does not measure the potential for fermentative metabolism in anaerobic microbial populations. Except for the indirect determination of nitrate reduction, there is little evidence that defines which kinds of microorganisms are responsible for the biological activity in the transition zone above the pycnocline of the Orca Basin brines.

While ATP levels were relatively high in the transition zone, they were extremely high in the brine at 62 m above the sediment interface (15.4 $ng \cdot l^{-1}$). Uridine uptake rates (normalized to brine volume) were slightly higher in the deep brine than in the transition zone. Uridine uptake rates of the deep brines normalized to ATP concentrations or cell counts were 30% and 50%, respectively, of the highest rates measured in the transition zone (Table 12.3). In fact, uridine uptake rates per ng ATP, a measure of nucleic acid synthesis, are fairly similar in the oxic seawater and anoxic brines except near boundary zones at the bottom of each layer. These data are some of the very few published that compare such metabolic activity of natural populations in neighboring seawater and brine habitats.

The anoxic brines of the Orca Basin have high levels of nutrients and hydrocarbons (Table 12.2). CH_4 concentrations are as high as 10.2–12.8 mg $C \cdot l^{-1}$ or three times the DOC values (Sackett et al., 1979). These concentrations of methane are about 100 times higher than those found in other anoxic brines except for the petroleum-associated East Flower Garden bank pool (Wiesenburg et al., 1985). While Sackett et al. (1979) measured $\delta^{13}CH_4$ values in the Orca Basin brines (-72 to -74.5‰), even lighter values (-85 to -105‰) were recorded in an earlier investigation (Sackett and Bernard, 1977). Relatively light $\delta^{13}CH_4$ values (-97‰) were found in the sediments, and the differences in $\delta^{13}C$ values between interstitial and water column brine were attributed to anaerobic consumption of CH_4 in near-surface sediments (Wiesenburg et al., 1985). Reid et al. (1977) noted that the $\delta^{13}C$ of the ΣCO_2 in the brine was light (-16‰), indicating a significant pool of biogenic CO_2.

The isotope signatures and the ratio of C_1:C_2+C_3 hydrocarbons (Table 12.2) indicate that the methane is biogenic and not a thermal alteration product of older petroleum. Sackett and Bernard (1977) and Sackett et al. (1979) suggested that some of the methane may be a petroleum-related component introduced with the influx of the brine. The latter investigators rejected *in situ* methanogenesis as the major source of methane because of the generally known inhibition of methanogens in sulfate-rich environments. Because we lack data demonstrating the relative roles of these two groups of microorganisms in hypersaline marine sediments, it is premature to assume that *in situ* methanogenesis is quantitatively unimportant in Orca Basin sediments.

In addition to methane measurements, Sackett et al. (1979) measured

Table 12.3 Radiolabeled uridine uptake in Orca Basin brines[a]

Zone	Depth, m	dpm uridine per ng ATP^{-1}	dpm uridine per 10^7 cells
Aerobic	1650	5430	114
Aerobic	1850	6680	115
Aerobic	2120	6870	308
Transition	2221	36,000	10,600
Anoxic	2260	>5380	1076
Anoxic	2292	>4900	1190
Anoxic	2338	10,310	5481

dpm, disintegrations per minute.
[a]Data of LaRock et al. (1979).

Table 12.4 Dissolved and particulate organic carbon, and inorganic carbon in the Orca Basin[a].

Carbon pool	Source	mg C·l^{-1}	$\delta^{13}C$, ‰
DOC	Overlying seawater	26.8–28.3	+0.2 to +0.4
	brine	51.7–56.9	−15.0 to −19.0
POC	Overlying seawater	0.5–0.6	−19.8 to −20.3
	brine	3.3–3.7	−23.0 to −27.0
Inorganic carbon	Overlying seawater	0.005–0.017	–
	brine	0.058–0.086	−17.0 to −22.6

DOC, dissolved organic carbon; POC, particulate organic carbon.
[a]Data of Sackett et al. (1979).

concentrations and $\delta^{13}C$ values of the dissolved and particulate organic carbon and the total inorganic carbon pools. In all cases the pool sizes were larger in the brines than in the overlying seawater (Table 12.4). While the $\delta^{13}C$ values of the particulate organic fraction of the brines are not unusual, the relatively light values of the DOC and inorganic carbon pools are indicative of reworking of buried organic matter. The extremely light $\delta^{13}C$ values of the methane are theoretically possible if the CO_2 source is extremely light, which is the case in the Orca Basin brines. Like the problem of determining the methane source, the source of isotopically light CO_2 (diffusion from lower sediments, lateral migration with brines, or *in situ* production) remains enigmatic.

12.2 East Flower Garden Bank

The East Flower Garden bank brine pool, located in the northwestern Gulf of Mexico (27° 53′N, 93° 38′W) is a small, shallow, hypersaline pond (30 cm deep, 30 m in diameter) located in the euphotic zone at 72 m depth (Brooks et al., 1979). For a "deep sea" hypersaline basin, it is unique in that biogeochemical processes apparently involve *in situ* photosynthesis. Brine flows down through the pool, over a sill, and down the axis of a submarine canyon. Although brine remains in the pool due to density stratification, it is not stagnant.

Specific organisms were not identified at the bottom of the pool, but the presence of purple patches (phototrophic bacteria?), a thin white layer of presumably elemental sulfur, and the presence of H_2S were noted. Colorless (non-phototrophic) sulfide-oxidizing bacteria were isolated. These data suggest that a complete sulfur cycle might be operating in this brine (18.8–19.5% salinity). The data in Table 12.2 show that nutrient levels are not particularly high. However, the flow of brine through the pool (estimated to be 2 days turnover time) may replenish phosphorus and nitrogen and stimulate biological productivity, a situation analogous to solar salterns. The relatively high concentration of microorganisms in the brine (83.3 ng ATP·l^{-1}) and 20 cm above the brine (43.5 ng ATP·l^{-1}) also indicate the presence of very active microbial communities. These concentrations are significantly higher than those measured by LaRock et al. (1979) in the Orca Basin brines (maximum = 15.4 ng ATP·l^{-1}) and in the transition zone above those brines (maximum = 4–6 ng ATP·l^{-1}).

Sulfur-producing, heterotrophic bacteria have been isolated and cultured from the brine pool (P.A. LaRock, pers. comm. in Brooks et al., 1979), but no details of the methods nor salinities used were given. Brooks et al. (1979) suggested that such microorganisms may live in the water column immediately above the brine, where they would oxidize sulfide diffusing from the brines. They could account for the high level of ATP found at this level.

Although substantial concentrations of hydrocarbons were measured from the East Flower Garden bank brines, the low C_1:C_2+C_3 ratio and the $\delta^{13}C$ isotope signatures (Table 12.2) suggest that the bulk of these gases were thermocatalytically produced in the petroleum fields of the region. However, Bernard et al. (1976) and Brooks et al. (1979) believe that some *in situ* methanogenesis may also be occurring. A comparison of CH_4 concentrations in the main pool and a smaller pool a few meters downstream, showed a 3.26-fold greater concentration in the second pool while the C_2 and C_3 hydrocarbons remained essentially unchanged. Brooks et al. (1979) concluded that the difference in CH_4 concentrations may either reflect organic degradation in the sediments that diffuse upward or an artifact caused by disruption of the sediment by the sampling vehicle.

The $\delta^{13}CH_4$ of the second smaller pool was lighter (−47.1‰) than that of the first pool (−40.0‰). To account for this difference, the excess CH_4 in the second pool would have to have a $\delta^{13}C$ value of −50‰, which is heavier than most biogenic CH_4. The authors argued that (anaerobic) methane consumption or substrate depletion could account for the accumulation of relatively heavy CH_4 produced *in situ*. Neither argument is compelling. CO_2 "substrate" was abundant (48.0 mg CO_2·l^{-1}) and the authors offered no evidence to prove that other substrates were limiting. The investigators inferred that molecular and isotopic changes away from the center of the main brine pool suggested *in situ* sulfate reduction and methanogenesis but some of these comparisons were made between strong and weak brines.

Although the East Flower Garden bank pools are small, their location in the euphotic zone with a presumably active phototrophic microbial community makes them unique among the oceanic brine pools so far documented. The East Flower Garden bank and the Orca Basin are only two of many salt diapirs in the northwest Gulf of Mexico. Further investigations of the brines and associated sediments comparing the microbial processes at these sites would demonstrate the potential for petroleum diagenesis in different kinds of hypersaline communities in the same region. They would also show whether such anaerobic processes in deep sea brines are qualitatively and quantitatively different from those in normal seawater environments of the same region. Dow (1984) has suggested that salt-controlled submarine topography in the Gulf of Mexico (due to subsurface flow of rock salt, forming diapirs, ridges, and sills) may be responsible for the formation of semi-isolated basins that became anoxic during periods of global warming and high sea level. Such basins are believed to be the sites of oil source bed accumulation during Pliocene and Miocene times. However, it remains to be demonstrated whether those anoxic basins were hypersaline or of normal marine composition at the time of organic matter accumulation.

12.3 Red Sea Hot Brines

The hot brines of the Red Sea were the focus of numerous studies during the 1960's, and most of that work was published in a volume edited by Degens and Ross (1969). While detailed analyses of the geophysical, hydrological, and chemical nature of the environment were presented, only two studies devoted to the microbiology of the hot brines were reported in that volume (Trüper, 1969; Watson and Waterbury, 1969). Those investigators recognized that their limited number of samples put constraints on conclusions concerning microbial processes in the brines. Subsequent studies by Heitzer and Ottow (1976) and Burke et al. (1981) have shed some more light on this subject. The following discussion will focus on the actual and inferred evidence of microbial activity in the brines, and the chemical and physical factors that may limit life in this deep sea environment. Comparisons with the brine basins of the Gulf of Mexico further define some of the unique attributes of the Red Sea brines that may restrict or preclude life in some parts of those basins.

The Red Sea brines are found in some 20 basins or "deeps" at over 2000 m depth in an active fault zone (Bäcker and Schoell, 1972). The Red Sea brines are thought to have evolved from seawater dissolving buried halite from evaporites in the geothermally active Red Sea Rift zone. Heated brines, which are less dense than cold dissolution brines, are believed to rise in the sediments and form pools in the deeps (Brooks et al., 1969). The major ion composition of one basin, the Discovery Deep, is given in Table 12.1.

Although Na^+ and Cl^- comprise the major ions (97% w/w) of the brines, differences exist between temperatures, size, and depth of the brine layers, as well as salinity gradients, total dissolved solids, and heavy metal concentrations of the deeps (Brewer and Spencer, 1969; Brewer et al., 1969; Bäcker and Schoell, 1972). The Atlantis II Deep has two brine layers of different salinities and temperatures (Brewer et al., 1969b). There is about a 30-m transition zone in which the salinity rises from 4.32% to 13.5%, while the temperature rises from 22°C to 44°C. Below the transition zone there is a 30-m zone of stable salinity and temperature (13.5%, 44°C), which overlies the densest and hottest brines (up to 15.65%, 60°C) (Bäcker and Schoell, 1972). In contrast, the Discovery Deep has a nearly 60-m salinity and temperature gradient to the bottom in which the salinity rises from 4.58% to 25.7%, and the temperatures increases to 44.8°C.

As a result of geothermal heating of the sediments, the brines emerge enriched in heavy metals. Table 12.5 summarizes trace metal concentrations and several other attributes of the Atlantis II Deep and the Discovery Deep. Other trace element analyses are given by Brooks et al. (1969). While the Discovery Deep and the 44°C layer of the Atlantis II brines share similar temperatures, the Discovery Deep brine has higher concentrations of the commonly toxic heavy metals (Zn, Cu, Co, Ni, and Pb). The hotter bottom brines of the Atlantis II Deep have even higher concentrations of the heavy metals.

The earliest studies of life in the Red Sea brines demonstrated they were largely sterile. Watson and Waterbury (1969) could not detect bacteria in the Atlantis II Deep brine and sediments, but obligate and facultative anaerobes were found in the overlying transition layer and in 22°C seawater. Positive enrichments for bacteria were made from Discovery Deep sediments but not from the brines. A variety of enrichments were made for aerobes and anaerobes using different media with seawater or brine as the base, and at 22°–56°C. While relatively lower heavy metal concentrations could possibly permit life in the Discovery Deep sediments, a variety of factors were cited which may generally limit and inhibit microbial life in the Red Sea brines: high salinity, high temperatures, high pressures, low carbon (DOC = 1.03–2.15 mg $C \cdot l^{-1}$, sediment carbon = 0.224–0.664%), little or no O_2, and high concentrations of certain heavy metals. In addition, the authors cited the need for more sampling to look for bacteria.

All enrichments for sulfate-reducing bacteria and nitrate-reducing bacteria from the brines and core muds of the two deeps were negative (Trüper,

Table 12.5 Chloride and trace metal concentrations of some Red Sea hot brines[a]

	Atlantis II Deep		Discovery Deep
	44°C	56°C	44.8°C
Cl^-	80.0	156	155
Fe	2.00×10^{-4}	8.1×10^{-2}	2.7×10^{-4}
Mn	8.2×10^{-2}	8.2×10^{-2}	5.46×10^{-2}
Zn	1.52×10^{-4}	5.4×10^{-3}	7.7×10^{-4}
Cu	1.72×10^{-5}	2.6×10^{-4}	7.5×10^{-5}
Co	8.0×10^{-7}	1.6×10^{-4}	1.29×10^{-4}
Ni	1.2×10^{-6}	nd	3.42×10^{-4}
Pb	8.8×10^{-6}	6.3×10^{-4}	1.65×10^{-4}
Brine area	55 km^2		11.5 km^2
Brine depth	126–178 m		175–209 m

[a]Data of Bäcker and Schoell (1972) and Brooks et al. (1969).
All data are in $g \cdot kg^{-1}$.
nd, no data.

1969). Enrichments were conducted at room temperature and at the respective brine temperatures. One of five sediment samples taken outside the brine area had sulfate-reducing bacteria while the brine-seawater transition zone of the Atlantis II Deep yielded positive enrichments from which three strains were isolated. All the isolated bacteria were strains of *Desulfovibrio* which had temperature optima of 35°–40°C. The strains from the Atlantis II transition zone grew well at 44°C in 10% NaCl medium and at 35°C in 0.5× transition zone brine. No denitrifying bacteria were isolated from any enrichment. Trüper cited many of the same reasons given by Watson and Waterbury (1969) to explain the limited success of anaerobic bacteria in the hot brines. In addition, the following specific factors that might limit sulfate-reducing bacteria were noted: no extremely halophilic sulfate-reducing bacteria have yet been found; the sulfate concentration of the Red Sea brines is much lower than that of seawater; and salt-tolerant sulfate-reducing bacteria may not have had enough evolutionary time to adapt in these isolated basins. The pH of the environment is suitable for these bacteria (pH 5.0–5.5), but even the non-brine muds yielded low numbers of bacteria. The nutrient status of the brines and sediments, which have not been determined, may also inhibit bacterial development.

Although circumstantial evidence for free sulfide in the brines was given (brass messengers returned blackened), no free sulfide could be measured in the Atlantis II Deep. The high concentration of Fe^{2+} in the brine would prevent sulfide accumulation, even in the transition zone from which sulfate-reducing bacteria were isolated (Trüper, 1969; Watson and Waterbury, 1969). Sulfate depletion was observed in the Atlantis II Deep while it was not found in most other deeps (Bäcker and Schoell, 1972). A smaller hypersaline basin, the Kebrit Deep (15.33% Cl^-, 23.3°C) was reported to have large amounts of CO_2 and H_2S as well as relatively high concentration of sulfate (2.2 $g \cdot kg^{-1}$) (Bäcker and Schoell, 1972). This basin should be sampled in any further studies to detect sulfate-reducing bacteria in the Red Sea brines.

Although Trüper (1969) had no success in finding denitrifying bacteria in the Red Sea brines, Heitzer and Ottow (1976) obtained positive enrichments using 10% NaCl and 28° or 37°C with sediments from several basins, including the Discovery, Atlantis II, and Kebrit Deeps. They found up to 10^6 bacteria·g $sediment^{-1}$ at the sediment surface with decreasing numbers of bacteria downward. Fifteen isolates of *Pseudomonas* spp. and one coccus were subjected to extensive biochemical testing. Of the several strains that could grow and denitrify in 20% salt, all could grow in $\leq$3% salt as well. Assays to test the effects of environmental factors (i.e., high temperature, pressure, and heavy metal concentrations) were not conducted. Although several kinds of bacteria have now been isolated from the Red Sea brines, their potential activity under *in situ* conditions is still unknown.

Burke et al. (1981) analyzed light hydrocarbons in Red Sea brines and sediments. By comparing C_1:C_2+C_3 ratios, they interpreted whether methane was of unaltered biogenic origin or a thermally altered product of older petroleum. Hydrocarbon concentrations were considerably lower than those from the Gulf of Mexico (Table 12.2). In the 44°C Atlantis II brine, the CH_4 concentration was about 155 $\mu l \cdot l^{-1}$ and the C_1:C_2+C_3 ratio was high (about 1100). Such a high ratio is indicative of biogenic origin of methane. In the lower brine body, about 120 $\mu l \cdot l^{-1}$ CH_4 was measured, but the C_1:C_2+C_3 ratio was only about 50, indicating a thermogenic origin for that methane. The CH_4 concentrations in the underlying sediments were only about half those measured in the brines. The authors concluded that the CH_4 was produced *in situ* or at the seawater-brine interface. The C_1:C_2 ratios of the Valdivia Deep brine and the Nereus Deep brine were 40–130 and 130–470, respectively, indicating a thermal or mixed biogenic-thermal origin for the CH_4.

Burke et al (1981) suggested that the absence of olefins, ethene, and propene in the brines could be explained by their consumption by bacteria in the surface layers of the sediment. Both ethene and propene were found in Red Sea sediments, with concentrations of 119 $nl \cdot l^{-1}$ and 11 $nl \cdot l^{-1}$, respectively, analyzed in the Atlantis II Deep sediments. Although this evidence is somewhat circumstantial, it appears valid to suggest that both methanogenesis and processes of anaerobic hydrocarbon biodegradation are operative to a degree in the Red Sea hot brines. These interpretations should be confirmed by: 1) detection of methanogens and methanogenesis; 2) determination of $\delta^{13}C$ signatures of methane at different levels in the brines and sediments; and 3) analyses to detect changes with time of low-molecular-weight alkenes in surficial sediments.

Stable isotope analyses of the inorganic carbon and sulfate pools do not provide clear-cut evidence of biological activity, but rather reflect the complicated history of brines in the Red Sea deeps. By measuring the ΣCO_2 concentration (125 ppm) and the $\delta^{13}CO_2$ (−1‰) in Red Sea water and comparing these values to those obtained in the brines, Craig (1969) interpreted that the excess CO_2 in the Atlantis II Deep (total CO_2 = 184.7 ppm, $\delta^{13}C$ = −5.6‰) resulted from a mixture of isotopically heavy carbonate and light organic carbon. The deficiency in CO_2 in the Discovery Deep brine (total CO_2 = 39.4 ppm, $\delta^{13}C$ = −16.8‰) could be attributed to the precipitation of isotopically heavy carbon.

The $\delta^{34}S$ values for sulfate in the brines and in Red Sea water were +20.3‰. The $\delta^{34}S$ values for sulfate in the brine were interpeted to reflect the fractionation of the sulfate of the original seawater that formed the brine in combination with processes of sulfate removal that do not fractionate it. The $\delta^{18}O$ values of sulfate showed some isotope fractionation between seawater (+9.5‰) and brine (+7.3‰), a reflection of some isotopic equilibrium at higher brine temperatures or fractionation processes during precipitation. Kaplan et al. (1969) analyzed total $\delta^{34}S$ fractionation and found several categories, including values of greater than +25‰ and less than −25‰. They concluded that the brine sulfate was derived from marine evaporites while the isotope signatures of sulfide in the Atlantis II Deep reflected a mixture of hydrothermal processes and brine introduction, or sulfate-reducing bacteria. The discussion by Craig (1969) summarizes some of the factors that could have influenced the final isotopic fractionation signatures of sulfates and sulfides, and they include: possible sulfate-reducing bacteria (recent), cycles of biological activity over time, brine-shale reactions, and the influence of geothermal brines.

Deep sea brines are rather uncommon, but they have been documented in the Gulf of Mexico, the Red Sea, and in the Mediterranean Sea. In the Tyro Basin near Greece, at $>$3300 m depth, 15 °C brines have been discovered with 157 g $Cl^{-1} \cdot kg^{-1}$, high organic carbon, and a strong smell of H_2S (de Lange and ten Haven, 1983; Jongsma et al., 1983; A. Yayanos, pers. comm.). Because subterranean evaporite deposits are similarly under pressure, deep sea brine systems may provide the most accessible *in situ* evidence of microbial processes involved in the diagenesis of organic matter under pressure in evaporite environments.

References

Addy, S.K. and **Behrens, E.W.** 1980. Time of accumulation of hypersaline anoxic brine in Orca Basin (Gulf of Mexico). *Marine Geology* 37: 241–252.

Bäcker, H. and **Schoell, M.** 1972. New deeps with brines and metalliferous sediments in the Red Sea. *Nature Physical Science* 240: 153–158.

Bernard, B.B., Brooks, J.M., and **Sackett, W.M.** 1976. Natural gas seepage in the Gulf of Mexico. *Earth and Planetary Science Letters* 31: 48–54.

Brewer, P.G. and **Spencer, D.W.** 1969. A note on the chemical composition of the Red Sea brines, pp. 174–179 in Degens, E.T. and Ross, D.A. (editors), *Hot Brines and Recent Heavy Metal Deposits in the Red Sea,* Springer-Verlag, New York.

Brewer, P.G., Densmore, C.D., Munns, R., and **Stanley, R.J.** 1969. Hydrography of the Red Sea brines, pp. 138–147 in Degens, E.T. and Ross, D.A. (editors), *Hot Brines and Recent Heavy Metal Deposits in the Red Sea,* Springer-Verlag, New York.

Brooks, J.M., Bright, T.J., Bernard, B.B., and **Schwab, C.R.** 1979. Chemical aspects of a brine pool in the East Flower Garden bank, northwestern Gulf of Mexico. *Limnology and Oceanography* 24: 735–745.

Brooks, R.R., Kaplan, I.R., and **Peterson, M.N.A.** 1969. Trace metal composition of the Red Sea geothermal brine and interstitial water, pp. 180–203 in Degens, E.T. and Ross, D.A. (editors), *Hot Brines and Recent Heavy Metal Deposits in the Red Sea,* Springer-Verlag, New York.

Burke, Jr., R.A., Brooks, J.M., and **Sackett, W.M.** 1981. Light hydrocarbons in Red Sea brines and sediments. *Geochimica et Cosmochimica Acta* 45: 627–634.

Craig, H. 1969. Geochemistry and origin of the Red Sea brines, pp. 208–242 in Degens, E.T. and Ross, D.A. (editors), *Hot Brines and Recent Heavy Metal Deposits in the Red Sea,* Springer-Verlag, New York.

Degens, E.T. and **Ross, D.A.** (editors). 1969. *Hot Brines and Recent Heavy Metal Deposits in the Red Sea.* Springer-Verlag. New York. 633 pp.

de Lange, G.J. and **ten Haven, H.L.** 1983. Recent sapropel formation in the eastern Mediterranean. *Nature (London)* 305: 797–798.

Dow, W.G. 1984. Oil source beds and oil prospect definition in the Upper Tertiary of the Gulf Coast. *Transactions—Gulf Coast Association of Geological Societies* 34: 329–339.

Heitzer, R.D. and **Ottow, J.C.G.** 1976. New denitrifying bacteria isolated from Red Sea sediments. *Marine Biology* 37: 1–10.

Jongsma, D., Fortuin, A.R., Huson, W., Troelstra, S.R., Klaver, G.T., Peters, J.M., van Harten, D., de Lange, G.J., and **ten Haven, L.** 1983. Discovery of an anoxic basin within the Strabo Trench, eastern Mediterranean. *Nature (London)* 305: 795–797.

Kaplan, I.R., Sweeney, R.E., and **Nissenbaum, A.** 1969. Sulfur-isotope studies on Red Sea geothermal brines and sediments, pp. 474–498 in Degens, E.T. and Ross, D.A. (editors), *Hot Brines and Recent Heavy Metal Deposits in the Red Sea,* Springer-Verlag, New York.

Kennett, J.P. and **Penrose, N.L.** 1978. Fossil Holocene seaweed and attached calcareous polychaetes in an anoxic basin, Gulf of Mexico. *Nature (London)* 276: 172–173.

LaRock, P.A., Lauer, R.D., Schwarz, J.R., Watanabe, K.K., and **Wiesenburg, D.A.** 1979. Microbial biomass and activity distribution in an anoxic, hypersaline basin. *Applied and Environmental Microbiology* 37: 466–470.

Reid, D.F., Brooks, J.M., Bernard, B.B., and **Sackett, W.M.** 1977. Chemical and isotopic analysis of the Orca Basin and East Flower Garden brines. *EOS Transactions, American Geophysical Union* 58: 1175.

Sackett, W.M. and **Bernard, B.B.** 1977. Chemical measurements of Orca Basin sediments. *EOS Transactions, American Geophysical Union* 58: 1175.

Sackett, W.M., Brooks, J.M., Bernard, B.B., Schwab, C.R., Chung, H., and **Parker, R.A.** 1979. A carbon inventory for Orca Basin brines and sediments. *Earth and Planetary Science Letters* 44: 73–81.

Sheu, D-D. and **Presley, B.J.** 1986a. Formation of hematite in the euxinic Orca Basin, northern Gulf of Mexico. *Marine Geology* 69: 309–321.

Sheu, D-D. and **Presley, B.J.** 1986b. Variations of calcium carbonate, organic carbon and iron sulfides in anoxic sediment from the Orca Basin, Gulf of Mexico. *Marine Geology* 70: 103–118.

Shokes, R.F., Trabant, P.K., Presley, B.J., and **Reid, D.F.** 1977. Anoxic hypersaline basin in the northern Gulf of Mexico. *Science* 196: 1443–1446.

Trefry, J.H. and **Presley, B.J.** 1977. Transition metal chemistry in the Orca Basin. *EOS Transactions, American Geophysical Union* 58: 1175.

Trüper, H.G. 1969. Bacterial sulfate reduction in the Red Sea hot brines, pp. 263–271 in Degens, E.T. and Ross, D.A. (editors), *Hot Brines and Recent Heavy Metal Deposits in the Red Sea,* Springer-Verlag, New York.

Watson, S.W. and **Waterbury, J.B.** 1969. The sterile hot brines of the Red Sea, pp. 272–281 in Degens, E.T. and Ross, D.A. (editors), *Hot Brines and Recent Heavy Metal Deposits in the Red Sea,* Springer-Verlag, New York.

Wiesenburg, D.A., Barnard, L.A., and **Schink, D.R.** 1977. Nitrogen in the Orca Basin: effects of denitrification at the brine-sea water interface. *EOS Transactions, American Geophysical Union* 58: 1175.

Wiesenburg, D.A., Brooks, J.M., and **Bernard, B.B.** 1985. Biogenic hydrocarbon gases and sulfate reduction in the Orca Basin brine. *Geochimica et Cosmochimica Acta* 49: 2069–2080.

13

Solar Salterns

13.1 Introduction

Solar salterns are convenient systems for analyzing and interpreting biogeochemical trends through salinity gradients or between comparable salinities of different localities. Most commercial solar salterns (salinas) consist of a series of shallow ponds connected in a sequence of increasingly saline brines (Figure 13.1). Salt companies maintain records of weather, brine flow, evaporation rates, and salinity. In seawater-fed systems, Ca^{2+} precipitates as carbonate and sulfate salts in the concentrating ponds (see locations marked in Figure 13.2), while the brines of the salt crystallizers are saturated with NaCl and are nearly devoid of Ca^{2+}. In some systems, potash salts containing various combinations of Na^+, Mg^{2+}, K^+, Cl^-, and SO_4^{2-} reach saturation in the most concentrated brines. These brines, called bitterns, are discarded or are processed for harvesting potash or other salts.

In many salterns, NaCl of at least 99.8% purity is harvested throughout the year. To protect the salt floors of the crystallizer ponds from dissolution due to rain, some salt companies harvest NaCl only during the dry season. The optimal product of a solar saltern is pure NaCl in the form of large crystals that are free of brine inclusions and insoluble particulate matter. The large crystals are easier to process (which includes washing with brine to remove contaminants) and market. The presence of brine inclusions (which contain other ions besides Na^+ and Cl^-) makes the crystals mechanically weak and lowers the purity of the salt. Due to a combination of physical and chemical effects, NaCl crystals produced in salinas in the summer tend to be larger than those that grow during the winter. Some of the factors that promote the formation of large crystals in commercial salinas include warm temperatures

Figure 13.1 Aerial view of a solar saltern, San Diego, California.

(with relatively little change between day and night), low humidity (and no rain), winds, dust-free air, and sufficient brine, in combination with proper design of crystallizer ponds that takes into consideration brine depth, convective currents, brine flow, and maintenance of proper brine salinities.

Solar salterns provide ideal model systems for studying evaporative sequences of minerals, biological composition and activity over a broad range of salinities, and detailed chemical partitioning between sediments and superficial and interstitial brines. Several different factors can cause significant deviations from the calculated succession of precipitates and ion composition during seawater evaporation in a saltern, including the composition of the seawater entering the system, leakage and reflux through pond floors and dikes, and flow patterns imposed by the management. A typical distribution of major ions in saltern ponds was given in Figure 1.3 and Table 1.1.

Biogeochemical processes in salterns not only provide a framework for assessing the interplay between biology and geochemistry across salinities, but also for comparing seasonal and temperature effects on rates of processes within a narrow range of salinities. Seasonal differences in both salt production and biological productivity are notable. In the summer, increased brine temperatures and longer day length, both due to increased solar radiation,

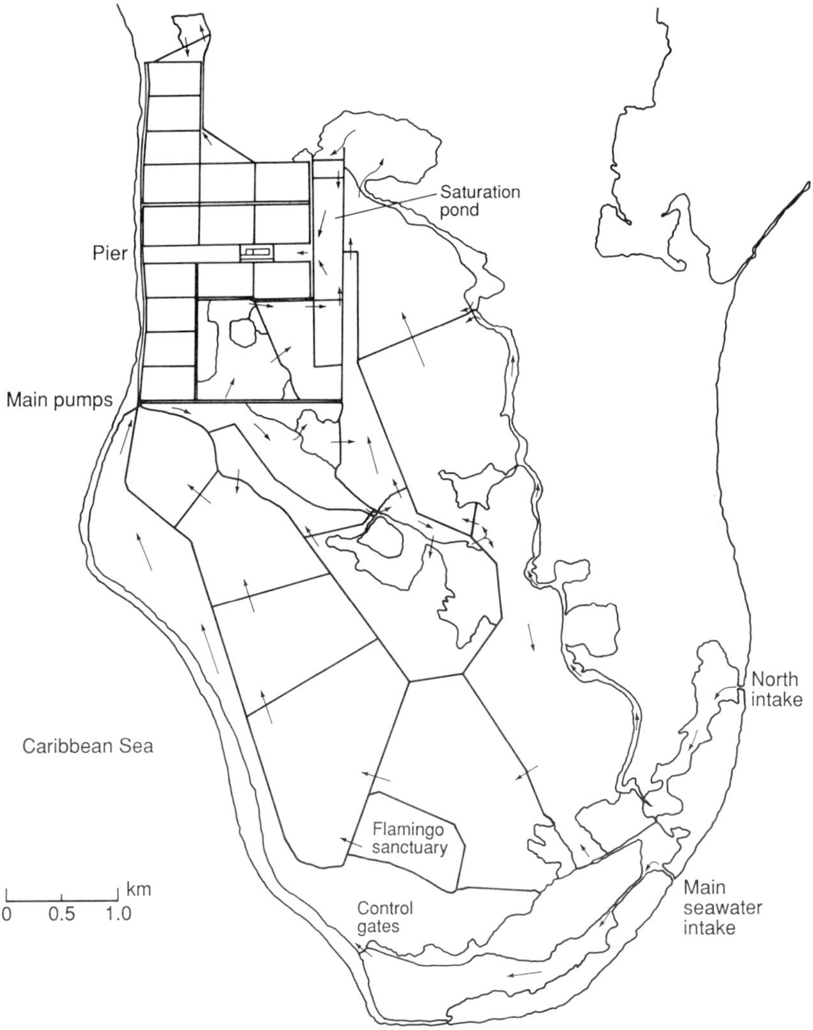

Figure 13.2 General layout of a saltern of the Antilles International Salt Company, Bonaire, Netherlands Antilles.

apparently promote biological productivity in solar salt ponds. While species diversity in general decreases with increasing salinity, organic productivity or standing crop does not necessarily decrease and may even increase, especially during the summer.

The following discussion examines some of the major chemical and physical attributes of salterns, including the presence or absence of $CaCO_3$ mineral precipitation, the fate of Sr^{2+} and other trace elements, the stable isotopes of brines, and nutrient concentrations. The published accounts of the microbiology of these systems are largely descriptive although some data on photosynthetic rates and sediment degradation have been published. The following account of the microbiology describes the similarities and differences between salinas. Even though little is known about *in situ* activity, enrichment culture studies and experiments with isolates have provided a wealth of information concerning potential microbial activity in these environments.

13.2 Chemistry

The pH and O_2 content of saltern brines are dependent on salinity and biological productivity. Brines typically range in pH between 7.5 and 8.5. Extremely concentrated brines have even lower pH (see Chapter 1). O_2 concentrations can fluctuate diurnally. Kinsman et al. (1974) determined O_2 saturation values for concentrated seawater. Although NaCl-saturated brines are saturated with about 2 mg $O_2 \cdot l^{-1}$, little or no O_2 has been detected in crystallizer brines (Sammy, 1983; Javor, unpublished data). This situation probably occurs because respiration exceeds primary productivity in extremely strong brines (see below).

When seawater of normal composition is allowed to evaporate, the minerals predicted to precipitate follow the order of calcium carbonate, gypsum, and halite (see Chapter 1). Under experimental conditions in the laboratory, significant $CaCO_3$ does precipitate (Bassegio, 1974). Based on calculations of these data, about 2.5–6.3 millimoles per liter $CaCO_3$ precipitates at the point that seawater is concentrated to nearly halite saturation. Calcium carbonates are the dominant minerals precipitated in sabkhas and ponds such as Solar Lake, but they appear to be minor precipitates in solar salterns (Javor, 1983a, 1983b; Gouleau, 1982; Landry and Jaccard, 1982; McCaffrey et al., 1987). White precipitates (whitings) composed of $CaCO_3$ and $CaSO_4$ particles (1.5–2.5 μm diameter) sometimes occur when concentrator brines at pH values above 8.5 suddenly mix with inflowing brines and there is a 2°–3°C temperature difference between the two brines (Sammy, 1983). Other reports of $CaCO_3$ precipitation in a saltern described Ca^{2+} concentration with respect to pond number, not brine concentration (Herrmann et al., 1973; Schneider and Herrmann, 1980). A re-evaluation of those data based on the data in Figure 1.1 and those of McCaffrey et al. (1987) indicates that little $CaCO_3$ precipitation occurs before gypsum reaches saturation.

The relative lack of $CaCO_3$ precipitation in solar salterns may be related to: 1) the relative homogeneity of brine salinities at any one local pond area at any one time; 2) the continuously subaqueous environment of the pond floors; 3) the relatively high DOC content of saltern brines; and/or 4) inhibition by increasing concentrations of Mg^{2+}. The whitings reported by Sammy (1983) indicate that the mixture of brines of different compositions or temperatures may be a prerequisite for large-scale $CaCO_3$ precipitation in these systems. $CaCO_3$ precipitation may occur in environments such as Solar Lake and sabkha flats because normal salinity seawater is concentrated in a single body of brine. The general lack of reflux or mixing in the unidirectional flow of brines in salterns apparently makes them unique hypersaline environments. A detailed chemical study of the dynamics of calcium carbonate saturation and precipitation in saltern systems would complement the large body of literature on carbonates in other marine and hypersaline environments.

Strontium (Sr^{2+}) typically co-precipitates with gypsum in a sabkha environment, but there is controversy as to whether most strontium precipitates as $SrSO_4$ (celestite) with halite (Braitsch, 1971; Butler et al., 1973). In solar salterns, Sr^{2+} concentration does not change with salinity up to gypsum saturation, where a little Sr^{2+} precipitates (Figure 13.3). However, mass Sr^{2+} precipitation does not occur until the brines are saturated with halite (Nadler and Magaritz, 1980; Javor, 1983a). Landry and Jaccard (1982) found that the total concentration and the loss of Sr^{2+} in brine solution could vary significantly in two different years in a saltern. Javor (1983a) showed that celestite and strontianite ($SrCO_3$) crystal morphology was influenced by dissolved organic matter. Kinsman (1969) suggested that high Sr^{2+}:Ca^{2+} ratios in Bahaman and Persian Gulf aragonites were related to organic complexing. Similar ob-

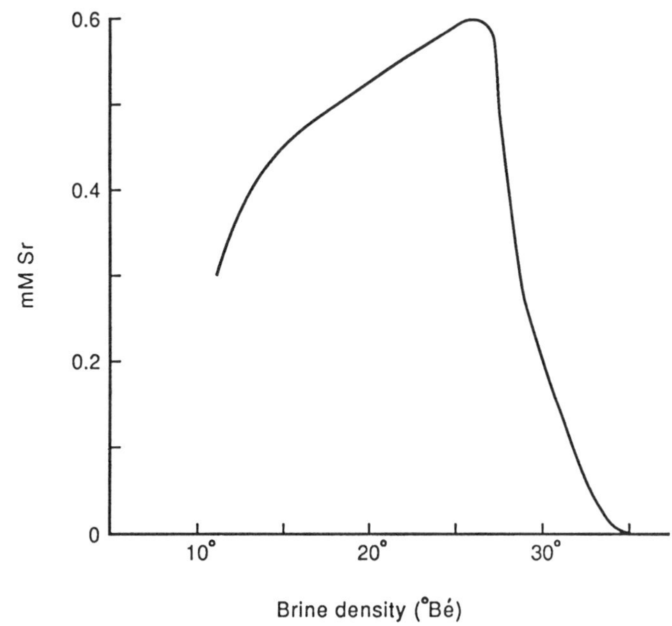

Figure 13.3 Sr^{2+} concentrations in the brines of the Exportadora de Sal saltern (from Javor, 1983a).

servations have been made in normal seawater for $CaCO_3$, and in brines for gypsum (see Chapter 4).

Bromide has been noted to accumulate in brines and to precipitate partially with NaCl. The partitioning of Br^- between halite and interstitial or trapped fluids has been used to interpret the amount of brine recycling in evaporite systems (Holser, 1979). Herrmann et al. (1973) and McCaffrey et al. (1987) measured Br^- in saltern brines and halite. While the data of Herrmann et al. (1973) indicate that Br^- concentration is not conservative as chlorinity changes in brines that are undersaturated with respect to NaCl, the data of McCaffrey et al. (1987), measured by a more precise technique (ion chromatography), indicate that Br^- is a conservative element. Because microorganisms and presumably DOC sequester Br^-, the possible effects of biological processes on Br^- concentration cannot be ignored entirely.

Trace elements in solar salterns have not been well characterized. Landry and Jaccard (1982) found Pb concentrations in saltern brines ranged from ≤ 1 to 10 $\mu g \cdot l^{-1}$. The higher concentrations were associated with brines saturated or nearly saturated with NaCl, but not all such dense brines had a high lead content. Fe concentrations mostly were in the range of 13–86 $\mu g \cdot l^{-1}$, but two samples (about 2.5× concentrated seawater) had Fe concentrations of 129 and 790 $\mu g \cdot l^{-1}$. The latter brine had an exceptionally high sugar content (24.5 $mg \cdot l^{-1}$ vs. 0.70–11.0 $mg \cdot l^{-1}$ in the other brines) and may have been associated with a bloom of microorganisms. Single analyses of such components cannot be properly evaluated without more detailed knowledge of the biological and organic components of the system.

13.3 Stable isotopes

Stable isotope distributions in solar salterns remain largely unstudied. In small ponds used to evaporate Mediterranean seawater, Nadler and Magaritz (1980) found that both $\delta^{18}O$ and δD (deuterium) values of brine water increased until seawater was 4× concentrated, and then decreased in more concentrated brines. Normal seawater H_2O has $\delta^{18}O$ values of +1.62 to +2.09‰ and δD values of +10.6 to +21.2‰. In 4× seawater, $\delta^{18}O$ was +3.99 to +5.59‰, and δD was +24.9 to +26.2‰. By the time the brines reached saturation with respect to halite, $\delta^{18}O$ of water decreased to +1.91‰ and δD decreased to +14.1‰. In 32-fold concentrated seawater, $\delta^{18}O$ was +0.08‰ and δD was +3.4‰. These isotope fractionations occurred in the absence of observable biological activity.

In the Salin-du-Giraud (France) saltern, the $\delta^{18}O$ content of H_2O was enriched in the summer (due to evaporation) but it decreased in the winter (due to rain dilution) (Pierre and Fontes, 1983). The same effects were recorded in the hydration water of gypsum ($CaSO_4 \cdot 2\,H_2O$).

The stable isotopes of the sulfate ion also showed fractionation effects due to salinity and biological activity (Pierre, 1985). The $\delta^{18}O$ of dissolved SO_4^{2-} increased from about +9.5‰ in seawater salinity to about +11‰ at gypsum saturation, and to about +12‰ in NaCl-saturated brines. The $\delta^{34}S$ of dissolved sulfate increased from about +20‰ to about +20.5‰ at gypsum saturation, and about +21‰ at halite saturation. Interstitial solutions in the salina sediments (no salinity given) showed evidence of bacterial sulfate reduction. The $\delta^{34}S$ values of sulfide were −2.4‰ and −15.0‰ in two samples. Porewater sulfate had $\delta^{34}S$ values of +21.7‰ and +30.7‰, and $\delta^{18}O$ values of +11‰ and +16.9‰.

13.4 Nutrient effects on the biota

The nutrient concentrations found in solar salterns depend on a variety of parameters. Geographic factors influencing nutrients include proximity to rivers, urban centers and pollution, the nutrient status of incoming seawater, and the climate. The nature and extent of the fauna and flora, the season of productivity, and management practices also influence nutrient concentration. In some cases, salt production can be improved by fertilization or inoculation with particular organisms (Davis, 1978; Jones et al., 1981). Davis (1978) showed that microbial mats could prevent brine leakage and that pigmented phytoplankton or bacterioplankton could affect water temperature by increasing solar absorption in the crystallizers. Mucilage production by *Aphanothece*-type cyanobacteria has been demonstrated to be detrimental to the growth of commercial-grade NaCl crystals (Baha Al-Deen and Baha Al-Deen, 1972). Jones et al. (1981) found that populations of mucilage-producing *Coccochloris elabens* (*Aphanothece halophytica*) could be checked by controlling algal populations with the inoculation of grazing brine shrimp.

Among unamended salterns, both oligotrophic (low nutrient) and eutrophic (high nutrient) systems have been described (Table 13.1). In concentrator ponds with about 2–4× seawater salinity, oligotrophic systems are characterized by benthic microbial mats with sparse phytoplankton. *Artemia salina*, which is primarily a filter-feeder, is never very abundant in oligotrophic salinas such as the Exportadora de Sal saltern. In moderately eutrophic systems, phytoplankton productivity exceeds brine shrimp grazing and both phytoplankton and benthic mats thrive (California salt ponds). In very eutrophic systems, shading by the phytoplankton can limit benthic mat productivity. From the standpoint of salt production, such a situation is undesirable since a benthic mat is needed to retard brine leakage. The low-nutrient/low-phytoplankton system also introduces less particulate organic carbon into ponds where salt crystallizes. Although it may be desirable to maintain brines colored by plankton in the crystallizers to aid in heat absorption and evaporation, there remains a trade-off of increased DOC contamination of the salt and inhibition of NaCl crystallization (see Chapter 3). To overcome these prob-

Table 13.1 Characteristics of solar salterns

Saltern	Salinity, %	Nutrient conc., μM			Nutrient status	Reference[a]
		PO_4^{3-}	NO_3^-	NH_4^+		
Alviso (San Francisco, California, U.S.A.)	3.36–9.35	0.3–13.0	0.5–32.5	8.0–34.0	Eutrophic	1
Dampier (Western Australia)	3.5–24.7	0.03–0.26	ND[b]	ND	Oligotrophic	5
	26.3–29.3	0	ND	ND		
Salin-de-Giraud (France)	3.5–saturated	<0.01–0.07	0.4–14.3	11–630	Moderately eutrophic	4
	>saturated[c]	<0.01–0.02	<2.0–25.0			
Exportadora de Sal (Mexico)	3.5–saturated	0	0–2.0	0–5.0	Oligotrophic	3
	>saturated[c]	0–10.0	2.0–32	5.0–48		
Western Salt (Chula Vista, California, U.S.A.)	3.5–saturated	0–4.0	0–37	0–120	Eutrophic	2
	>saturated[c]	2.0–5.0	2.0–26	2.0–60		

[a] (1) Carpelan (1957); (2) Javor (1983a); (3) Javor (1983b); (4) Landry and Jaccard (1972); (5) Sammy (1983).
[b] Not determined.
[c] Crystallizer and bitterns brines.

lems, salt company managers sometimes employ various "tricks" to optimize salt quality, including chemical and ultraviolet treatment to kill microorganisms, and purification of the salt by dissolution and recrystallization.

It has been popularly accepted by saltern managers that red-colored brines (due to *Dunaliella* and/or halobacteria) increase solar absorption and cause the temperature of the brines to rise, but there are no published data documenting this effect. An experiment designed to show the effect of halobacterial coloration of brines on solar heating was described in Figure 1.5. The maximum temperature attained by the densely-colored brine was 46°C while the clear brine reached only 39°C. In that saltern, mid-morning brine temperatures typically varied between 19° and 30°C (Javor, 1983b), although temperatures up to about 35°C were occasionally encountered during summer (Javor, unpublished data). Temperatures up to 41°C were recorded in a Spanish saltern (Rodriguez-Valera et al., 1985). Others working in salt ponds have found that density differences that inhibit vertical mixing (a layer of fresher brine overlying dense brine) can cause bottom brines to be too hot for the bare feet (B. Soderberg, pers. comm.).

13.5 Microbiology and primary productivity

The effect of salinity on different microbial types has been discussed in Chapter 2 and Chapters 5–11. The microbial species composition of solar salterns tends to be similar in different systems. After seawater has been concentrated about two-fold, the photosynthetic community is largely restricted to cyanobacteria, phototrophic bacteria, some diatoms, and the green algae *Dunaliella viridis* and *D. salina* (Davis, 1978). In brines concentrated to above gypsum saturation (about 4× seawater), the photosynthetic community is largely restricted to *Dunaliella* and *Aphanothece halophytica* (or related taxa). Only *Dunaliella,* halotolerant and moderately halophilic eubacteria, and halobacteria thrive in NaCl-saturated seawater (see Chapters 7–9). No living organisms appear in bitterns brines which have a natural blue-green (low humic content) or yellow (high humic content) color. The water activity of bitterns brines drops below a_w 0.70 (see Chapter 1) and the Mg^{2+} content increases to one molar and greater. These factors, as well as relatively low Na^+ concentrations or other chemical factors, probably preclude life in bitterns or potash brines (Javor, 1983b; 1984).

Although halobacteria are known to color salt ponds red, no one has demonstrated how active they are *in situ.* Using natural saltern brines supplemented with peptone, Javor (1984) showed that isolates of halobacteria grew poorly in saltern brines saturated with NaCl and that moderately halophilic eubacteria had little or no detectable growth in those brines. Further experiments demonstrated that the high concentrations of Na^+ and Mg^{2+} in the strong brines limited bacterial growth. The high numbers of bacteria found in crystallizer brines may in part be due to passive concentration with evaporation. Wais and Daniels (1985) described how bacteriophage could also help control the population size of halobacteria in a salt pond.

During the time of the studies by Javor (1983a, 1983b, 1984), the Exportadora de Sal saltern was essentially devoid of *Dunaliella*. The absence of this alga was attributed to the oligotrophic nutrient status of the ponds (see Table 13.1). With subsequent changes in brine flow and composition imposed by the management, low populations of *Dunaliella* were regularly found in the crystallizers in that saltern (H. Estrada, pers. comm.). Davis (1978) noted the quantitative increase in species and biomass in a Bahaman saltern after the addition of N and P fertilizers. After fertilization, 13 species of cyanobacteria, 2 species of chlorophytes, 4 species of pyrrophytes, one species of cryptomonad, and 4 genera of diatoms were recognized in seawater at $\geq$5.5 salinity. A qualitative change in species composition was also noted in the Western Salt (California, U.S.A.) saltern after a winter of heavy rains flooded the ponds. Some of the concentrator ponds developed thick mats where only thin, sparse algal felts had been noted before (Javor, 1983a, unpublished data).

Protozoa are significant members of the microbiotas of saltern concentrator ponds (see Chapter 10), but little is known of their biology in these habitats. In ponds designed for outdoor culture of *Dunaliella* (presumably rather eutrophic), 14 ciliates, 10 zooflagellates, and 4 sarcodines were frequently observed in $\geq$15% seawater (Post et al., 1983) (see Table 10.1). Davis (1978) counted 11 species of protozoans in a Bahaman saltern. In Exportadora de Sal brines (about 6–10% salinity), the ciliate *Fabrea salina* occasionally formed such blooms that the brine appeared filled with a suspension of fine grains of black pepper (Javor, 1983b, unpublished data).

Primary productivity has been measured in several saltern systems. In the Alviso salt ponds (San Francisco, California, U.S.A.), Carpelan (1957) found that rates of photosynthesis in ponds of 4.7–9.4% salinity in August, 1951, were 0.65–1.25 ml $O_2 \cdot l^{-1} \cdot h^{-1}$. When these values were calculated in terms of biomass (as optical density, wavelength not given), there was an inverse relationship between salinity and photosynthetic rates. The greatest productivity was in the 4.7% salinity pond (14.3 ml $O_2 \cdot l^{-1} \cdot h^{-1} \cdot$O.D. unit^{-1}) while the lowest rate was at the outlet of the 9.4% salinity pond (2.2 ml $O_2 \cdot l^{-1} \cdot h^{-1} \cdot$O.D. unit^{-1}). Primary productivity at higher salinities was not determined.

Carpelan (1957) noted that microbial activity could greatly influence the concentration of dissolved oxygen in the brines. The greatest difference in morning and afternoon oxygen concentrations was noted in the 6.4% salinity pond, where O_2 increased from 2.3 to 6.8 ml $O_2 \cdot l^{-1}$ between morning and afternoon (exact hours not given).

In Salina Fortuna (Puerto Rico), Copeland (1967) also noted a rise in O_2 concentrations along with an increase in pH (from 7.78 to 7.95 in 15% salinity). Copeland and Jones (1965) measured primary productivity in the plankton of Salina Fortuna in 15, 19.5, and 22.5% salinity brines in terms of O_2 and CO_2 metabolism. The respiratory quotient (moles CO_2 produced per mole O_2 consumed) in all three salinities was comparable (1.5–1.7). The photosynthetic quotient (moles O_2 produced per mole CO_2 consumed) increased with salinity from 0.3 to 1.2. The biomass in each brine was not given but areal rates and depth were tabulated. Those rates were recalculated in terms of brine volume and compared to planktonic productivity rates in Laguna Madre (Gulf of Mexico) in Table 13.2.

In contrast to the apparently eutrophic Salina Fortuna saltern, the Exportadora de Sal saltern had very low or "negative" rates of photosynthesis in the plankton (Javor, 1983b). Investigations using microelectrodes in the microbial mats in those ponds showed that most of the primary productivity apparently occurred in the benthos (Jørgensen et al., 1987). Competition for sulfide by phototrophic bacteria and colorless sulfur bacteria was also demonstrated in those microbial mats (Jørgensen and Des Marais, 1986).

13.6 Microbial processes in the sediments

The activity of sulfate-reducing bacteria was detected in saltern sediments at salinities as high as 30% (Klug et al., 1985) (see Figure 4.4). Sulfide concentrations as high as 18 mM were detected at 13 cm depth in 9% salinity pond

Table 13.2 Community metabolism of plankton in the Salina Fortuna saltern (Puerto Rico) and Laguna Madre (Gulf of Mexico)[a]

Salinity, %	Temperature, °C	O_2, $mg \cdot l^{-1}$	O_2, $g \cdot m^{-3} \cdot d^{-1}$ Photosynthesis	O_2, $g \cdot m^{-3} \cdot d^{-1}$ Respiration	Photosynthetic quotient	Site[b]
Salina Fortuna						
15.0	27.6–34.9	2.5–3.5	3.68	9.94	0.37	
19.5	26.7–36.5	2.6–3.0	2.67	4.33	0.62	
22.5	26.1–37.0	1.7–2.6	5.25	10.67	0.50	
Laguna Madre						
12.1	18.2–20.5	0–0.16	0.50	50.4	0.01	LC
12.1	18.2–19.5	0–0.10	0.48	47.1	0.01	LC
5.0	24.8–28.7	5.71–6.47	1.07	0.93	1.67	C1
9.5	20.1–24.1	4.41–5.43	4.09	9.61	0.43	C1
11.4	7.8–16.4	4.11–6.51	4.00	9.07	0.44	C1
9.4	20.1–26.6	4.66–6.24	4.74	9.78	0.48	C2
11.5	9.0–14.2	3.06–4.97	11.4	17.0	0.67	C2
11.1	20.1–24.0	4.82–6.54	3.56	5.89	0.60	C2
9.3	20.2–24.0	4.76–5.72	1.87	8.00	0.25	C3
11.5	26.2–30.8	2.95–5.15	12.4	16.4	0.75	C3
11.3	7.7–15.1	4.82–6.79	4.30	7.55	0.57	C3
11.8	27.2–30.8	2.95–5.65	8.16	7.43	1.10	C4
12.0	11.8–36.0	4.83–5.78	2.81	2.81	1.00	C4

[a] From Copeland and Jones (1965).
[b] LC, La Capia station; C1, Carvajal station 1; C2, Carvajal station 2; C3, Carvajal station 3; C4, Carvajal station 4.

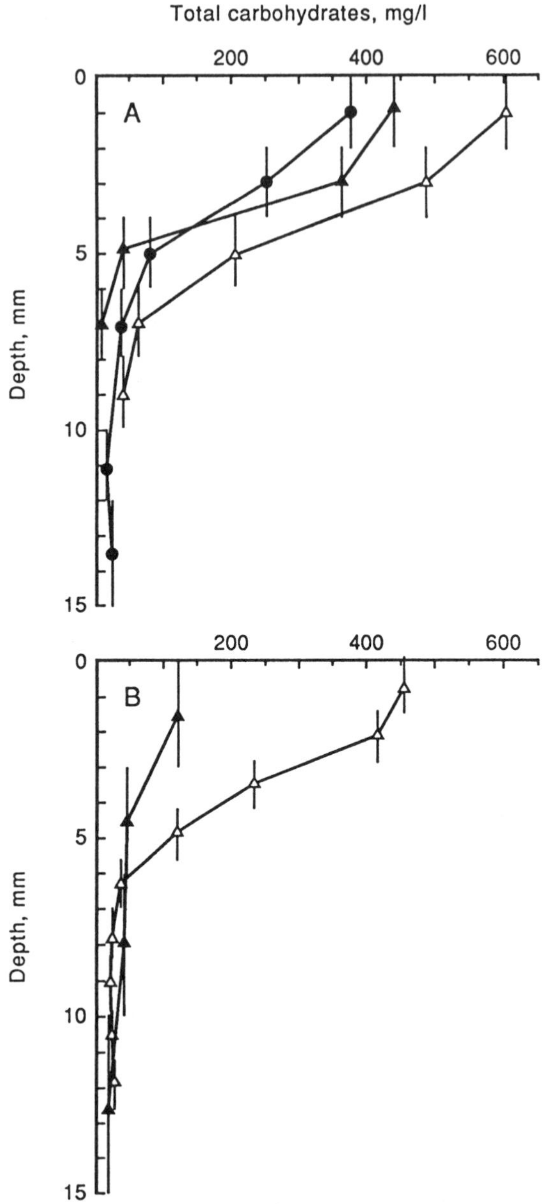

Figure 13.4 Total carbohydrates (determined with the phenol-hydrazine sulfate method) of microbial mats. A: *Aphanothece* mats (13.2–17.8% salinity during the course of the study) in 1987 from the Israel Salt Company (Eilat) saltern; circles, 4 January; closed triangles, 23, April; open triangles, 6 June. B: Solar Lake; closed triangles, 15 March (9.0% salinity); open triangles, 21 May (13.0% salinity). The concentrations of total carbohydrates in the brines overlying the microbial mat are indicated with an arrow on the abscissa.

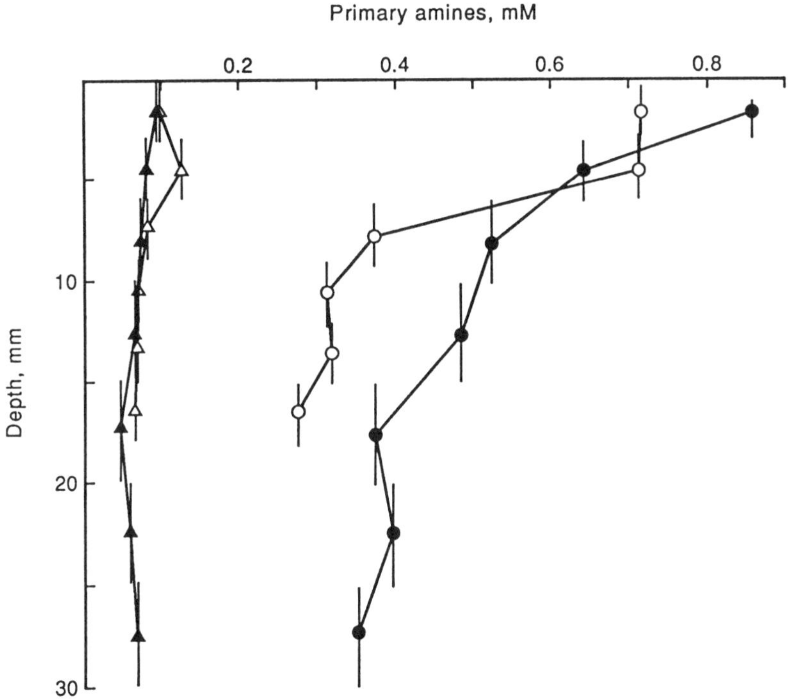

Figure 13.5 Dissolved free amino acids (triangles) and total amino acids (circles) measured as primary amines in porewaters of microbial mats (see Figure 13.4). Open symbols: *Aphanothece* mat, 2 March. Closed symbols: Solar Lake, 15 March. Primary amines were determined in unhydrolyzed (free amino acids) and hydrolyzed (total amino acids) samples with fluorescamine against glycine standards. Brines overlying the *Aphanothece* mats had 9–31 μM free primary amines and 35–69 μM total primary amines.

sediments. The greatest rates of sulfate reduction were found in the top centimeter of the less saline sediments. At 13–15 cm depth, sulfate reduction was still detectable in the hypersaline muds but it was negligible in the 3.3% salinity muds due to sulfate depletion.

Although fermentation rates in saltern sediments were not measured by Klug et al.(1985), fermentation products (low-molecular-weight fatty acids) were detected and quantified. There was a general increase in the concentration of these fatty acids, especially acetate, with salinity (see Figure 4.1). The results suggest that microbial fermentation greatly exceeds sulfate reduction in extremely hypersaline sediments. CH_4 (but not methanogenesis) was detected in 4.2% salinity sediments by Klug et al. and in 15% salinity sediments of the Eilat, Israel, saltern (A. Oren, pers. comm.).

Concentrations of dissolved total carbohydrates and total amino acids in the surface brines of a eutrophic and an oligotrophic saltern generally increased with salinity (see Figures 4.2 and 4.3). Concentrations of dissolved carbohydrates (Figure 13.4) and amino acids (Figure 13.5) in the interstitial brines of *Aphanothece* mats (about 15% salinity) of the Eilat, Israel, saltern were highest in the top 4 mm of the mats (the photosynthetic zone), and they greatly exceeded the concentrations found in the overlying brines (Javor, unpublished data). Summer values were higher than winter values, probably due to warmer ambient temperatures and increased irradiance and daylength. The concentrations of these DOC pools are similar to those measured in Solar Lake mats.

Because of their worldwide distribution and their accessibility for sampling, studies of biogeochemical processes in salterns can provide models for understanding the interactions between microorganisms and the geochemistry of evaporite environments in general. For example, salinas provide the opportunity to compare primary productivity and degradation rates under both aerobic and anaerobic conditions across a wide range of salinties. The effects of salinity and DOC on ion activity coefficients could be compared between surface brines and interstitial brines of the sediments. Likewise, the stable isotopic signatures of carbon and sulfur could be compared between salinities, between high and low redox potentials, and between systems of high and low organic productivity.

References

Baha Al-Deen, B. and **Baha Al-Deen, A.H.** 1972. Posible efecto de microalgas en la forma de cristalización de cloruro de sodio en la Salina de Araya. *Boletín del Instituto Oceanográfico Universidad de Oriente* 11: 35–38.

Baseggio, G. 1974. The composition of sea water and its concentrates, pp. 351–358 in Coogan, A.H. (editor), *Fourth Symposium on Salt*, vol. 2, Northern Ohio Geological Society, Cleveland.

Braitsch, O. 1971. *Salt Deposits. Their Origin and Composition.* Springer-Verlag. Berlin. 297 pp.

Butler, G.P., Krouse, R.H., and **Mitchell, R.** 1973. Sulphur-isotope geochemistry of an arid, supratidal evaporite environment, Trucial Coast, pp. 453–462 in Purser, B.H. (editor), *The Persian Gulf. Holocene Carbonate Sedimentation and Diagenesis in a Shallow Epicontinental Sea*, Springer-Verlag, New York.

Carpelan, L.H. 1957. Hydrobiology of the Alviso salt ponds. *Ecology* 38: 375–390.

Copeland, B.J. 1967. Environmental characteristics of hypersaline lagoons. *Texas University Contributions to Marine Science* 12: 207–218.

Copeland, B.J. and **Jones, R.S.** 1965. Community metabolism in some hypersaline waters. *Texas Journal of Science* 17: 188–205.

Davis, J.S. 1978. Biological communities of a nutrient enriched salina. *Aquatic Botany* 4: 23–42.

Gouleau, D. 1983. Chemistry of interstitial water in sediments at the Salin-de-Giraud solar salt works (southern France). *Géologie Méditerranéenne* 9: 349–367.

Herrmann, A.G., Knake, D., Schneider, J., and **Peters, H.** 1973. Geochemistry of modern seawater and brines from salt pans: main components and bromine distribution. *Contributions to Mineralogy and Petrology* 40: 1–24.

Holser, W.T. 1979. Trace elements and isotopes in evaporites, pp. 295–346 in Burns, R.G. (editor), *Marine Minerals.* Mineralogical Society of America Short Course Notes, vol. 6, Washington, D.C.

Javor, B.J. 1983a. Nutrients and ecology of the Western Salt and Exportadora de Sal saltern brines, pp. 195–205 in Schreiber, B.C. and Harner, H.L. (editors), *Sixth Symposium on Salt,* vol. 1, The Salt Institute, Toronto.

Javor, B.J. 1983b. Planktonic standing crop and nutrients in a saltern ecosystem. *Limnology and Oceanography* 28: 153–159.

Javor, B.J. 1984. Growth potential of halophilic bacteria isolated from solar salt environments: carbon sources and salt requirements. *Applied and Environmental Microbiology* 48: 352–360.

Jones, A.G., Ewing, C.M., and **Melvin, M.V.** 1981. Biotechnology of solar salt fields. *Hydrobiologia* 82: 391–406.

Jørgensen, B.B. and **Des Marais, D.J.** 1986. Competition for sulfide among colorless and purple sulfur bacteria in cyanobacterial mats. *FEMS Microbiology Ecology* 38: 179–186.

Jørgensen, B.B., Cohen, Y., and **Des Marais, D.J.** 1987. Photosynthetic action spectra and adaptation to spectral light distribution in a benthic cyanobacterial mat. *Applied and Environmental Microbiology* 53: 879–886.

Kinsman, D.J.J. 1969. Interpretation of Sr^{2+} concentrations in carbonate minerals and rocks. *Journal of Sedimentary Petrology* 39: 486–508.

Kinsman, D.J.J., Boardman, M., and **Borcsik, M.** 1974. An experimental determination of the solubility of oxygen in marine brines, pp. 325–327 in Coogan, A.H. (editor), *Fourth Symposium on Salt,* vol. 1, Northern Ohio Geological Society, Cleveland.

Klug, M., Boston, P., Francois, R., Gyure, R., Javor, B., Tribble, G., and **Vairavamurthy, A.** 1985. Sulfur reduction in sediments of marine and evaporite environments, pp. 128–157 in Sagen, D. (editor), *The Global Sulfur Cycle,* National Aeronautics and Space Administration Technical Memoir 87570, Washington, D.C.

Landry, J.-C. and **Jaccard, J.** 1982. Chimie des eaux libres dans le marais salant de Salin-de-Giraud (Sud de la France). *Géologie Méditerranéenne* 9: 329–348.

McCaffrey, M.A., Lazar, B., and **Holland, H.D.** 1987. The evaporation path of seawater and the coprecipitation of Br^- and K^+ with halite. *Journal of Sedimentary Petrology* 5: 928–937.

Nadler, A. and **Magaritz, M.** 1980. Studies of marine solution basins—isotopic and compositional changes during evaporation, pp. 115–129 in Nissenbaum, A. (editor), *Hypersaline Brines and Evaporitic Environments,* Developments in Sedimentology 28, Elsevier Scientific, New York.

Pierre, C. 1985. Isotopic evidence for the dynamic redox cycle of dissolved sulphur compounds between free and interstitial solutions in marine salt pans. *Chemical Geology* 53: 191–196.

Pierre, C. and **Fontes, J.C.** 1983. Isotopic study of brines and gypsum from the Salin-de-Giraud solar salt works (southern France). *Géologie Méditerranéenne* 9: 479–486.

Post, F.J., Borowitzka, L.J., Borowitzka, M.A., Mackay, B., and **Moulton, T.** 1983. The protozoa of a Western Australian hypersaline lagoon. *Hydrobiologia* 105: 95–113.

Sammy, N. 1983. Biological systems in north-western Australian solar salt fields, pp. 207–215 in Schreiber, B.C. and Harner, H.L. (editors), *Sixth Symposium on Salt,* vol. 1, The Salt Institute, Toronto

Schneider, J. and **Herrmann, A.G.** 1980. Saltworks--natural laboratories during the

evaporation of seawater, p. 371–381 in Bersticker, A.C., Hoekstra, K.E. and Hall, J.F. (editors), *Symposium on Salt*, Northern Ohio Geological Society, Cleveland.
Wais, A.C. and **Daniels, L.L.** 1985. Populations of bacteriophage infecting *Halobacterium* in a transient brine pool. *FEMS Microbiology Ecology* 31: 323–326.

14

Solar Lake

14.1 Introduction

Solar Lake, on the Sinai Peninsula (Gulf of Aqaba), is an extremely small (50 by 140 m), shallow (4–6 m deep), sea-marginal pond (Figure 14.1). Solar Lake can probably claim an unofficial record of the most number of scientific investigations for such a small body of water. Some preliminary data on this pond were first reported by Por (1972), and the dynamic physical, chemical, and microbiological cycles in the lake were described in detail by Cohen and co-workers (Cohen et al., 1977a, 1977b, 1977c). Cohen's studies were followed by investigations by others to further define the microorganisms that comprise the biota, their activities, and their inorganic and organic geochemical remains. The small size, shallow depth, and the relative ease of access to the lake in the 1970s facilitated sample collection and measurements of *in situ* microbial activities. From the standpoint of understanding the biogeochemical dynamics of modern hypersaline environments, Solar Lake is the best known system in the world. Although it should not be considered a model for all marine-derived hypersaline systems, it can serve as a guide for similar investigations in other hypersaline habitats.

Solar Lake (Figure 14.2) is a monomictic body of water subject to density stratification, intense solar heating in the lower layers (hence, the name), high evaporation rates, and complex and intense microbial interactions in the water column, on the benthos, and in the sediments (Cohen et al., 1977a, 1977b, 1977c; Krumbein et al., 1977; Jørgensen and Cohen, 1977). Some of the limnological attributes are given in Table 14.1. The lake is separated from the Red Sea by a 60-m wide gravel bar. It is fed by occasional rain showers and by seawater seepage from the Red Sea. This type of sea-marginal lake has

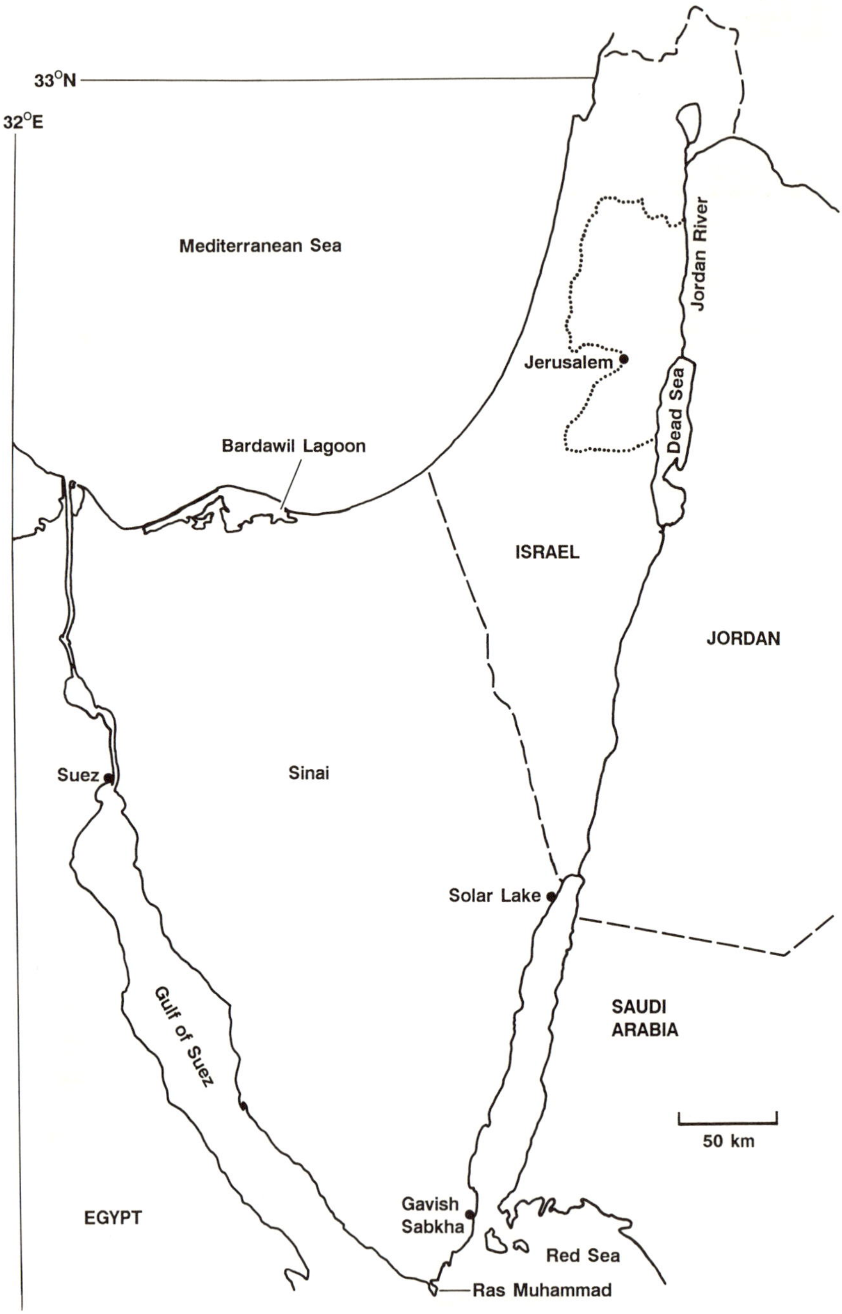

Figure 14.1 Location of hypersaline environments in the Sinai (Solar Lake, Gavish Sabkha, Ras Muhammad pool, and Bardawil Lagoon) and in Israel-Jordan (Dead Sea).

Figure 14.2 Solar Lake, Sinai Peninsula.

Table 14.1 Some limnological characteristics of Solar Lake, Sinai peninsula

Size: 50 × 140 m
Age: 4000 years total sediment accumulation. The oldest algal mats are 2400 years old.
Annual evaporation rate: about 3.0 meters
Annual precipitation: 6–98.2 mm
Sediments: carbonates and gypsum
Residence time of water: 5.5 months
Mixing period: 4–13 weeks in summer
Salinity: 15–18%
pH: 8.2–8.8
Temperature: 27°C
O_2: oxic at bottom
Sulfide: absent in water
Light penetration: relatively clear; 25.7% of surface light reaches bottom
Stratification period: September to July
Salinity gradient: 6.8% at top, 16–18% at bottom
pH gradient: 8.1–8.2 in water column, 6.9 at bottom
Temperature gradient: 27°C at surface, up to 60.5°C at 2.5–3 m depth, 40°C at bottom
O_2: anoxic at bottom
Redox gradient: +390 mV at surface, −185 mV near bottom
Sulfide: 39 $mg \cdot l^{-1}$ maximum in water near bottom
Light penetration: very turbid

From Cohen et al. (1977a, 1977c)

been termed an anchialine pool by Por (1985), meaning it has no surface connection to the sea. Rainwater runoff, which can form a less dense layer floating on the denser brine of the lake, can contribute significantly to temperature stratification (Cohen et al., 1977a). While seawater does seep into the pond, brine loss back through the gravel bar is retarded by the presence of beachrock, gypsum crusts, and laminated microbial mats. The oldest mats in the sediments are about 2400 years old, indicating the limnological and climatic conditions of this environment have been stable during this period.

Density stratification begins in the autumn (Table 14.1) and continues through the winter, spring, and early part of summer. With increasing solar radiation and evaporation, the epilimnion (0–1.0 m) becomes increasingly more saline until the top and bottom of the lake have the same density at the beginning of the mixing period (holomixis) in July or August. Holomixis lasts 4–13 weeks. During holomixis, parameters such as salinity, pH, and Eh, and temperature are constant throughout the lake while, during stratification, extremely steep physical and chemical gradients occur. These annual cycles impose physical and chemical boundaries that delimit microbial populations in the water column and benthos.

14.2 Microbiology and primary productivity in the water column

The phototrophic flora of the water column and some rates of primary productivity were presented by Cohen et al. (1977b) and are summarized in Table 14.2. The maximal carbon fixation rate, 8015 mg $C \cdot m^{-2} \cdot d^{-1}$ is one of the highest rates of primary productivity ever recorded in a non-polluted body of water. The actual rates of carbon fixation ($^{14}C\text{-}HCO_3^-$ assimilation) were calculated from alkalinity measurements but were not corrected for the carbonate alkalinity of the brines (see Chapter 1). Because of the unmeasured inorganic carbon concentrations, it is impossible to compare CO_2 assimilation rates in other brines with those of Solar Lake. Empirically-derived apparent dissociation constants of the carbonate system of this body of water have not been calculated. The values given by Cohen et al. (1977b, 1977c) should be considered as estimates of unknown precision, especially when they are compared to productivity rates in other hypersaline systems. Monitoring oxygen evolution, which is sometimes used as an alternative measure of primary productivity rates, would not assess the role of phototrophic bacteria nor would it give an accurate estimate of the photosynthetic potential of the cyanobacteria capable of anoxygenic photosynthesis. The very high standing crops of planktonic organisms that develop each year during stratification in Solar Lake suggest that primary productivity rates are high or that grazers are few below the epilimnion.

During stratification, the greatest primary productivity rates (in terms of

Table 14.2 Phototrophic microorganisms, microbial biomass, and primary productivity in the Solar Lake water column

Summer mixing period

Microorganisms: Cyanobacteria (*Aphanothece halophytica**, *A. littoralis, Oscillatoria salina, Microcoleus* sp., *Spirulina* sp., *S. labyrinthiformis*), total number $2 \times 10^3 \cdot ml^{-1}$

Diatoms (*Amphora coffeaeformis**, *Nitzschia thermalis, N.* sp.), total number $\leq 2 \times 10^4 \cdot ml^{-1}$

Heterotrophic bacteria, total number $6.7 \times 10^5 \cdot ml^{-1}$ near bottom, somewhat less near top

Biomass: 3 g $C \cdot m^{-3}$

Primary productivity: <100 mg $C \cdot m^{-3} \cdot d^{-1}$ for all depths

Stratification period

Epilimnion (0–1.0 m)

Microorganisms: Similar to summer, but total number around $10^4 \cdot ml^{-1}$, *Dactylococcopsis salina** dominates

Heterotrophic bacteria, total number $2 \times 10^5 \cdot ml^{-1}$

Biomass: 1–8 g $C \cdot m^{-3}$

Primary productivity: 50–100 mg $C \cdot m^{-3} \cdot d^{-1}$

Metalimnion (1.0–2.5 m)

Microorganisms: Purple phototrophic bacteria (*Chromatium violescens*,* up to $10^6 \cdot ml^{-1}$, *Lamprocystis* sp.)

Filamentous cyanobacteria (*O. salina, O. limnetica, Microcoleus* sp.), up to 2×10^3 filaments$\cdot ml^{-1}$

Heterotrophic bacteria, up to $9.5 \times 10^6 \cdot ml^{-1}$

Biomass: 4–18 g $C \cdot m^{-3}$

Primary productivity: 100–<1000 mg $C \cdot m^{-3} \cdot d^{-1}$

Hypolimnion (2.5 m–bottom)

Microorganisms: Upper hypolimnion has green phototrophic bacteria (*Prosthecochloris* sp.*) up to $10^6 \cdot ml^{-1}$, purple phototrophic bacteria, and filamentous cyanobacteria.

Lower hypolimnion has floccose benthic mat with fewer green phototrophic bacteria ($2 \times 10^5 \cdot ml^{-1}$) but a dense bloom of filamentous cyanobacteria (same species as in the metalimnion).

Heterotrophic bacteria, $1.8 \times 10^6 \cdot ml^{-1}$

Biomass: 4–28 $g \cdot m^{-3}$

Primary productivity: up to 4960 mg $C \cdot m^{-3} \cdot d^{-1}$

Maximum primary productivity in lake: 8015 mg $C \cdot m^{-2} \cdot d^{-1}$, 91% in the metalimnion + hypolimnion

Annual primary productivity: 59.09 g $C \cdot m^{-2} \cdot yr^{-1}$

From Cohen et al. (1977b, 1977c); Walsby et al. (1983).
* Dominant species

lake area) were in the metaliminion and hypolimnion, where over 8 g of carbon was fixed per $m^2 \cdot d$. The high standing crop of microorganisms greatly attenuates light in the lower part of the water column during stratification. Low rates of photosynthesis in the epilimnion may be partially due to grazing pressure by the brine shrimp *Artemia salina*. The experimental evidence of Walsby et al. (1983) showed that high temperatures (100% mortality after 16 hours at 37° C) and anoxia restrict the brine shrimp to the epilimnion. Extremely high grazing rates (up to 10^6 cyanobacterial cells consumed each hour by each adult brine shrimp) indicate that photosynthesis must be fairly great in the epilimnion (unless grazing rates are much slower in dilute suspensions of cells and/or the brine shrimp consume other bacteria as well). An evaluation of photoassimilation rates per mg chlorophyll *a* indicated the epilimnetic population has a relatively low potential for photosynthesis (Cohen et al., 1977b).

A detailed survey of the Solar Lake biota during stratification was made by Hirsch (1980). A total of 149 morphotypes were isolated in culture and their taxonomic affinities and their distribution within the lake were tabulated. Of the total isolates, 81% were procaryotes. The eucaryotes were represented by diatoms, flagellates (green or golden, and colorless), non-flagellated algae, and ciliates. Fungi were absent. All the eucaryotes except for one flagellate were confined to the oxic layers of the lake, and 21 of the total 28 eucaryotes were confined to the water surface or shore mat. Among the procaryotes, only 28% were confined to the lake surface or shore mat.

One-third of the morphotypes identified by Hirsch were primary producers, including cyanobacteria, phototrophic bacteria, and eucaryotic algae. Hirsch found 20 different morphotypes of cyanobacteria in Solar Lake while Potts (1980) counted 24 different species. Hirsch (1980) and Cohen et al. (1977b) noted that the phototrophic bacteria and some of the cyanobacteria were restricted to the hypolimnion. A great variety of heterotrophic bacteria were recognized by Hirsch (1980) but could not be identified with cultivated, known genera. While some experiments testing the effects of growth substrates and temperatures were done, measurements of growth at different salinities were not described.

The study of Hirsch (1980) complemented an investigation by Cohen et al. (1977c) of bacterial activity in the water column of Solar Lake. Six different bacterial layers (plates) were observed during stratification. The presence and activity of these bacteria were confirmed by several techniques: direct and viable cells counts, light and dark $^{14}CO_2$ incorporation, measurement of respiratory activity, and determinations of chlorophyll *a*, protein, and ATP. During holomixis higher viable counts were obtained in 12% salinity medium in comparison to 3% salinity medium, and greater counts were obtained at lower incubation temperatures (20° > 35° ≫ 48° C). During stratification, the greatest viable counts of the bacteria in the epilimnion were obtained in 3% salinity at 35°C. At 2.5 m depth, low counts were obtained in both media at all temperatures. At 4.5 m depth, greater viable counts were obtained in 3% salinity medium than in 12% salinity medium although the lake has a salinity of 18% at this depth.

Enrichments for a variety of aerobic and anaerobic bacteria have been made. While Hirsch (1980) obtained enrichments of the sulfate-reducing bacterium *Desulfovibrio* from hypolimnetic water, Cohen et al. (1977c) could not isolate any sulfate-reducing bacteria from any depth in summer or winter. They did find putrefactive bacteria that produced H_2S from cysteine in bottom water samples. Jørgensen et al. (1979b) detected low rates of sulfate reduction near the bottom of the hypolimnion during stratification (4.1–6.9 μmol $SO_4^{2-} \cdot l^{-1} \cdot d^{-1}$).

Sulfur-oxidizing bacteria (thiobacilli) were detected just above the thermocline where they presumably thrive near the O_2/H_2S transition zone. These bacteria typically grow as chemoautotrophs, reducing CO_2 at the expense of energy obtained by the oxidation of sulfide. Dark CO_2 uptake at the top of the metalimnion may be in part due to these bacteria. In spring, peaks in dark CO_2 fixation were measured at 1.0, 2.0, and 4.0 m depth. When the results from those measurements were expressed as CO_2 fixed·mg protein^{-1}, the greatest dark CO_2 assimilation rates were found at the lake surface, with lower rates at 2 and 4 m. The lowest efficiency of dark CO_2 assimilation was measured at 1 m depth. Because of the extreme salinity stratification in the lake during the spring, it is uncertain whether the different CO_2 fixation rates at each depth were more due to biological processes or to salinity effects on the carbonate equilibria.

Relatively high dark CO_2 fixation rates in the hypolimnion were thought to be due to non-reductive CO_2 assimilation by the mixed phototrophic and non-phototrophic communities. CO_2 fixation in the dark by photosynthetic microorganisms is typically about 6% of that measured in the light. The highest dark CO_2 incorporation rate measured (1014 mg C·m^{-3}·d^{-1}) was about one-fifth of the maximum rate of photosynthetic uptake of $^{14}CO_2$ (4960 mg C·m^{-3}·d^{-1}) (Cohen et al., 1977c). The carbon necessary to support such high heterotrophic activity in the plankton was believed to have been produced in the benthic cyanobacterial mats. These mats produce 5–12 g C·m^{-2}·d^{-1} (Krumbein et al., 1977). Plates of actively growing bacteria were also detected through ATP measurements (relatively high concentrations of ATP were found, nearly 100 mg·m^{-3} at 1.0 m depth, with lower peaks at 3.0–3.5 m and at 4.5 m) and O_2 uptake measurements (high rates of respiration were found, 1700 m mole O_2 consumed·m^{-3}·h^{-1} at 4.5 m, with lower activity peaks at 1.0–2.0 m and at 3.0 m). Based on a wide variety of estimates, Cohen et al. (1977c) tabulated the major types of organisms that dominate the different layers of the lake during stratification (Table 14.3).

The benthos of Solar Lake was characterized by four different types of microbial mats distinguished by their general morphology, species composition, and location (Krumbein et al., 1977; Jørgensen et al., 1983): shallow

Table 14.3 Dominant metabolic types of communities in the water column of Solar Lake during stratification

Depth, m	Oxic	Community
0–1.0	+	Chemolithotrophs
1.0	+	Chemoorganotrophs
2.0	±	Mixed chemolithotrophs and phototrophs
2.5	−	Phototrophs
3.0–3.5	−	Low biomass and low CO_2 uptake
4.0	−	Chemoorganotrophs and phototrophs

From Cohen et al. (1977c).
+, present; ±, O_2 variable, −, O_2 absent.

flat mat, deep flat mat, blister or pinnacle mat, and gelatinous or floccose mat. The shallow flat mat, covered by very little water, consisted of a surface film of diatoms and the bacterium *Achromatium* (0–0.1 mm depth), underlain by 0.3 mm of a blue-green layer of the cyanobacterium *Microcoleus*. Layers of carbonate grains, filamentous and coccoid cyanobacteria, and flexibacteria were found between 0.4 and 1.2 mm depth. Below 2.5 mm, all the layers repeated themselves. The active photic zone was 0.8 mm deep. Photosynthetic rates were maximal in the *Microcoleus* layer (50 μmol $O_2 \cdot cm^{-3} \cdot h^{-1}$ produced, 1.5 mg $C \cdot cm^{-2} \cdot d^{-1}$ assimilated as $^{14}CO_2$).

The deep flat mat (covered with about 50 cm water) had essentially the same composition as the shallow flat mat except that it lacks *Achromatium*. The exact metabolic nature of *Achromatium* is unknown since it has not been isolated in pure culture. Photosynthetic rates up to 25 μmol $O_2 \cdot cm^{-3} \cdot h^{-1}$ were measured in the *Microcoleus* zone of this mat (Jørgensen et al., 1983). The active photic zone was 2.5 mm deep. Extreme diurnal variations in O_2, H_2S, pH, and Eh were found in the photic zone of these mats using microelectrodes (Jørgensen et al., 1979a; Revsbech et al., 1983; Cohen, 1984): 0–1400 μM O_2, 0–50 μM H_2S, pH 7.7–9.6, and −50mV to +280 mV at the mat surface.

The *Microcoleus* layer of the mat was subjected to oxic conditions during the day and to about 200 μM sulfide at night in a mat investigated by Jørgensen et al. (1986). The investigators found that the cyanobacteria in the mat shifted between oxygenic photosynthesis under low sulfide conditions and predominantly anoxygenic photosynthesis under high sulfide conditions. Both types of photosynthesis occurred concurrently in the mats in moderate sulfide concentrations (around 300–500 μM). The diatoms on the mat were inhibited by these sulfide concentrations.

Nitrogenase activity (measured as acetylene reduction to ethylene) was detected in *Microcoleus* mats of Solar Lake (Potts, 1980). High background levels of spontaneous ethylene production were found throughout the water column (Potts, 1979). The background rates of spontaneous ethylene production increased toward the bottom and were not affected by formaldehyde and 0.1 M HCl. The background ethylene production was attenuated when the lake water was first filtered or when the headspace of the incubation vials was first flushed with N_2 or H_2. No potential source or agent of ethylene production was suggested. It is uncertain whether the acetylene reduction (nitrogenase) activity measured in the *Microcoleus* mat represents N_2 fixation potential or activity related to ethylene production.

The blister mat or pinnacle mat occurred at 1.0–2.5 m depth. The surface 0.4 mm consisted of diatoms and the coccoid cyanobacterium *Synechococcus*. It was underlain by an orange zone (1.1 mm thick) of *Phormidium* and flexibacteria. Below a 0.5-mm thick layer of $CaCO_3$ grains were mm-thick layers of filamentous cyanobacteria and flexibacteria. At 2.5–3 mm depth (*Phormidium*-flexibacteria), photosynthetic rates up to 6 μmol $O_2 \cdot cm^{-3} \cdot h^{-1}$ were measured (Jørgensen et al., 1983). The photic zone extended down to 4.0–4.5 mm depth.

In the deepest part of the lake, a gelatinous or floccose mat develops. During stratification the photic zone of this mat was 10 mm deep. It consisted of a variety of cyanobacteria including coccoid (*Aphanothece, Synechococcus*) and filamentous forms (*Oscillatoria limnetica, O. salina, Phormidium*) as well as flexibacteria. Maximum photosynthetic rates were 0.4–0.5 μmol $O_2 \cdot cm^{-3} \cdot h^{-1}$ (Jørgensen et al., 1983). During holomixis, oxygenic photosynthesis in the mats causes them to accumulate oxygen bubbles and float. The development of non-phototrophic bacteria (including myxobacteria and non-phototrophic sulfur bacteria) accompany degradation and breakdown of the mat during the mixing period (Krumbein et al., 1977).

The luxuriant growth of cyanobacteria that developed under anaerobic conditions and in high sulfide concentrations on the bottom sediments of the deepest part of Solar Lake led to the discovery of anoxygenic photosynthesis in *O. limnetica, O. salina, A. halophytica,* and *Microcoleus* sp. (Cohen, 1975, 1984; Cohen et al., 1975; Garlick et al., 1977; Oren et al., 1977; Oren and Shilo, 1979). These species use sulfide instead of water as an electron donor for photosynthesis, while some strains have been reported capable of using ferrous iron as electron donor (Cohen, 1984). *O. limnetica,* the best characterized species, is the most tolerant of sulfide. It can perform anoxygenic photosynthesis in 1–8.5 mM sulfide, with optimal rates (1600 μmol CO_2 fixed$\cdot$mg protein$^{-1} \cdot h^{-1}$) in 3.5 mM sulfide (Garlick et al., 1977). During sulfide oxidation, elemental sulfur globules accumulate on the outside of the cells (Cohen et al., 1975). In the dark under anaerobic conditions, the elemental sulfur serves as an electron sink for cell maintenance or the cells can ferment endogenous polyglucose (Oren and Shilo, 1979). A more detailed description of the physiology of these cyanobacteria is presented in Chapter 8.

The dynamics of the sulfur cycle of the littoral zone mats were described by Jørgensen and Cohen (1977), Jørgensen et al. (1979a), and Cohen (1984). Very high rates of sulfate reduction (5400 nmol reduced$\cdot cm^{-3} \cdot d^{-1}$) were determined near the mat surface, while negligible rates ($<$1 nmol reduced$\cdot cm^{-3} \cdot d^{-1}$) were found at one-meter sediment depth (Jørgensen and Cohen, 1977). The high rates of sulfate reduction near the surface were accompanied by low SO_4^{2-}/Cl^- ratios (with respect to Gulf of Aqaba water), high concentrations of free sulfide (about 1 mM), and the absence of gypsum. Gypsum occurs in deeper parts of Solar Lake and its absence in the shallow mats is probably due to sulfate depletion by bacteria (Krumbein et al., 1977). Based on measured rates and sulfate pools, Jørgensen and Cohen (1977) calculated sulfate turnover times of 22 days near the mat surface, and 160 years at 50 cm depth. In the mat sediment profile, 50% of the sulfate-reducing activity was found in the top 0.5 cm, and 90% within the top 3 cm.

Viable counts of anaerobic bacteria at different depths in the sediments demonstrated that the total number of anaerobic heterotrophs was always 3–

100 times greater than the number of sulfate-reducing bacteria. In the top 0–2 mm of sediment, 2.5×10^6 sulfate-reducing bacteria and 6×10^6 anaerobic heterotrophs per cm^3 sediment were detected. In a comparison of sulfate reduction rates with viable counts of sulfate-reducing bacteria according to depth, the viable counts decreased faster than the activity rates. The viable counts may not accurately represent the active population of sulfate-reducing bacteria at depth. However, it is likely that fermentation rates exceeded sulfate reduction rates, since the products of fermentation (e.g., low molecular weight acids) are typical substrates used by sulfate-reducing bacteria.

Evidence that methanogens co-exist with sulfate-reducing bacteria in the littoral zone mats was presented by Giani et al. (1984). In the top cm of sediment, 3×10^4 to 3×10^5 methanogens$\cdot ml^{-1}$ were enumerated by viable counts, with decreasing numbers at depth. The predominant methanogen produced from enrichments was a *Methanosarcina* sp. It grew better on methylated amines than on $CO_2 + H_2$ or acetate. This is significant because Boon (1984) reported the presence of trimethylamine only in the upper layers of mat.

While Hirsch (1980) noted that one eucaryote (a flagellate) was found in the anaerobic zone of the lake, Wilbert and Kahan (1981) noted that none of the 16 species of ciliates they found in the lake were associated with the hot, anaerobic zones. The ciliates were associated with flat mats and pinnacle mats down to the lower boundary of the epilimnion. Most of them were of marine origin. The ciliates included two predators, five grazers of algae, and eight bacteriovorous species. Illustrations were shown for most of these species. In addition to protozoans, an acoelan flatworm of unknown distribution or species was noted by Por (1972). A turbellarian worm, *Macrostomum* sp., was identified by Gerdes et al. (1985). Other metazoans found in Solar lake associated with the blister mat are a copepod (*Robertsonia salsa*) and larvae and adults of two insects (*Ochthebius* sp. and *Paraberosus* sp.). Eucaryotes may be generally excluded from the sediments and brines below the pycnocline due to their lack of tolerance of sulfide and anoxic conditions, as well as their inability to survive at elevated temperatures.

14.3 Sediment and porewater chemistry

Low concentrations of FeS have been detected in Solar Lake sediments (Jørgensen and Cohen, 1977), but pyrite is essentially absent due to limited concentrations of iron. Iron is brought in with detritus during floods. Cohen et al. (1980) and Aizenshtat et al. (1983) found that the water column during stratification and the sediments were rich in polysulfides due to a lack of iron and the presence high concentrations of NH_4^+ at pH 8.0. The polysulfides led to secondary enrichment of buried organic matter with organically-bonded sulfur. Stable accumulations of elemental sulfur were also detected (Jørgensen and Cohen, 1977; Aizenshtat et al., 1983).

Trace metals in Solar Lake sediments were measured by Gaudette and Lyons (1984) and compared to a variety of measurements in other sediments by Long et al. (1985). The following average concentrations in shallow Solar Lake sediments in and below mats were recorded by Gaudette and Lyons (1984) (in $\mu g \cdot g$ dry sediment^{-1}): Fe, 8000; Cd, <0.36; Cr, 10; Cu, 9.4; Pb, 3.8; and Zn, 27. The vertical profiles of Fe, Cu, and Zn suggested they were associated with sulfides, while Pb was probably associated with carbonates. These trace metal concentrations overlap those measured by Nissenbaum (1974) in Dead Sea sediments, and are close to the range of concentrations found between average carbonates and average shales (Long et al., 1985). Gaudette and Lyons (1984) concluded that Solar Lake-type sediments are not more enriched in metals than are nearshore sediments, but that the microbial mats in the lake may serve as a locus for the initial concentration of the heavy metals.

Concentrations of porewater phosphate (0.9–3.5 μM) in 5–35 cm deep Solar Lake sediment (Lyons et al., 1984) were similar to concentrations measured in laminated solar saltern mats in Mexico (0–7.4 μM) and mats plus sediments in the adjacent carbonate-gypsum sabkha (3.8–8.3 μM) (Javor, unpublished data). Total phosphorus in Solar Lake sediments (550–883 μg P per g sediment) (Lyons et al., 1984) was similar to total P in the solar saltern mat sediments in Mexico (87–1586 μg $P \cdot g$ sediment^{-1}, average $= 584$ μg $P \cdot g$ sediment^{-1}) and the adjacent sabkha sediments (198–980 μg $P \cdot g$ sediment^{-1}, average $= 624$ μg $P \cdot g$ sediment^{-1}) (Javor, unpublished data). Lyons et al. (1984) compared dissolved phosphate and total P with hypersaline sediments in Bonaire (with and without microbial mats) and found similar concentrations as those listed above. Phosphorus concentrations do not appear to fluctuate with intensity of sulfate reduction (Lyons et al., 1984), nor with maintenance of continually anaerobic conditions (lakes and salterns vs. sabkhas). Phosphorus may accumulate with certain classes of organic material, or more importantly, as complexes or salts of Ca^{2+} and Mg^{2+}.

Organic geochemistry In spite of high primary productivity in the shallow mats, they only accrete about 30–40 g $C \cdot m^{-2} \cdot yr^{-1}$, or about 10% of the rate calculated from primary productivity measurements (Cohen et al., 1980). Degradation by a variety of bacteria accounts for the loss of organic carbon with depth. Surveys of the organic geochemistry of Solar Lake sediments have provided a wealth of information concerning the degradation and preservation of organic matter in such a hypersaline environment (Lyons et al., 1982; Aizenshtat et al., 1983, 1984; Boon et al., 1983; Boon, 1984; Edmunds and Eglinton, 1984; Hines and Burlingame, 1984; Klok et al., 1984). Some of these data have been summarized in Tables 14.4 and 14.5 and Figures 13.2 and 13.3.

Table 14.4 Concentrations of major lipids (in ppm) of Solar Lake[a]

	Depth (in mm)				
Lipid	0–3	3–10	10–20	268–300	568–620
Fatty acids n-$C_{16:0}$	1184	85	34	1.4	0.9
Fatty acids n-$C_{18:0}$	100	36	17	0.8	0.5
Alcohols n-$C_{18:0}$	50	67	5	12	24
Alcohols n-$C_{14:0}$	41	38	1.9	5.5	15
Alcohols n-$C_{16:0}$	21	17	1.6	3.1	16
Hydrocarbons n-$C_{16:1}$	270	8.2	0.1	0	0
Hydrocarbons n-$C_{16:0}$	30	0.9	0.1	1.2	0.2
Hydrocarbons n-$C_{17:0}$	135	5.9	0.2	1.2	1.2
Hydrocarbons n-$C_{18:0}$	62	2.3	0.3	1.2	0.2
Hydrocarbons n-$C_{20:0}$	low	5.1	0.5	0.9	0.3
Steroids: individual compounds as high as 15 ppm					
Hopanoic acids	0	36	70	2.5	3.5
Hopanoid ketones	0	0.5	0.8	1.0	3.1[b]
Hopanoid alcohols	0.6	10	2.6	6.3	16[b]
Hopanoid hydrocarbons	1.1	2.3	0.6	1.9	1.5[b]
Acyclic isoprenoids					
Phytol	576	335	23	97	54
Carotenoids	136	–	–	–	26
Phyt-1-ene	128	1.6	0.1	0.3	0.4
Squalene	15.3	6.7	1.4	0.6	0

[a] From Edmunds and Eglinton (1984). Some values were estimated from figures.
[b] From a depth of 650–690 mm.

Dissolved organic carbon (DOC) values are very high in Solar Lake sediments. Six horizons (5–81 cm depth) in four cores taken from 0.5–1.0 m lake depth had DOC values ranging from 119 to 818 mg DOC·l^{-1} (Lyons et al., 1982). The average of 24 samples was 273 mg·l^{-1}. The profile at 15–53 cm core depth suggests that DOC production from POC is greater than its oxidation and removal. The authors stated that these are the highest DOC values ever recorded from recent sediments. Aizenshtat et al. (1984) suspected that the DOC was largely carbohydrates. They found both free and bound amino acids were abundant as well (Table 14.5).

Total organic carbon (TOC) values decreased with depth. Aizenshtat et al. (1984) reported TOC decreased from 15–17% at the mat surface to 5–6% at 80 cm depth. Klok et al. (1984) found 15.2–26.6% organic carbon at the mat surface and 4.5% organic carbon at 60–70 cm depth. Much lower values of TOC reported by Boon et al. (1983) could reflect the incomplete oxidation of organic carbon in their method of analysis (Aizenshtat et al., 1984). Carbohydrates constituted the major proportion of the TOC (Boon et al., 1983; Klok et al., 1984; Table 14.5). In the top 0.5 mm of mat, carbohydrates (predominantly glucose with smaller amounts of other monosaccharides) constituted 34.1% of the organic matter (Klok et al., 1984). Carbohydrates (and especially glucose) decreased with depth, with only 7.5% of the residual organic matter of carbohydrate composition at 60–70 cm sediment depth. The

Table 14.5 Major organic components of Solar Lake sediments

Component	Characterization	References[a]
Carbohydrates	Dominate throughout; major component of buried cell sheath material; glucose, major carbohydrate down to 145 mm depth; xylose, major carbohydrate at 265–658 mm depth; other sugars, rhamnose, fucose, arabinose, ribose, mannose, galactose	2, 3, 5
Amino acids	Free amino acids around 50 mM, decreasing with depth; bound amino acids around 30% of organic matter near mat surface	1
Humic acids	Not present	1
Fatty acids	n-$C_{16:0}$ acids dominate, followed by n-C_{18} acids; no components $>C_{30}$; unsaturated:saturated fatty acid ratio decreases with depth; rapid decrease in straight-chain and monomethyl acids with depth; degradation to n-alcohols; fatty acids largely from cyanobacteria; iso- and anteiso-fatty acids from heterotrophs	2, 3, 4
Alcohols	n-$C_{18:0}$ alcohols dominate; C_{14} and C_{16} alcohols present in significant concentrations	4
Hydrocarbons	n-$C_{16:1}$ dominates surface layers; unsaturated C_{16}, C_{17}, C_{18}, and C_{20} hydrocarbons dominate lower layers; no hydrocarbons $>C_{30}$ (no higher plant input); cyanobacteria are the source of surface hydrocarbons; deeper hydrocarbons may be due to reworking by bacteria or reactions with protokerogen	1, 2, 4
Acyclic isoprenoids	Phytol dominates and is the major lipid (by weight) at depth; diagenetic products of phytol important, with all components decreasing with depth; squalene and carotenoids significant near mat surface; no archaebacterial isoprenoids nor diatom carotenoids detected	4
Steroids	Low concentrations of a diverse array of sterols; probably of cyanobacterial origin; show evidence of diagenesis with depth; ergosterol (indicator of fungi) not present	2, 4
Hopanoids	Largely absent in surface mat; maximum concentrations at 3–20 mm depth with acid fraction dominant; C_{32}-$\beta\beta$ is the major hopanoic acid; eubacterial origin	4

[a] (1) Aizenshtat et al. (1984); (2) Boon et al. (1983); (3) Boon (1984); (4) Edmunds and Eglinton (1984); (5) Klok et al. (1984).

remnants of microbial mats in the lower sediments include abundant empty sheaths of cyanobacteria. Dissolved carbohydrates and amino acids in the porewaters of the surface of the mats were shown in Figures 13.2 and 13.3.

Table 14.4 summarizes the concentrations of the major lipids in Solar Lake sediments while Table 14.5 outlines the major components and some of their attributes. A number of novel and unidentified compounds have been reported as well. The general picture of the profile of organic compounds of the mats and sediments shows that most of the marker molecules of the top mats are completely transformed or degraded after 2400 years of burial in spite of the fact that the sediments are hypersaline and anaerobic. Although such environments have been thought to promote the preservation of organic matter, it is obvious that selective degradation is characteristic of such a habitat. Further experimentation would determine whether rates of degradation and transformations could be detected in the sediments.

Stable isotopes The profile of stable carbon isotopes reveals very heavy values of organic carbon in Solar Lake (Schidlowski et al., 1984; Aizenshtat et al., 1984). Schidlowski et al. (1984) found that organic carbon in the mat was the heaviest encountered in the biosphere, with a $\delta^{13}C$ of $-5.7 \pm 1.4‰$ versus the standard (Peedee Belemnite). Average organic carbon has $\delta^{13}C$ values of -20 to $-30‰$. The authors attributed the heavy values to CO_2 limitation in Solar Lake, resulting in the ineffective discrimination of ^{12}C and ^{13}C by the phototrophic microorganisms. Six horizons in a 100-cm long sediment core had organic carbon $\delta^{13}C$ values of -4.4 to $-8.4‰$. A more detailed core profile measured by Aizenshtat et al. (1984) showed a slight trend of increasingly lighter values from the mat top 2.2 mm ($-6.65‰$) to 19.5–21.0 mm depth ($-8.63‰$). Values in the top 2 cm averaged $-6.87‰$. They attributed the ^{12}C enrichment at the surface to discrimination by the enzyme ribulose 1,5-bisphosphate carboxylase (the first enzyme in the CO_2 fixation pathway in photosynthesis). Slightly lighter carbon below was thought to be controlled by activities of sulfate-reducing bacteria which degrade organic carbon. A similar phenomenon was noted in microbial mats in Baja California, Mexico (D. DesMarais, pers. comm. in Aizenshtat et al., 1984).

Limited availability of CO_2 has also been suggested to explain the stable carbon isotope distribution in carbonates in Solar Lake (Aharon et al., 1977; Aizenshtat et al., 1984). Aharon et al. (1977) demonstrated that the average $\delta^{13}C$ value for both aragonite and dolomite was $+4‰$ and for Mg-calcite was -8.7 to $-9.7‰$. They concluded that the dolomite does not originate in the algal layers, but rather in association with aragonite mixed with Mg^{2+}-rich brines. Mg-calcite was interpreted to have formed in a closed system, driven by gypsum dissolution by bacteria. Ca^{2+}-enriched brines mixed with HCO_3^- derived from organic decay would result in the precipitation of Mg-calcite. The $\delta^{18}O$ values of the carbonates support this model. Aizenshtat et al. (1984) found the $\delta^{13}C$ values of carbonates varied with depth and were a result of a complex system of precipitation. At the bottom of the cores, where oxygenated seawater mixed with the sediments, both light ($\delta^{13}C = -10‰$) and heavy ($\delta^{13}C = +4‰$) carbonates were found, probably as a result of biological activity and chemical deposition, respectively.

The stable isotope profiles of various sulfur species confirm the high degree of activity of microbial sulfur cycling found in other biological and chemical studies. There was no fractionation of dissolved sulfate in the Solar Lake water column ($\delta^{34}S = +20.1$ to $20.9‰$) in comparison to Red Sea water ($\delta^{34}S = +20.7‰$), which concurs with the finding that little or no sulfate reduction was detected in the water column (Aizenshtat et al., 1983). Sediment sulfate becomes heavier between the surface ($\delta^{34}S = +22.9‰$) and 45–80 cm depth ($\delta^{34}S$ = +24.8‰), but lighter values were found at 80–99 cm depth ($\delta^{34}S$ = +23.3 to 23.4‰) (Aizenshtat et al., 1983, 1984). In the same profile, S-bound organic matter (protokerogen) became lighter between the surface and 80 cm ($\delta^{34}S$ = −12.2 to −26.3‰) but was slightly heavier ($\delta^{34}S$ = −24.1‰) at 90 cm depth. Dissolved sulfide and elemental sulfur were light ($\delta^{34}S$ = −16.3‰ and as low as −21‰, respectively) while gypsum sulfate at the bottom of the core was heavy ($\delta^{34}S$ = +22.2‰). The progressive enrichment of light organic sulfur along with the increase in total S-bound protokerogen with depth (from 1.4 to 8.2%) were interpreted to be a result of polysulfide reactions with protokerogen material (Aizenshtat et al., 1983). Isotopic signatures of polysulfide should be similar to those of sulfide and elemental sulfur.

The variety of biogeochemical studies performed on Solar Lake samples have provided a complicated but still incomplete picture of the range and types of microbial activities in that system. Primary productivity is high, but there is uncertainty about the measured rates due to ignorance about actual total inorganic carbon concentrations. Organic matter produced *in situ* is not well preserved under anaerobic, hypersaline conditions, although salinities greater than 18% were not encountered in Solar Lake. Few comparative studies of more hypersaline environments have been published to support or contradict the popular notion that only minimal degradation occurs in the most extremely saline habitats. The findings at Solar Lake should serve as a springboard to challenge these concepts.

References

Aharon, P., Kolodny, Y., and **Sass, E.** 1977. Recent hot brine dolomitization in the "Solar Lake," Gulf of Elat, isotopic, chemical, and mineralogical study. *Journal of Geology* 85: 27–48.

Aizenshtat, Z., Stoler, A., Cohen, Y., and **Nielsen, H.** 1983. The geochemical sulphur enrichment of Recent organic matter by polysulfides in the Solar-Lake, pp. 279–288 in Bjørnoy, M., et al. (editors), *Advances in Organic Geochemistry 1981,* John Wiley and Sons, New York.

Aizenshtat, Z., Lipner, G., and **Cohen, Y.** 1984. Biogeochemistry of carbon and sulfur cycles in the microbial mats of the Solar Lake (Sinai), pp. 281–312 in Cohen, Y., Castenholz, R.W. and Halvorson, H.O. (editors), *Microbial Mats: Stromatolites,* MBL Lectures in Biology, vol. 3, Alan R. Liss, New York.

Boon, J.J. 1984. Tracing the origin of chemical fossils in microbial mats: biogeochemical investigations of Solar Lake cyanobacterial mats using analytical pyrolysis methods, pp. 313–342 in Cohen, Y., Castenholz, R.W. and Halvorson, H.O. (editors), *Microbial Mats: Stromatolites,* MBL Lectures in Biology, vol. 3, Alan R. Liss, New York.

Boon, J.J., Hines, H., Burlingame, A.L., Klok, J., Rijpstra, W.I., de Leeuw, J.W., Edmunds, K.E., and **Eglinton, G.** 1983. Organic geochemical studies of Solar Lake laminated cyanobacterial mats, pp. 207–227 in Bjornoy, M. et al. (editors), *Advances in Organic Geochemistry 1981,* John Wiley and Sons, New York.

Cohen, Y. 1984. The Solar Lake cyanobacterial mats: strategies of photosynthetic life under sulfide, pp. 133–148 in Cohen, Y., Castenholz, R.W. and Halvorson, H.O. (editors), *Microbial Mats: Stromatolites,* MBL Lectures in Biology, vol. 3, Alan R. Liss, New York.

Cohen, Y., Padan, E., and **Shilo, M.** 1975. Facultative anoxygenic photosynthesis in the cyanobacterium *Oscillatoria limnetica. Journal of Bacteriology* 123: 855–861.

Cohen, Y., Goldberg, M., Krumbein, W.E., and **Shilo, M.** 1977a. Solar Lake (Sinai). 1. Physical and chemical limnology. *Limnology and Oceanography* 22: 597–608.

Cohen, Y., Krumbein, W.E. and **Shilo, M.** 1977b. Solar Lake (Sinai). 2. Distribution of photosynthetic microorganisms and primary production. *Limnology and Oceanography* 22: 609–620.

Cohen, Y., Krumbein, W.E., and **Shilo, M.** 1977c. Solar Lake (Sinai). 3. Bacterial distribution and production. *Limnology and Oceanography* 22: 621–634.

Cohen, Y., Aizenshtat, Z., Stoler, A., and **Jørgensen, B.B.** 1980. The microbial geochemistry of Solar Lake, Sinai, pp. 167–172 in Trudinger, P.A. and Walter, M.R. (editors), *Biogeochemistry of Ancient and Modern Environments,* Springer-Verlag, New York.

Edmunds, K.L.H. and **Eglinton, G.** 1984. Microbial lipids and carotenoids and their early diagenesis in the Solar Lake laminated microbial mat sequence, pp. 343–389 in Cohen, Y., Castenholz, R.W. and Halvorson, H.O. (editors), *Microbial Mats: Stromatolites,* MBL Lectures in Biology, vol. 3, Alan R. Liss, New York.

Garlick, S., Oren, A., and **Padan, E.** 1977. Occurrence of facultative anoxygenic photosynthesis among filamentous and unicellular cyanobacteria. *Journal of Bacteriology* 129: 623–629.

Gaudette, H.E. and **Lyons, W.B.** 1984. Trace metal concentrations in modern sabkha environments, pp. 425–434 in Cohen, Y., Castenholz, R.W. and Halvorson, H.O. (editors), *Microbial Mats: Stromatolites,* MBL Lectures in Biology, vol. 3, Alan R. Liss, New York.

Gerdes, G., Spira, J., and **Dimentan, C.** 1985. The fauna of the Gavish Sabkha and the Solar Lake—a comparative study, pp. 322–345 in Friedman, G.M. and Krumbein, W.E. (editors), *Hypersaline Ecosystems. The Gavish Sabkha.* Ecological Studies 53, Springer-Verlag, New York.

Giani, D., Giani, L., Cohen, Y., and **Krumbein, W.E.** 1984. Methanogenesis in the hypersaline Solar Lake (Sinai). *FEMS Microbiology Letters* 25: 219–224.

Hines, H.B. and **Burlingame, A.L.** 1984. Chemical degradations of residual organic matter from laminated cyanobacterial mats from Solar Lake, Israel, pp. 391–410 in Cohen, Y., Castenholz, R.W. and Halvorson, H.O. (editors), *Microbial Mats: Stromatolites,* MBL Lectures in Biology, vol. 3, Alan R. Liss, New York.

Hirsch, P. 1980. Distribution and pure culture studies of morphologically distinct Solar Lake microorganisms, pp. 41–60 in Nissenbaum, A. (editor), *Hypersaline Brines and Evaporitic Environments,* Developments in Sedimentology 28, Elsevier, New York.

Jørgensen, B.B. and **Cohen, Y.** 1977. Solar Lake (Sinai). 5. The sulfur cycle of the benthic cyanobacterial mats. *Limnology and Oceanography* 22: 657–666.

Jørgensen, B.B., Revsvech, N.P., Blackburn, H., and **Cohen, Y.** 1979a. Diurnal cycle of oxygen and sulfide microgradients and microbial photosynthesis in a cyanobacterial mat sediment. *Applied and Environmental Microbiology* 38: 46–58.

Jørgensen, B.B., Kuenen, J.G., and **Cohen, Y.** 1979b. Microbial transformations of sulfur compounds in a stratified lake (Solar Lake, Sinai). *Limnology and Oceanography* 24: 799–822.

Jørgensen, B.B., Revsbech, N.P., and **Cohen, Y.** 1983. Photosynthesis and structure of benthic microbial mats: microelectrode and SEM studies of four cyanobacterial communities. *Limnology and Oceanography* 28: 1075–1093.

Jørgensen, B.B., Cohen, Y., and **Revsbech, N.P.** 1986. Transition from anoxygenic to oxygenic photosynthesis in a *Microcoleus chthonoplastes* cyanobacterial mat. *Applied and Environmental Microbiology* 51: 408–417.

Klok, J., Cox, H.C., Baas, M., de Leeuw, J.W., and **Schenck, P.A.** 1984. Carbohydrates in recent marine sediments. II. Occurrence and fate of carbohydrates in a recent stromatolitic deposit: Solar Lake, Sinai. *Organic Geochemistry* 7: 101–109.

Krumbein, W.E., Cohen, Y., and **Shilo, M.** 1977. Solar Lake (Sinai). 4. Stromatolitic cyanobacterial mats. *Limnology and Oceanography* 22: 635–656.

Long, D.T., Lyons, W.B., and **Gaudette, H.E.** 1985. Trace metal concentrations in modern marine sabkhas. *Chemical Geology* 53: 185–189.

Lyons, W.B., Gaudette, H.W., and **Gustafson, H.C.** 1982. Dissolved organic carbon in pore waters from a hypersaline environment. *Organic Geochemistry* 3: 133–135.

Lyons, W.B., Hines, M.E., and **Gaudette, H.E.** 1984. Major and minor element pore water geochemistry of modern marine sabkhas: the influence of cyanobacterial mats, pp. 411–423 in Cohen, Y., Castenholz, R.W. and Halvorson, H.O., *Microbial Mats: Stromatolites,* MBL Lectures in Biology, vol. 3, Alan R. Liss, New York.

Nissenbaum, A. 1974. Trace elements in Dead Sea sediments. *Israel Journal of Earth-Sciences* 23: 111–116.

Oren, A., Padan, E., and **Avron, M.** 1977. Quantum yields for oxygenic and anoxygenic photosynthesis in the cyanobacterium *Oscillatoria limnetica. Proceedings of the National Academy of Science U.S.A.* 74: 2152–2156.

Oren, A. and **Shilo, M.** 1979. Anaerobic heterotrophic dark metabolism in the cyanobacterium *Oscillatoria limnetica*: sulfur respiration and lactate fermentation. *Archives of Microbiology* 122: 77–84.

Por, F.D. 1972. Hydrobiological notes on the high-salinity waters of the Sinai Peninsula. *Marine Biology.* 14: 111–119.

Por, F.D. 1985. Anchialine pools—comparative hydrobiology, pp. 136–144 in Friedman, G.M. and Krumbein, W.E. (editors), *Hypersaline Ecosystems. The Gavish Sabkha.* Ecological Studies 53, Springer-Verlag, New York.

Potts, M. 1979. Ethylene production in a hot brine environment. *Archiv für Hydrobiologie* 87: 198–204.

Potts, M. 1980. Blue-green algae (Cyanophyta) in marine coastal environments of the Sinai Peninsula; distribution, zonation, stratification and taxonomic diversity. *Phycologia* 19: 60–73.

Revsbech, N.P., Jørgensen, B.B., Blackburn, T.H., and **Cohen, Y.** 1983. Microelectrode studies of the photosynthesis and O_2, H_2S, and pH profiles of a microbial mat. *Limnology and Oceanography* 28: 1062–1074.

Schidlowski, M., Matzigkeit, U., and **Krumbein, W.E.** 1984. Superheavy organic carbon from hypersaline microbial mats. *Naturwissenschaften* 71: 303–308.

Walsby, A.E., van Rijn, J., and **Cohen, Y.** 1983. The biology of a new gas-vacuolate cyanobacterium, *Dactylococcopsis salina,* sp. nov., in Solar Lake. *Proceedings of the Royal Society of London* B 217: 417–447.

Wilbert, N. and **Kahan, D.** 1981. Ciliates of Solar Lake on the Red Sea shore. *Archiv für Protistenkunde* 124: 70–95.

15

Gavish Sabkha and Other Hypersaline Marine Sabkhas, Pools, and Lagoons

Sea-marginal environments subject to hypersalinity include a variety of tidal flats and depressions, salt marshes, lagoons, and other physiographic features in which microorganisms thrive. Some of these environments are subject to extreme salinity variations and sometimes complete desiccation. In some cases, they offer the opportunity to study in a single basin the precipitation and diagenesis of a variety of evaporite minerals, and the potential biological and organic chemical interactions with those minerals. In other cases, they offer the opportunity to profile the biogeochemical attributes of a stage in the evolution of an evaporite basin. Both of these approaches are necessary for accurate interpretations of ancient marine evaporite deposits.

15.1 Gavish Sabkha

Besides Solar Lake, the Red Sea coast of the Sinai is noted for several other hypersaline embayments, including the Gavish Sabkha and Ras Muhammad pool (see Figure 14.1). A variety of largely descriptive papers covering biological, geological, and chemical aspects of these environments was published in a volume edited by Friedman and Krumbein (1985). Aspects of those studies are discussed in comparison with the results in the Solar Lake investigations and with investigations of hypersaline habitats in the Gulf of Mexico, the Pacific coast of Baja California, Mexico, and Australia. Much of the work describing hypersaline environments in the Persian Gulf was presented in a volume by Purser (1973) and those studies will not be discussed here except where comparisons between the systems are noteworthy.

Krumbein (1985) and Friedman (1985) discussed various characteristics of sabkhas. They are smooth, flat plains or salt flats that gradually grade upward from a tropical sea or are separated from the sea by a low or permeable barrier. The mineralogy and reactions involved in evaporite precipitation in sabkhas has been well-documented for the Persian Gulf. The majority of evaporite salts in that environment precipitate from interstitial waters within the sabkha and are therefore of secondary origin (Purser, 1985). This situation makes it difficult to relate mineralogy with biological activity, since buried microbial communities may not be active in the geochemical regime at the time of an investigation.

Minerals and trace elements In the Gavish Sabkha, gypsum forms within microbial mats as small crystals (5 mm), and below the mats as larger crystals (up to 20 cm). The presence of intact layers of microbial mats within the large gypsum crystals indicates that the gypsum replaces carbonate sediments upon burial of the mats. Gypsum formation in this sabkha is similar to that found in the Persian Gulf (Purser, 1985).

In contrast to Solar Lake, Gavish Sabkha has only a small area of permanent water. This pool remains very shallow all year around (Gerdes and Krumbein, 1984). The exposed areas are subject to the precipitation of a variety of minerals besides gypsum, including carbonates, celestite, and halite. Gypsum is locally absent but those hypersaline muds have FeS and smell of H_2S (Gavish et al., 1985). The authors attributed the lack of gypsum in those sites to sulfate-reducing bacteria. The same phenomenon was observed in the shallow microbial mats of Solar Lake.

Pb, Zn, and Cr contents of Gavish Sabkha mats and sediments are similar to those of Solar Lake. Cu was about half as abundant and Fe about one-sixth as abundant in the sabkha (Gaudette and Lyons, 1984; Lyons and Gaudette, 1985; Long et al., 1985). The differences in Cu and Fe concentrations between the two sites may be related to differences of influx of detritus in the two systems.

Microbiology Microbial mats dominated by cyanobacteria develop in some of the hypersaline sites in the Gavish Sabkha (Gerdes et al., 1985a). In 6.5–15% seawater, nodular communities of procaryotes were described. The dominant species was *Pleurocapsa,* with lesser numbers of other coccoid and filamentous forms. The filamentous cyanobacteria were favored at higher salinity. Both purple sulfur and green phototrophic bacteria were present, along with the sulfur-oxidizing *Beggiatoa* and unidentified spirilla and budding bacteria.

In 18–25% seawater, small, dome-shaped, laminated mats occur, growing in thickness to 10–13 mm in the winter. The top layer consisted predominantly of slime-producing, coccoid cyanobacteria identified as *Synechocystis, Synechococcus,* and/or *Aphanocapsa,* although it may be difficult to distinguish these taxa from the *Aphanothece,* the dominant coccoid cyanobacterium in other hypersaline habitats (Golubic, 1980). Beneath the surface were four other zones: a layer of coccoid *Pleurocapsa* and *Gloeothece,* a layer of *Microcoleus*-type cyanobacteria, a pink layer of purple phototrophic bacteria (*Thiocapsa*), and another layer of filamentous cyanobacteria (*Phormidium*-type). Flexibacteria were also abundant. During periods of relatively low salinity, diatoms (*Nitzschia* and *Amphora*) were found associated with the mats. Between the mat and the black, reducing sediment, a zone of purple and green phototrophic bacteria developed.

Halobacteria predominate in the zone of salt pans where brines are of 30–36% salinity. Beneath the salt crusts some filamentous and coccoid cyanobacteria were found. Many of the mat-forming cyanobacteria were isolated by Gerdes et al. (1985a) and were found to be capable of growth in a medium of 23% salinity.

In a zone of mats where the salinity varied between 18 and 20%, oxygen was present, being found at 10 mm and at 18 mm depth (3–4 mg $O_2 \cdot l^{-1}$), and coexisting with a sulfide peak at 4 mm depth (20 mg$\cdot l^{-1}$). The slime of the mats apparently hinders gas movement and allows oxygen to remain in solution below an apparent zone of sulfate reduction near the top of the mat. This reverse zonation of oxygen and sulfide has not been described before.

Gerdes et al. (1985a) compared concentrations of photosynthetic pigments of different microbial mats in Gavish Sabkha (Table 15.1). These values are compared to pigment concentrations in mats from several other environ-

Table 15.1 Chlorophyll, bacteriochlorophyll, and photosynthetic rates of microbial mats in sabkha and hypersaline marine environments

Site	Salinity, %	Chl *a*, $\mu g \cdot cm^{-3}$	Bchl *a*, $\mu g \cdot cm^{-3}$	Photosynthesis, mg $C \cdot m^{-2} \cdot h^{-1}$	Reference
Gavish Sabkha	5.5–7.0	24.6	12.1	150–520	Gerdes et al.
	6.5–7.0	45.6	3.14	(maximum, all salinities)	(1985a)
	10–18	55.1	3.28		
Spencer Gulf	ca. 6.4	8.9–28.8	—	30–613 (most 100–300)	Bauld (1984)
Shark Bay	Not given	2.5–51.2	—	17–113	Bauld (1984)
Laguna Guerrero Negro	4.5–9.6	70.2–70.7[a]	25.5–31.2[a]	—	Javor (1979)
Solar Lake	≤18				Jørgensen et al. (198
Shallow flat mat		—	—	211[b]	
Deep flat mat		—	—	160[b]	
Blister mat		—	—	157[b]	
Gelatinous mat		—	—	14[b]	

[a] Values expressed per cm^2.
[b] Conversion of O_2 produced to C assimilated.

ments. The greater abundance of chlorophylls in the Guerrero Negro mats probably reflects the lack of inorganic sediment and the differences in species composition (Javor, 1979; Javor and Castenholz, 1981). In Gavish Sabkha, the ratio of chlorophyll *a*:bacteriochlorophyll *a* appears to increase with salinity, perhaps reflecting the limited ability of phototrophic bacteria to succeed in these mat communities. In contrast, the same high salinities promote the development of planktonic phototrophic bacteria in Solar Lake and solar salterns (see Chapters 13 and 14).

Photosynthetic rates of Gavish Sabkha mats were 150–520 mg $C \cdot m^{-2} \cdot h^{-1}$ (Gerdes et al., 1985a), which are largely comparable to the rates measured in the mats of Solar Lake and Spencer Gulf (Table 15.1). Productivity in Gavish Sabkha apparently is not nutrient-limited. Both phosphate (1.8–8.9 μM) and nitrate plus nitrite (1.6–5.9 μM) were present in Gavish Sabkha (Gerdes et al., 1985a). Ammonia was not reported. Nutrient concentrations in Solar Lake have not been published.

Erlich and Dor (1985) made an extensive inventory of the phototrophic microorganisms in Gavish Sabkha. Diatoms were found at most collecting sites at salinities up to 20.5%. *Dunaliella* was usually found up to 15–20% salinity. Cyanobacteria were nearly ubiquitous. They dominated the microflora at salinities >10% and were found at up to 32.7% salinity. The authors identified 33 species of cyanobacteria and 28 species of diatoms, and recorded their abundances. Their illustrations and photomicrographs of all the species should be a useful guide for the identification of algae in other hypersaline habitats.

Among the cyanobacteria, most of the listed taxa tolerated at least 18% salinity, but only two species (*Aphanothece halophytica* and *Schizothrix arenaria*) were associated with 25–33% salinity brines. Among the diatoms, the most halotolerant were *Amphora coffeaeformis* and species of *Navicula* and *Nitzschia*. *Nitzschia lembiformis* was found almost exclusively at high salinities (around 15%). No halophilic diatom has ever been reported before (see Chapter 9). This diatom is not found in Solar Lake or Bardawil Lagoon (another hypersaline embayment in the Sinai), but it was identified in hypersaline ponds near the Dead Sea.

Extremely halophilic bacteria isolated from Gavish Sabkha include cocci, pleomorphic rods, and square bacteria (Kessel et al., 1985). About 10^7 $cells \cdot ml^{-1}$ were found in late summer and early fall (Stoeckenius et al., 1985). These studies showed the square bacteria have bacteriorhodopsin pigments. Other characteristics of these halophilic bacteria are discussed in Chapters 5 and 6. Stalked bacteria tentatively identified as *Pedobacterium* were also seen (Kessel et al., 1985).

A large variety of salt-tolerant metazoans, including copepods, ostracods, a nematode, a turbellarian worm, a gastropod, and insects were found at various sites in Gavish Sabkha (Gerdes and Krumbein, 1984; Gerdes et al., 1985b). The salinity tolerances of these animals are listed in Table 2.3. Gerdes et al. (1985b) noted that several of the same taxa were found in the winter in Solar Lake, where they were associated with the pinnacle mat (9–14% salinity). However, a richer variety of invertebrates was associated with the more exposed, Gavish Sabkha environment. *Artemia salina* was notably absent in Gavish Sabkha.

Organic geochemistry The organic geochemistry of the Gavish Sabkha sediments was described by De Leeuw et al. (1985) and Boon et al. (1985). The general character of the organic components of the sediments is very similar to that reported for Solar Lake, with a few notable exceptions. Glucose is the major carbohydrate of the top mat, but it is less abundant in the Gavish Sabkha mat (0.48% vs. 7.8% of the sediment dry weight) (Boon et al., 1983; de Leeuw et al., 1985). Inorganic sediment brought in by floods probably accounts for the difference.

Some similar acyclic isoprenoids that were intrepreted to be of eubacterial origin in Solar Lake, namely phytene and squalene, were thought to be possibly of archaebacterial origin in the Gavish Sabkha (Edmunds and Eglinton, 1984; de Leeuw et al., 1985). No definitive archaebacterial isoprenoids (i.e., C_{25} or C_{40}) were found in either environment. Among the lipids in both environments, phytol was the most abundant. It is believed to be a degradation product of chlorophyll. In the upper sediments of Gavish Sabkha, the concentration of alcohols was much greater than fatty acids, which in turn were about as abundant as sterols and hopanoids. Hydrocarbons were the least abundant (de Leeuw et al., 1985). In Solar Lake and other recent environments, fatty acids exceed fatty alcohols in abundance. The same is true for the abundance of these compounds in a sandy layer 50 cm below the Gavish Sabkha mat. The authors concluded that the high fatty alcohol content may represent a response of the microbiota to extremely hypersaline conditions or a relatively large contribution by phototrophic bacteria.

Stable isotopes The stable isotopes of carbon and oxygen of the Gavish Sabkha and Solar Lake sediments were compared by Schidlowski et al. (1985). The organic carbon of Gavish Sabkha was isotopically heavy, but not quite as heavy as that in Solar Lake mats. The average $\delta^{13}C_{org}$ for Solar Lake was −5.4‰. The Gavish Sabkha surface mat $\delta^{13}C_{org}$ was −8.1 to −9.8‰. Buried mats had $\delta^{13}C_{org}$ values of −8.4 to −11.7‰. Two clay-containing horizons had very light carbon (−16.8 to −17.0‰) which are probably associated with flood deposits.

The $\delta^{13}C$ values for inorganic carbon (−3.5 to +4.7‰ in Gavish Sabkha, −3.3‰ in Solar Lake) were reported to be typical for carbonates precipitated in closed environments (Schidlowski et al., 1985). The $\delta^{18}O$ values for carbonates (+32.1‰ average for Gavish Sabkha, +35.2‰ average for Solar Lake) are heavier than most marine carbonates (+26 to +32‰). ^{18}O enrichment of carbonates in the hypersaline habitats may reflect equilibration with isotopically heavy water that results from evaporation. The $\delta^{18}O$ content of the brines was not reported. The heavy organic carbon in both environments was attributed to CO_2 limitation.

15.2 The Ras Muhammad Pool

The Ras Muhammad pool in the southern Sinai undergoes yearly cycles of salinity that result in the alternate deposition of algal-carbonate sediments and gypsum. The result is an accumulation of varve-like sediments (Kushnir, 1981; Friedman et al., 1985). Kushnir (1981) found that interstitial brine composition of Ca^{2+} and SO_4^{2-} was anomalously high (up to two-fold supersaturated). In the lower layers of the sediment, supersaturation was relieved by subsurface gypsum growth that eventually disrupted the algal laminae.

Kushnir (1981) discounted any significant role of sulfate-reducing bacteria by citing that the $\delta^{34}S$ of sulfate in the gypsum (+22.8 ±0.2‰) was approximately that of normal marine gypsum. This value is somewhat higher than the $\delta^{34}S$ of sulfates from the sabkha of Laguna Ojo de Liebre, Baja California, Mexico, where Holser and Kaplan (1966) calculated an average value of +21.0‰. A closer examination of the Ca^{2+}, SO_4^{2-}, and Cl^- content of the interstitial brines of the Ras Muhammad pool measured by Kushnir (1981) suggests that significant sulfate depletion (high Ca^{2+}:SO_4^{2-} values) occurred in four varve cycles between layers 20 and 26 (Fig. 15.1). Minor fluctuations in Ca^{2+}:SO_4^{2-} content can be seen in other horizons. Because gypsum is the only sulfate mineral that is stable under this salinity regime, sulfate losses relative to Ca^{2+} must result from bacterial reduction.

Friedman et al. (1985) also concluded that, at least periodically, sulfate reduction affects the brine composition in Ras Muhammad pool. During 1970–1971, when surface salinities increased from 13 to 31.5%, Ca^{2+} increased from 1.48 $g{\cdot}l^{-1}$ to 5.30 $g{\cdot}l^{-1}$ while sulfate decreased from 9.70 $g{\cdot}l^{-1}$ to 4.60 $g{\cdot}l^{-1}$. The authors noted the smell of H_2S around the pool. During the 1977–1979 period, such a SO_4^{2-}:Ca^{2+} decrease was not observed. Friedman et al. (1985) did not comment on the results of Kushnir (1981).

The $\delta^{13}C$ and $\delta^{18}O$ values of carbonates show different trends with depth in the Ras Muhammad pool sediments (Friedman et al., 1985). Unfortunately, there is no way to accurately compare the chemical analyses of Kushnir (1981) and Friedman et al. (1985) from the published data. The $\delta^{13}C$ fractionation of carbonates show a trend for increasingly heavier values with depth between 2.2 and 47.2 cm (−0.85 to +4.68‰), and a slight decrease at 64.0 cm (+3.30‰). The heavier values at depth are comparable to those measured in Solar Lake. Friedman et al. (1985) attributed the increasingly heavier values due to a $\delta^{13}C$-enriched residuum of bicarbonate produced by organic decomposition. However, such decomposition typically results in ^{12}C enrichment.

The Ras Muhammad microbial mats were largely associated with ara-

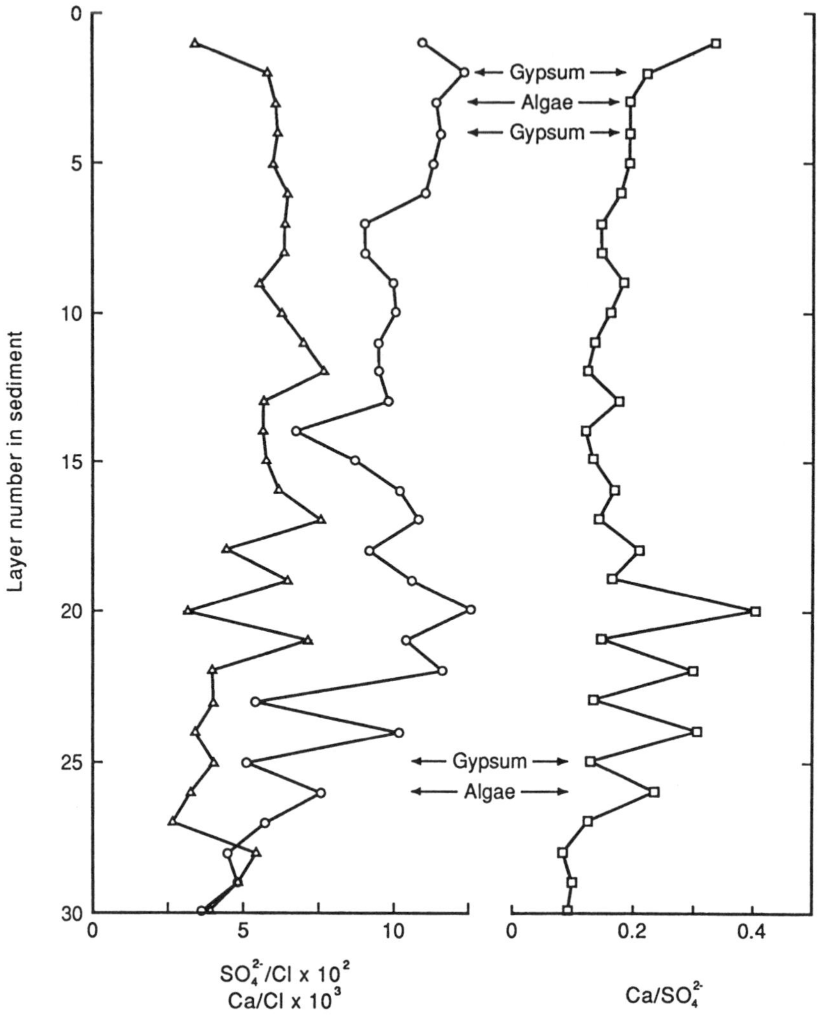

Figure 15.1 Ratios of calcium, sulfate, and chloride in porewaters of sediments from Ras Mohammad, Sinai (calculated from Kushnir, 1981). Thirty samples were taken from a core of 86 cm depth with no indication of the actual depth of each measured layer. Gypsum or algal layers mentioned by Kushnir are noted on the figure. Triangles = sulfate/chloride, circles = calcium/chloride, squares = calcium/sulfate.

gonite, although Mg-calcite was also noted (Friedman et al., 1985). Dolomite was largely associated with gypsum, where it was also found in Solar Lake (Aharon et al., 1977). In Solar Lake, C in Mg-calcite was isotopically light and aragonite and dolomite carbon was relatively heavy (see Chapter 14). The published details of carbonate distribution in the Ras Muhammad pool are insufficient to allow similar conclusions to be reached.

In the same depth range where $\delta^{13}C$ values of carbonate increased (2.2–47.2 cm), $\delta^{18}O$ values of carbonates remained nearly constant (+3.37 to +3.72‰) in Ras Muhammad pool. A heavier value (+5.20‰) was recorded at 64.0 cm depth. In Solar Lake, aragonite had $\delta^{18}O$ values of +3 to +4‰ (Aharon et al., 1977) while the $\delta^{18}O$ values of Mg-calcite were lighter and those for dolomite were heavier. Comparisons between the stable isotope values of carbonates of Solar Lake and Ras Muhammad pool show different trends. In Solar Lake, lighter carbon occurred with lighter oxygen in Mg-carbonate. In Ras Muhammad pool sediments, trends of increasing $\delta^{13}C$ values were not parallelled by any significant changes in $\delta^{18}O$ values. The Ras Muhammad sediment is subjected to more dramatic salinity changes and to wet/dry cycles. More detailed stable isotope studies of C, O, and S would clarify the interrelationships between biological and purely physico-chemical processes in that system.

15.3 Coastal lagoons

Some coastal lagoons which maintain a tidal connection to the sea are characterized by stagnation and high salinities in the farthest inland reaches.

Putrid Sea The Sivash, or Putrid Sea, a large (2700 km^2) arm of the Sea of Azov, is one such system (Zenkevitch, 1963). Salinities up to 12.4–16.6% have been measured in the southern part where the shallow waters (up to 3.2 m) undergo extreme temperature fluctuations between summer (30°–35°C) and winter (−1° to −3°C). The temperature fluctuations lead to mass mortality or migration of animals out of the more hypersaline waters. The hypersaline Sivash is characterized by low O_2, organic-rich muds, nutrient accumulation in the winter, and an impoverished fauna (*Artemia* and *Chironomus* only).

Gulf of Mexico Lagoons similar to that of the Putrid Sea are found along the Texas and Mexican coasts of the Gulf of Mexico (Copeland and Jones, 1965; Copeland, 1967). In Laguna Tamaulipas, Copeland (1967) showed a strong negative correlation between salinity and number of fish species. In 5.1% salinity, 29 species of fish were recorded. In 9.5% salinity, eight species of fish were found. In 11–12% salinity, only two species were seen (*Cyprinodon variegatus* and *Menidia beryllina*). No fish were observed in ≥14% salinity. A variety of environmental characteristics of lagoons and salterns were also compared in that study, including pH, alkalinity, and oxygen solubility.

Copeland and Jones (1965) measured community metabolism in hypersaline waters of Laguna Madre and in the Salina Fortuna solar saltern (see Chapter 13). Measurements of photosynthesis and respiration were done under different *in situ* temperatures, which makes it difficult to make correlations according to salinity. Photosynthetic quotients were mostly below 1.0. In 12.1% salinity at La Capia, they were extremely low (0.01), probably due to pollution.

Based on brine volume, the moderately hypersaline brines of Laguna Madre generally had high rates of photosynthesis. These rates were mostly comparable to rates similarly measured in the Salina Fortuna salterns (see Chapter 13). The highest rate of photosynthesis in Laguna Madre (12.35 g $O_2 \cdot m^{-3} \cdot d^{-1}$ at Carvajal station 3, 11.5% salinity) was nearly as high as the maximum rate measured in the hypolimnion of Solar Lake (see Chapter 14). Some of the high productivity rates in the Laguna Madre brines may be attributable to pollution.

Laguna Ojo de Liebre The Laguna Ojo de Liebre complex on the Pacific coast of Baja California, Mexico, consists of three lagoons. Laguna Ojo de Liebre is the largest (about 50 km long). The intertidal flats and supratidal sabkha that extend from Laguna Ojo de Liebre are associated with algal mats and carbonate, gypsum, halite, and diagenetic evaporite deposits. The high planktonic productivity in the lagoon itself (47.2 mg $C \cdot m^{-3} \cdot d^{-1}$) (Phleger and Ewing, 1962), along with the fairly heavy growth of the sea grass *Zostera* in the lagoon and microbial mats in the intertidal zone, are the source of much of the organic matter in the sabkha.

Both tidal and wind action cause the migration and accumulation of organic debris in the sabkha (Phleger and Ewing, 1962; Phleger, 1969, 1971). Phleger (1969) noted that fine particles of clay and silt can be deposited by wind action over the extensive, flat sabkhas of the region. Such fine sediments are usually associated with deep water, low energy systems or storm influxes. At Laguna Ojo de Liebre, the environmental setting of a flat, coastal plain situated between the cool Pacific Ocean on the northwest and the Vizcaino desert on the southeast creates a weather pattern of frequent, and often very strong, northwesterly winds blowing across the lagoon to and across the supratidal sabkha.

Biological productivity in the intertidal zone of Laguna Ojo de Liebre has hardly been studied. Cyanobacterial mats composed primarily of *Lyngbya* and *Calothrix* up to 1-cm thick cover broad expanses of the intertidal zone (Javor and Castenholz, 1981). Bacteriochlorophyll *a* was detected in those mats, but little bacterial photosynthesis and no anoxygenic photosynthesis by the cyanobacteria were detected (Javor and Castenholz, 1981, 1984).

It is likely that productivity rates of the Laguna Ojo de Liebre mats are similar to those of Laguna Guerrero Negro (in the same lagoon complex) (Javor and Castenholz, 1984), Solar Lake, Spencer Gulf (South Australia), and Shark Bay (Western Australia) (Bauld, 1984), since these environments share many of the same chemical and climatological attributes (Table 15.1). Javor and Castenholz (1984) found three mat types in Laguna Guerrero Negro with the following maximum photosynthetic rates (in mg C·µg chl a^{-1}·h^{-1}): *Microcoleus* mat, 0.282; *Lyngbya* mat, 0.122; and *Calothrix* mat, 0.100. All the mats had salinity optima slightly more saline than normal seawater. *Microcoleus* mats in Spencer Gulf had a somewhat higher salinity optimum (see below). Table 15.1 shows that different kinds of mats from different hypersaline lagoons or sabkhas demonstrate highly variable rates of oxygenic photosynthesis. The variability is probably related to a host of environmental factors (i.e., salinity, light and temperature) and species composition. Because of this variability, it is difficult to estimate annual productivity rates of microbial mat systems in these types of hypersaline environments.

Spencer Gulf and Shark Bay, Australia The open waters of Spencer Bay and Shark Bay are characterized by moderate hypersalinity: 3.6–4.8% in Spencer Gulf and 5.5–7.0% in Shark Bay (Bauld, 1984). Microbial mats colonize intertidal sediments in Spencer Gulf and both intertidal and subtidal sediments in Shark Bay. The intertidal microbial mats are subjected to widely fluctuating salinities and complete desiccation. Experiments to demonstrate photosynthetic potential of microbial mats of Spencer Gulf were carried out in artificial seawater of different salinities to which equal concentrations of $NaHCO_3$ and $Na_2{}^{14}CO_3$ were added (to avoid the problem of measuring HCO_3^- and CO_3^{2-} in natural brines). Maximum photosynthetic potential of mats dominated by the cyanobacterium *Microcoleus chthonoplastes* occurred in 7.0–10.5% salinity. Rates were high in 14.0% salinity and were still detectable in 17.5% salinity seawater.

Sulfate reduction in intertidal sediments of Spencer Gulf and Shark Bay was determined by Skyring (1984). *Microcoleus* mats plus sediments typically had the highest rates. In Spencer Gulf mats, reduction rates of 2–104 mmol SO_4^{2-}·m^{-2}·d^{-1} were measured with an average rate of 22 mmol SO_4^{2-}·m^{-2}·d^{-1}. In Shark Bay *Microcoleus* mats plus sediments, reduction rates of 5–10 mmol SO_4^{2-}·m^{-2}·d^{-1}. Spencer Gulf *Lyngbya* mat sediments showed higher sulfate reduction rates in winter than in summer (21 vs. 2 mmol SO_4^{-2}·m^{-2}·d^{-1}). *Entophysalis* mats in Shark Bay had negligible rates. In Shark Bay mat-associated sediments (mat type not given), most of the sulfate reduction measured in the top 20 cm of sediments occured in the top cm (1880 nmol·g^{-1}·d^{-1}). These rates are only one-third of those measured in Solar Lake mats (Jørgensen and Cohen, 1977). In the Shark Bay sediments, total organic carbon (0.90–1.70%) and organic nitrogen (0.10–0.72%) showed no trends with depth. Although the sediments are relatively organic-rich, much of that organic matter is apparently recalcitrant to anaerobic microbial degradation.

Stable isotopes The stable isotopes of S, C, and O of the Laguna Ojo de Liebre evaporites have been described (Holser and Kaplan, 1966; Holser et al., 1981; Pierre, 1982). Pierre (1982) also gave detailed ion analyses and the distribution of δD. The $\delta^{34}S$ fractionation of various sulfate sources are given in Table 15.2. The average $\delta^{34}S$ fractionation for all sulfate samples was +20.4±1.2‰ (Holser and Kaplan, 1966). Some pools showed substantial $\delta^{34}S$ depletion from the average, including some interstitial brines and water from the head of the lagoon (+18.6 to +18.8‰). Gypsum associated with black halite was slightly heavy ($\delta^{34}S$ = +21.0 to +21.3‰) but black gypsum associated with *Zostera* did not reflect isotope fractionation due to sulfate reduction (+20.4‰). A more detailed examination of the types and extent of organic degradation would probably show why local differences in sulfur isotope fractionation occur.

The $\delta^{13}C$ values of the carbonates show some carbonates are chemical precipitates under the control of ocean-atmosphere equilibrium ($\delta^{13}C$ = +2‰), while some carbonates have a relatively large input of biogenic CO_2 ($\delta^{13}C$ = −3.7 to −5.3‰) (Holser et al., 1981). The bulk carbonate has $\delta^{18}O$ values of +0.9 to +2.6‰. The higher values are associated with dolomite, suggesting that the crystallization process is associated with ^{18}O enrichment. Such fractionation apparently is not unusual. At one site, porewater H_2O was isotopically light ($\delta^{18}O$ = −1.2 to +0.9‰) and porewater SO_4^{2-} was heavy ($\delta^{18}O$ = +9.5 to +11.3‰). The authors believed these variations were due to the influence of meteoric waters in the supratidal regions of the sabkha. The same site yielded isotopically light sulfur in the sulfate pool ($\delta^{34}S$ = +17.3 to +19.0‰). Oxidizing conditions precluded the activity of sulfate-reducing bacteria. Pore fluid migration of a mixture of marine-derived brines and dissolution of gypsum by continental waters in addition to diagenetic, subsurface gypsum precipitation, make the interpretation of stable isotopic signatures in Ojo de Liebre and other sabkhas, such as the Abu Dhabi tidal flats in the Persian Gulf, a very difficult and complex problem (Butler et al., 1973).

Table 15.2 Distribution of $\delta^{34}S$ in sulfates of the Laguna Ojo de Liebre sabkha[a]

Sample description	$\delta^{34}S$, ‰
Water from the head of the lagoon	+18.6
Brown mud with diagenetic(?) gypsum, 1 m depth below gypsum muds and algal mats, on a tidal flat	+20.8–21.0
Brine in top 2 cm of white to red salt, at the edge of salt flats, 189 g $Cl^{-} \cdot l^{-1}$	+18.8
Gypsum sands, alternating light to dark greenish grey, underlying salt:	
2.5 cm depth, black halite with red organic specks	+21.0–21.3
5–6 cm depth, light halite sand	+20.8
22–24 cm depth, black gypsum with sea grass, no S^{2-}	+20.4
24–26 cm depth, light greyish green gypsum with sea grass	+19.6
Gypsum concentrated in 3 mm of a salt bed:	+21.0
Under the salt	+22.1
Brines	+18.7
Mean for all samples	+20.4±1.2

[a] From Holser and Kaplan (1966).

References

Aharon, P., Kolodny, Y., and **Sass, E.** 1977. Recent hot brine dolomitization in the "Solar Lake," Gulf of Elat, isotopic, chemical, and mineralogical study. *Journal of Geology* 85: 27–48.

Bauld, J. 1984. Microbial mats in marginal marine environments: Shark Bay, Western Australia, and Spencer Gulf, South Australia, pp. 39–58 in Cohen, Y., Castenholz, R.W. and Halvorson, H.O. (editors), *Microbial Mats: Stromatolites,* MBL Lectures in Biology, vol. 3, Alan R. Liss, New York.

Boon, J.J., Hines, H., Burlingame, A.L., Klok, J., Rijpstra, W.I., de Leeuw, J.W., Edmunds, K.E., and **Eglinton, G.** 1983. Organic geochemical studies of Solar Lake laminated cyanobacterial mats, pp. 207–227 in Bjørnoy, M. et al. (editors), *Advances in Organic Geochemistry 1981,* John Wiley and Sons, New York.

Boon, J.J., de Leeuw, J.W., and **Krumbein, W.E.** 1985. Biogeochemistry of Gavish Sabkha sediments. II. Pyrolysis mass spectrometry of the laminated microbial mat in the permanently water-covered zone before and after the desert sheetflood of 1979, pp. 368–380 in Friedman, G.M. and Krumbein, W.E. (editors), *Hypersaline Ecosystems. The Gavish Sabkha,* Ecological Studies 53, Springer-Verlag, New York.

Butler, G.P., Krouse, R.H., and **Mitchell, R.** 1973. Sulphur-isotope geochemistry of an arid, supratidal evaporite environment, Trucial Coast, pp. 453–462 in Purser, B.H. (editor), *The Persian Gulf. Holocene Carbonate Sedimentation and Diagenesis in a Shallow Epicontinental Sea,* Springer-Verlag, New York.

Copeland, B.J. 1967. Environmental characteristics of hypersaline lagoons. *Texas University Contributions to Marine Science* 12: 207–218.

Copeland, B.J. and **Jones, R.S.** 1965. Community metabolism in some hypersaline waters. *Texas Journal of Science* 17: 188–205.

de Leeuw, J.W., Sinninghe Damsté, J.S., Klok, J., Schenck, P.A., and **Boon, J.J.** 1985. Biogeochemistry of Gavish Sabkha sediments. I. Studies on neutral reducing sugars and lipid moieties by gas chromatography-mass spectrometry, pp. 350–367 in Friedman, G.M. and Krumbein, W.E. (editors) *Hypersaline Ecosystems. The Gavish Sabkha,* Ecological Studies 53, Springer-Verlag, New York.

Edmunds, K.L.H. and **Eglinton, G.** 1984. Microbial lipids and carotenoids and their early diagenesis in the Solar Lake laminated microbial mat sequence, pp. 343–389 in Cohen, Y., Castenholz, R.W. and Halvorson, H.O. (editors), *Microbial Mats: Stromatolites,* MBL Lectures in Biology, vol. 3, Alan R. Liss, New York.

Erlich, A. and **Dor, I.** 1985. Photosynthetic microorganisms of the Gavish sabkha, pp. 296–321 in Friedman, G.M. and Krumbein, W.E. (editors), *Hypersaline Ecosystems. The Gavish Sabkha,* Ecological Studies 53, Springer-Verlag, New York.

Friedman, G.M. 1985. Gulf of Elat (Aqaba). Geological and sedimentological framework, pp. 39–71 in Friedman, G.M. and Krumbein, W.E. (editors), *Hypersaline Ecosystems. The Gavish Sabkha,* Ecological Studies 53, Springer-Verlag, New York.

Friedman, G.M. and **Krumbein, W.E.** (editors). 1985. *Hypersaline Ecosystems. The Gavish Sabkha.* Ecological Studies 53. Springer-Verlag. New York. 484 pp.

Friedman, G.M., Sneh, A., and **Owen, R.W.** 1985. The Ras Muhammad Pool: implications for the Gavish Sabkha, pp. 218–237 in Friedman, G.M. and Krumbein, W.E. (editors), *Hypersaline Ecosystems. The Gavish Sabkha,* Ecological Studies 53, Springer-Verlag, New York.

Gaudette, H.E. and **Lyons, W.B.** 1984. Trace metal concentrations in modern marine sabkha sediments, pp. 425–434 in Cohen, Y., Castenholz, R.W. and Halvorson, H.O. (editors), *Microbial Mats: Stromatolites,* MBL Lectures in Biology, vol. 3, Alan R. Liss, New York.

Gavish, E., Krumbein, W.E., and **Halevy, J.** 1985. Geomorphology, mineralogy and groundwater geochemistry as factors of the hydrodynamic system of the Gavish system, pp. 186–217 in Friedman, G.M. and Krumbein, W.E. (editors), *Hypersaline Ecosystems. The Gavish Sabkha,* Ecological Studies 53, Springer-Verlag, New York.

Gerdes, G. and **Krumbein, W.E.** 1984. Animal communities in Recent potential stromatolites of hypersaline origin, pp. 59–83 in Cohen, Y., Castenholz, R.W. and Halvorson, H.O. (editors), *Microbial Mats: Stromatolites,* MBL Lectures in Biology, vol. 3, Alan R. Liss, New York.

Gerdes, G., Krumbein, W.E., and **Holtkamp, E.** 1985a. Salinity and water activity related zonation of microbial communities and potential stromatolites of the Gavish Sabkha, pp. 238–266 in Friedman, G.M. and Krumbein, W.E. (editors), *Hypersaline Ecosystems. The Gavish Sabkha,* Ecological Studies 53, Springer-Verlag, New York.

Gerdes, G., Spira, J., and **Dimentan, C.** 1985b. The fauna of the Gavish Sabkha and the Solar Lake--a comparative study, pp. 322–345 in Friedman, G.M. and Krumbein, W.E. (editors), *Hypersaline Ecosystems. The Gavish Sabkha,* Ecological Studies 53, Springer-Verlag, New York.

Golubic, S. 1980. Halophily and halotolerance in cyanophytes. *Origins of Life* 10: 169–183.

Holser, W.T. and **Kaplan, I.R.** 1966. Isotope geochemistry of sedimentary sulfates. *Chemical Geology* 1: 93–135.

Holser, W.T., Javor, B., and **Pierre, C.** 1981. *Geochemistry and ecology of salt pans at Guerrero Negro, Baja California.* Geological Society of America Cordilleran Section, 1981 Annual Meeting, Field Trip Number 1. 56 pp.

Javor, B.J. 1979. *Ecology, physiology and carbonate chemistry of blue-green algal mats, Laguna Guerrero Negro, Mexico.* Ph.D. Thesis, University of Oregon, Eugene. 260 pp.

Javor, B.J. and **Castenholz, R.W.** 1981. Laminated microbial mats, Laguna Guerrero Negro, Mexico. *Geomicrobiology Journal* 2: 237–273.

Javor, B.J. and **Castenholz, R.W.** 1984. Productivity studies of microbial mats, Laguna Guerrero Negro, Mexico, pp. 149–170 in Cohen, Y., Castenholz, R.W. and Halvorson, H.O. (editors), *Microbial Mats: Stromatolites,* MBL Lectures in Biology, vol. 3, Alan R. Liss, New York.

Jørgensen, B.B. and **Cohen, Y.** 1977. Solar Lake (Sinai). 5. The sulfur cycle of the benthic cyanobacterial mats. *Limnology and Oceanography* 22: 657–666.

Kessel, M., Cohen, Y., and **Walsby, A.E.** 1985. Structure and physiology of square-shaped and other halophilic bacteria from the Gavish Sabkha, pp. 267–287 in Friedman, G.M. and Krumbein, W.E. (editors), *Hypersaline Ecosystems. The Gavish Sabkha,* Ecological Studies 53, Springer-Verlag, New York.

Krumbein, W.E. 1985. Introduction and definitions, pp. 13–17 in Friedman, G.M. and Krumbein, W.E. (editors), *Hypersaline Ecosystems. The Gavish Sabkha,* Ecological Studies 53, Springer-Verlag, New York.

Kushnir, J. 1981. Formation and early diagenesis of varved evaporite sediments in a coastal hypersaline pool. *Journal of Sedimentary Petrology* 51: 1193–1204.

Long, D.T., Lyons, W.B., and **Gaudette, H.E.** 1985. Trace-metal concentrations in modern marine sabkhas. *Chemical Geology* 53: 185–189.

Lyons, W.B. and **Gaudette, H.E.** 1985. Trace metal concentrations in sediments from the Gavish Sabkha, pp. 346–349 in Friedman, G.M. and Krumbein, W.E. (editors), *Hypersaline Ecosystems. The Gavish Sabkha.* Ecological Studies 53, Springer-Verlag, New York.

Phleger, F.B. 1969. A modern evaporite deposit in Mexico. *Bulletin of the American Association of Petroleum Geologists* 53: 824–829.

Phleger, F.B. 1971. Reply. Ecology of a marine evaporite. *Bulletin of the American Association of Petroleum Geologists* 55: 2040–2042.

Phleger, F.B. and **Ewing, G.C.** 1962. Sedimentology and oceanography of coastal lagoons in Baja California, Mexico. *Geological Society of America Bulletin* 73: 145–182.

Pierre, C. 1982. Teneurs en Isotopes Stables (^{18}O, ^{2}H, ^{13}C, ^{34}S) et Conditions de Genèse des Évaporites Marines: Application à Quelques Milieux Actuels et au Messinien de la Méditerranée. Ph.D. Thesis. Université de Paris-Sud, Centre d'Orsay. 266 pp.

Purser, B.H. (editor). 1973. *The Persian Gulf. Holocene Carbonate Sedimentation and Diagenesis in a Shallow Epicontinental Sea.* Springer-Verlag. New York. 471 pp.

Purser, B.H. Coastal evaporite systems, pp. 72–102 in Friedman, G.M. and Krumbein, W.E. (editors), *Hypersaline Ecosystems. The Gavish Sabkha,* Ecological Studies 53, Springer-Verlag, New York.

Schidlowski, M., Matzigkeit, U., Mook, W.G., and **Krumbein, W.E.** 1985. Carbon isotope geochemistry and ^{14}C ages of microbial mats from Gavish Sabkha and the Solar Lake, pp. 381–401 in Friedman, G.M. and Krumbein, W.E. (editors), *Hypersaline Ecosystems. The Gavish Sabkha,* Ecological Studies 53, Springer-Verlag, New York.

Skyring, G.W. 1984. Sulfate reduction in marine sediments associated with cyanobacterial mats in Australia, pp. 265–275 in Cohen, Y., Castenholz, R.W. and Halvorson, H.O. (editors), *Microbial Mats: Stromatolites,* MBL Lectures in Biology, vol. 3, Alan R. Liss, New York.

Stoeckenius, W., Bivin, D., and **McGinnis, K.** 1985. Photoactive pigments in halobacteria from Gavish Sabkha, pp. 288–295 in Friedman, G.M. and Krumbein, W.E. (editors), *Hypersaline Ecosystems. The Gavish Sabkha,* Ecological Studies 53, Springer-Verlag, New York.

Zenkevitch, L. 1963. *Biology of the Seas of the U.S.S.R.* George Allen and Unwin. London. 955 pp.

16

Antarctic Lakes

16.1 Introduction

Several hypersaline Antarctic lakes have been investigated to determine their chemical constituents, the origin of their salts, and the composition and activity of their biological communities. These lakes are considered oases in desert valleys devoid of higher plants. The challenge of discovery of biological and chemical activities in such extreme environments, characterized by hypersalinity and sometimes sub-zero temperatures, is matched by the challenge of conducting research in such remote areas where logistics problems (i.e., coring through 4 m of ice, frequently inclement weather, and limited numbers of personnel) are compounded by the rather short field season. As a result, most of the relatively large body of literature on these lakes concerns chemical analyses or descriptive biology of samples collected and subsequently analyzed in the laboratory. However, several studies of *in situ* biological activities have also been made.

This survey examines some of the biogeochemical parameters of several hypersaline lakes in the Dry Valleys of Southern Victoria Land (Lake Vanda, Lake Bonney, and Don Juan Pond), the Vestfold Hills (Deep Lake), and the Syowa Oasis (Lake Hunazoko and Lake Suribati) (see map, Figure 1.1). It largely focuses on the processes found only in extremely hypersaline milieux which, in some cases, are found only near the bottom of the lakes. The apparent lack of certain common microbiological processes, or at least the inability to detect them under the restrictive conditions of the Antarctic, is compared and discussed in relation to these processes in other Antarctic environments. For more detailed discussions of general and specific aspects of the saline lakes of Antarctica, several reviews are available (Wilson, 1970,

1979; Torii et al., 1975; Priddle and Heywood, 1980; Burton, 1981; Torii and Yamagata, 1981; Wright and Burton, 1981; Heywood, 1984; Matsumoto et al., 1984a; and Tomiyama and Kitano, 1985).

The climate of the Dry Valleys is marked by low precipitation and high evaporation rates. The snow that does fall is usually lost by sublimation due to winds and low relative humidity. Inflow into the Dry Valley lakes is fed by groundwater, intermittent streams, or glacial meltwaters. The best studied hypersaline Antarctic lakes are Lake Vanda and Lake Bonney. These meromictic lakes were first described by Wilson and Wellman (1962) and Angino and Armitage (1963). Both lakes have a permanent ice-cover about 4 m thick, and are chemically stratified with a dense saline layer near the lake bottoms. Like other hypersaline Antarctic lakes, they are devoid of multicellular organisms, except in the very shallow, dilute moats around the lakes (Parker and Simmons, 1985).

16.2 Lake Vanda

The major limnological characteristics of Lake Vanda are shown in Figure 16.1 and Table 1.1. Located in the Wright Valley, it is a deep lake (68.8 m) and rather large (8.5 km long and up to 2.4 km wide). It is characterized by inverse temperature stratification with temperatures up to 23.5°C (Vincent et al., 1981) or 25°C (Wilson and Wellman, 1962) at the bottom as a result of solar heating (Wilson et al., 1974). The pH of the surface waters (3.9–48.4 m depth) ranged from 6.72 to 7.99, while the very hypersaline, deep waters (58.9 to 64.6 m depth) were lower (5.45–5.80) (Torii et al., 1975). The major ions are similar to those of seawater.

Chemistry The origins of the salts in Lake Vanda have been assessed from studies on stable isotope ratios, chemical composition, and hydrological evolution. One school of thought interprets the data as showing that the salts are a remnant of an evaporated seawater fjord covered by fresh surface water (Nakai et al., 1975a, 1975b; Nakai and Mizutani, 1977; Tomiyama and Kitano, 1985). These interpretations are based on values of $\delta^{13}C$, $\delta^{18}O$, and $\delta^{34}S$ of the water, salts and sediments. Morikawa et al. (1975) also hypothesized a seawater origin from studies on Mg^{2+}:K^{+} ratios of the water. Another school of thought believes that the salts of the lake bottom are of non-marine origin, derived mainly from the chemical weathering of bedrock (Jones and Faure, 1967; Matsubaya et al., 1979). These interpretations are based on strontium isotope values and chemical composition (high Ca^{2+} relative to Mg^{2+} and Na^{+}).

The monimolimnion of Lake Vanda increases in temperature below 40 m and remains saturated or supersaturated with O_2 down to 55 m (Torii et al., 1975). Below 55 m, dissolved salts increase from 4.7% to 11.4% at the

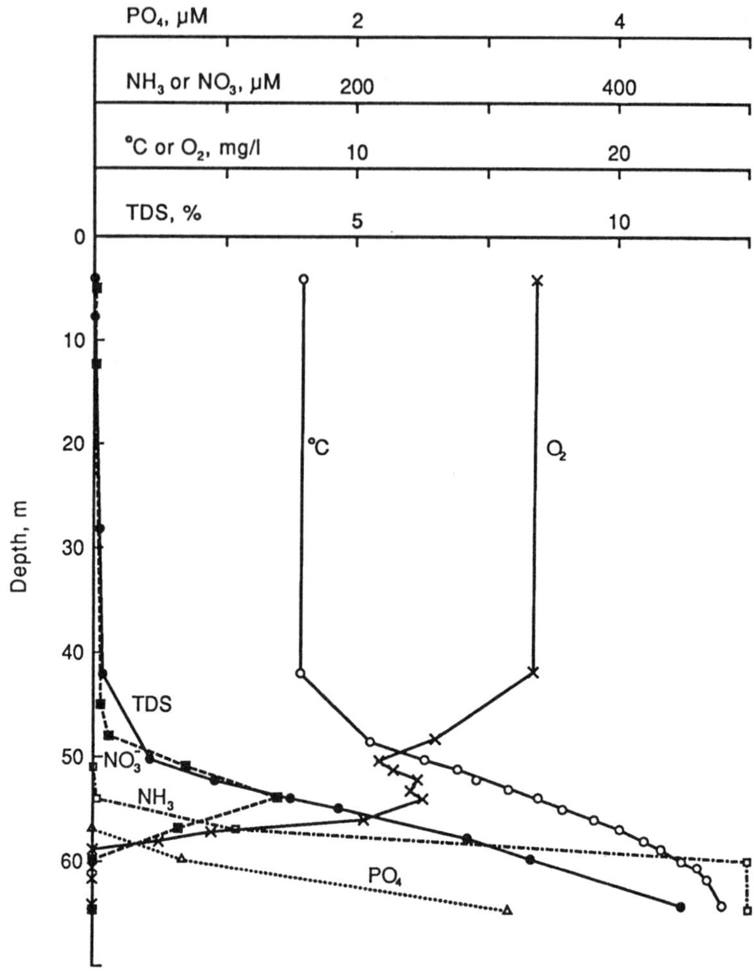

Figure 16.1 Some limnological features of Lake Vanda. TDS (closed circles), temperature (open circles), O_2 (crosses), ammonia (open squares), nitrate (closed squares), reactive phosphate (triangles). The data were taken from Torii et al. (1975) and Canfield and Green (1985).

bottom while dissolved O_2 is zero at 59 m and below. The bottom waters of Lake Vanda contain 1.18 mM sulfide (Torii et al., 1975).

The lakes and ponds of the Wright Valley tend to have high nitrate concentrations, believed to be derived from the weathering of the surrounding evaporites and soils (Torii et al., 1975). Because different investigators have used different methods to measure nutrients, and salt interference with the chemical assays may lead to erroneous results, the concentrations of nutrients in Lake Vanda reported in several studies differ by more than ten-fold in some cases (Matsumoto et al., 1982). The values reported by Canfield and Green (1985) demonstrate the trends in nutrient concentrations with respect to lake depth. PO_4^{3-}-P was less than 0.04 μM in the 5–57 m depth range, and it increased at 60 and 65 m to 0.66 μM and 3.20 μM, respectively. Total phosphate values were 2–4 times those of reactive phosphate. Ammonia showed a similar but more dramatic increase with depth, from 4.14 μM at 54 m to 507 μM in the 60–65 m depth range. Nitrate and nitrite values were low at the lake surface, but they increased to maxima of 143 μM and 0.76 μM, respectively, at 54 m (just above the chemocline). In the 60–65 m depth range, no nitrate or nitrite was detected. The nitrate and nitrite peaks were ascribed to the activities of a layer of nitrifying bacteria and were accompanied by a peak in nitrous oxide concentration (Vincent et al., 1981).

Wada et al. (1984) found high nitrate concentrations (179–209 μM) at 54–56 m depth (15.9–24.6 g $Cl^-\cdot kg^{-1}$, or about seawater salinity). This nitrate was significantly rich in $\delta^{15}N$ (+10.3 to +13.4‰) in comparison to sedimentary organic nitrogen (−4.6‰) probably as a result of isotope fractionation due to nitrate assimilation by algae. This would suggest that nitrate assimilation by the phytoplankton at this depth occurs under light-limited conditions. No discussion of nitrifying bacteria at these depths was given in that paper.

Microbiology Lake Vanda can be broadly defined as a two-tiered system: a cold, dilute, oxygenated, lighted zone where primary productivity and aerobic decomposition occur, and a temperate, hypersaline, anoxic, dark zone where probably only decomposition processes occur. The transition zone has an active community of bacteria (nitrifiers and possibly sulfur-oxidizing and other bacteria). A variety of microbial communities, including well-developed microbial mats, have been described from the non-hypersaline depths of Lake Vanda (down to 31 m or about 0.1% salinity) (Love et al., 1983; Wharton et al., 1983). From the standpoint of understanding biogeochemical processes in extremely hypersaline environments, only the deepest water (8.4–11.4% salinity) and the associated sediments are of interest. Very few data have been published for these depths. Goldman et al. (1967) reported the plankton at 60 m had 10^6 cells$\cdot l^{-1}$ of coccoid cyanobacteria (probably *Synechocystis*) and 10^5 filaments$\cdot l^{-1}$ of cyanobacteria (*Phormidium*-type), and was devoid of the phytoflagellates and heliozoans observed in shallower and more dilute water. Parker et al. (1982) found the dominant phytoplankter was a cryptophyte, *Chroomonas lacustris*. Chlorophyll *a* concentrations throughout the water column showed little variation (0.1 $mg\cdot m^{-3}$), although values for samples taken between 50 and 65 m (the chemocline) were not shown. Osnitskaya and Chudina (1978) described the isolation of a *Chromatium* (probably *C. vinosum*) from the lake.

Goldman et al. (1967) found that only 1% of the surface light energy reaches waters at 30–40 m depth in the summer. While Parker and Simmons (1985) stated that 18% of the photosynthetically active radiation (PAR) penetrated Lake Vanda's ice-cover, Parker et al. (1982) measured only 5.2% PAR penetration at 5 m, and only 0.84% PAR at 25 m depth. Most measurements of photosynthetic productivity have shown significant rates only in the rather dilute parts of the lake (Goldman, 1964; Parker et al., 1982; Seaburg et al., 1983). Reports of high photosynthetic activity in deep, essentially dark waters of Lake Vanda (Goldman et al., 1967; Vincent et al., 1981) should be reevaluated.

Vincent et al. (1981) discovered several layers of distinct microbiological activity in the plankton. The 52.5–55 m-depth range (about 2.4–4.7% salinity) had a layer of nitrifying bacteria, at 57 m (about 7.0% salinity) they measured a peak in DNA synthesis (from tritiated thymidine uptake measurements) coincident with a peak in photosynthetic activity, while the 59.5–62.5 m-depth range (about 8.0–10.0% salinity) was associated with denitrification activity.

Monimolimnion and sediments The data of Vincent et al. (1981), coupled with the data showing high concentrations of sulfide in bottom waters led Canfield and Green (1985) to conclude that below 60 m depth, the only microbial processes which are likely to be quantitatively significant for the oxidation of organic matter are denitrification and sulfate reduction. The presence of sulfate-reducing bacteria has only been implied from sulfide measurements since *in situ* measurements of activity are lacking. Barghoorn and Nichols (1961) suggested that the black, sulfidic muds in a hypersaline (13.3% salinity) kettlehole pond in the Wright Valley resulted from the activities of sulfate-reducing bacteria. Sulfate-reducing bacterial activity can also be implied from the relative depletion of sulfate with respect to calcium at depth. But, since it is not known whether the salts are marine or non-marine, sulfate depletion may not necessarily be entirely controlled by biological activity.

The $\delta^{34}S$ values for the gypsum in the lake sediments are low (+20.4–22.4‰) compared to sediment porewater sulfate (+39.1–48.8‰) (Nakai et al., 1975a). The $\delta^{18}O$ and $\delta^{34}S$ values of dissolved sulfate in the deepest parts of the lake (65–68 m) are also high (+6.0 to +6.2‰ and +42.1 to +46.0‰, respectively) (Nakai et al., 1975b). The $\delta^{34}S$ values for H_2S at 65 and 68 m are light (+7.7‰ and +10.5‰, respectively), which led the investigators to conclude that these stable isotope fractionations were the result of the activities of sulfate-reducing bacteria. Because the sulfide and the heavy sulfur isotopes of the brines in the monimolimnion are trapped, there is no intuitive way to estimate whether the activities of the sulfate-reducing bacteria are actually detectable, or whether they are immeasurably slow over very long periods of time.

Fermentation processes have not been measured nor even mentioned in reports of Antarctic hypersaline lakes, although they have been observed in more dilute Antarctic environments (Burton and Barker, 1979; Yarrington and Wynn-Williams, 1985). They may be an important link in the degradation process, producing the low molecular weight substrates utilized during anaerobic respiration by denitrifying and sulfate-reducing bacteria. Sugiyama et al. (1967) enriched for and isolated fungi and yeasts from the bottom waters and sediments of Lake Vanda. Benoit et al. (1971) did not observe any yeasts or molds at 57 m, nor could they isolate any bacteria from this depth. They used three-fourths-strength seawater medium under aerobic conditions. Neither anaerobic or hypersaline isolations were attempted. It is possible that both bacteria and yeasts are important agents of fermentations in the moderate temperatures of the bottom of the lake, but *in situ* measurements have yet to be made. The presence of methane in the bottom water of the lake suggests that methanogens may be active as well (Waguri, 1976).

The sediments of the deepest parts of the lake, which emit a strong odor of sulfide, have been described as black to brown sands and silts (Nakai et al., 1975b) and as grey-green, and green sandy muds (Nelson and Wilson, 1972). The latter investigators found that they are rich in finely disseminated organic matter (average of 6% by weight) and variable (up to 10% by weight) concentrations of authigenic calcite. Gypsum laminae are also observed in sediment cores. The activities of microorganisms in these sediments apparently have not been studied.

Within the hypersaline monimolimnion, trace metals show various distributions (Figure 16.2). Aluminum remains fairly constant throughout the water column while Zn, Ni, and Co vary with depth but are rather conservative with respect to salinity. Fe is greatly enriched in the anaerobic brine, but it is apparently oxidized rapidly in the deepest oxic zones of the lake. Cu is also greatly enriched in the deep zones. The values shown in Figure 16.2 are actually low with respect to Cu values in the shallower parts of the lake, since Cu is already enriched at 50 m (74.8 $\mu g \cdot l^{-1}$) relative to the average concentration measured in the 4–45 m zone (24.8 $\mu g \cdot l^{-1}$). Mn shows the greatest enrichment of the measured trace elements in the anaerobic brines. Masuda et al. (1982, 1984) concluded that the trace metals are largely derived from airborne particles. The depth distribution follows expected patterns of solution in a low Eh-pH environment that typically results from intense activity of fermenting and sulfate-reducing bacteria. Lake Vanda thus provides an example of how evaporitic environments may be a source of heavy metal sedimentary deposits.

Organic geochemistry Significant analyses have been performed on the organic geochemical constituents of Antarctic lakes, including Lake Vanda. Beginning with the early observations that the deep brines are brown in color (Wilson and Wellman, 1962; Goldman et al., 1967), subsequent reports by Matsumoto and co-workers describe the distribution of the major organic constituents. Matsumoto and Hanya (1977) measured the increase of total

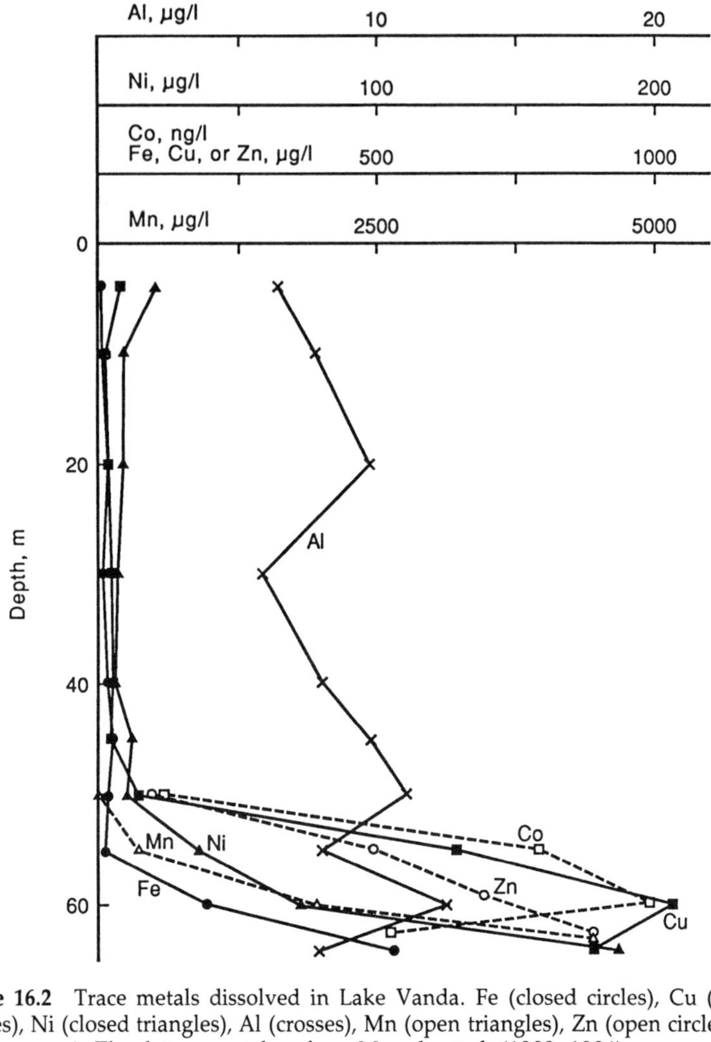

Figure 16.2 Trace metals dissolved in Lake Vanda. Fe (closed circles), Cu (closed squares), Ni (closed triangles), Al (crosses), Mn (open triangles), Zn (open circles), Co (open squares). The data were taken from Masuda et al. (1982, 1984).

organic carbon (TOC) and ethyl acetate-extractable organic carbon (EOC) from above the dense brine (55 m) to the bottom of Lake Vanda (66 m). TOC increases from 1.9 to 63.8 mg·l^{-1} while EOC increases from 0.2 to 5.1 mg·l^{-1}. Surficial waters and influent waters have much lower values. Sediment values are also low, with a TOC content of 1.9 mg·l^{-1} and an EOC content of 0.077 mg·l^{-1} (Matsumoto et al., 1984a). EOC:TOC ratios increase with depth in the 5.4–50.4 m-depth range (up to 0.30), but then decrease to the bottom of the lake (0.06) and in the sediments (0.04). The higher EOC:TOC ratios in the more dilute layers of the lake suggest the presence of fresh organic matter, including living organisms.

Fatty acids constitute the majority of identified compounds except at the bottom and in the sediments where hydrocarbons and sterols comprise the greatest proportion (Matsumoto et al., 1984b). Fatty acids are relatively scarce in the 5.4–50 m-range, and show the greatest concentration (60 $\mu g \cdot l^{-1}$) at 55.4 m. This depth is nearly coincident with the layers of nitrifying bacteria (52.5–55 m) and DNA synthesis/photosynthetic activity (57 m) described by Vincent et al. (1981). At 66 m the fatty acids are characterized by C_8–C_{32} compounds, with acids of even carbon numbers more common than odd. Saturated acids predominate over unsaturated acids, except $C_{16:1}$ and $C_{18:1}$ acids are more common than the saturated C_{16} and C_{18} fatty acids (Matsumoto and Hanya, 1977). At 55 m, the C_{12}–C_{28} fatty acids are predominantly unsaturated.

Hydrocarbons are only found below 60 m in Lake Vanda, with concentrations of 6.5 $\mu g \cdot l^{-1}$ at 65.9 m and 7.4 $\mu g \cdot g^{-1}$ dry sediment in the bottom sediments (Matsumoto et al., 1984b). The $C_{18:0}$ branched compound, 2,6-dimethyl hexadecane, is the major hydrocarbon in the brines (up to 75% of the total hydrocarbons) and an important component (42% of the total hydrocarbons) in the sediments. Its presence has not yet been recorded in living organisms and it may arise from microorganisms of unknown composition, perhaps unique to this environment (Matsumoto et al., 1984a). Sterols increase from the bottom (1.4 $\mu g \cdot l^{-1}$) to the muds (4.1 $\mu g \cdot l^{-1}$). Phytol concentration in the sediments is relatively low (0.39 $\mu g \cdot g^{-1}$ dry sediment (Matsumoto et al., 1984b). The high ratios of C_{29}:C_{27} sterols (1.1–5.3), which are normally indicative of a relatively large contribution by vascular plants, require other interpretations in the desert environment of the Wright Valley (Matsumoto et al., 1984a). The lack of any contribution by higher plants is confirmed by the very low content of phenolic acids (0–0.6 $\mu g \cdot l^{-1}$).

16.3 Lake Bonney

Much of the work on Lake Bonney has been done in comparative studies with Lake Vanda, since both lakes are located in the same region and they are both perennially ice-covered and meromictic. Lake Bonney, which lies in Taylor Valley, consists of two connected hypersaline basins. The chemistry and limnological characteristics of the two basins are quite different (Table 1.1 and Figure 16.3). In the west basin, the dilute mixolimnion (0–12 m) is oxic, whereas the monimolimnion is anoxic from 15 m to the bottom (29.5 m). No H_2S has been detected (Torii et al., 1975). The salinity increases from 8.5 m (4.36%) to the bottom (12.64%). The entire lake is cold, with the highest temperature at 8.5 m (1.9°C) and the lowest temperature at the bottom (−4.6°C). Alkalinity values are about 20 times greater in the west lobe brines than in Lake Vanda brines (Torii et al., 1975). Hoehn et al. (1977) found a

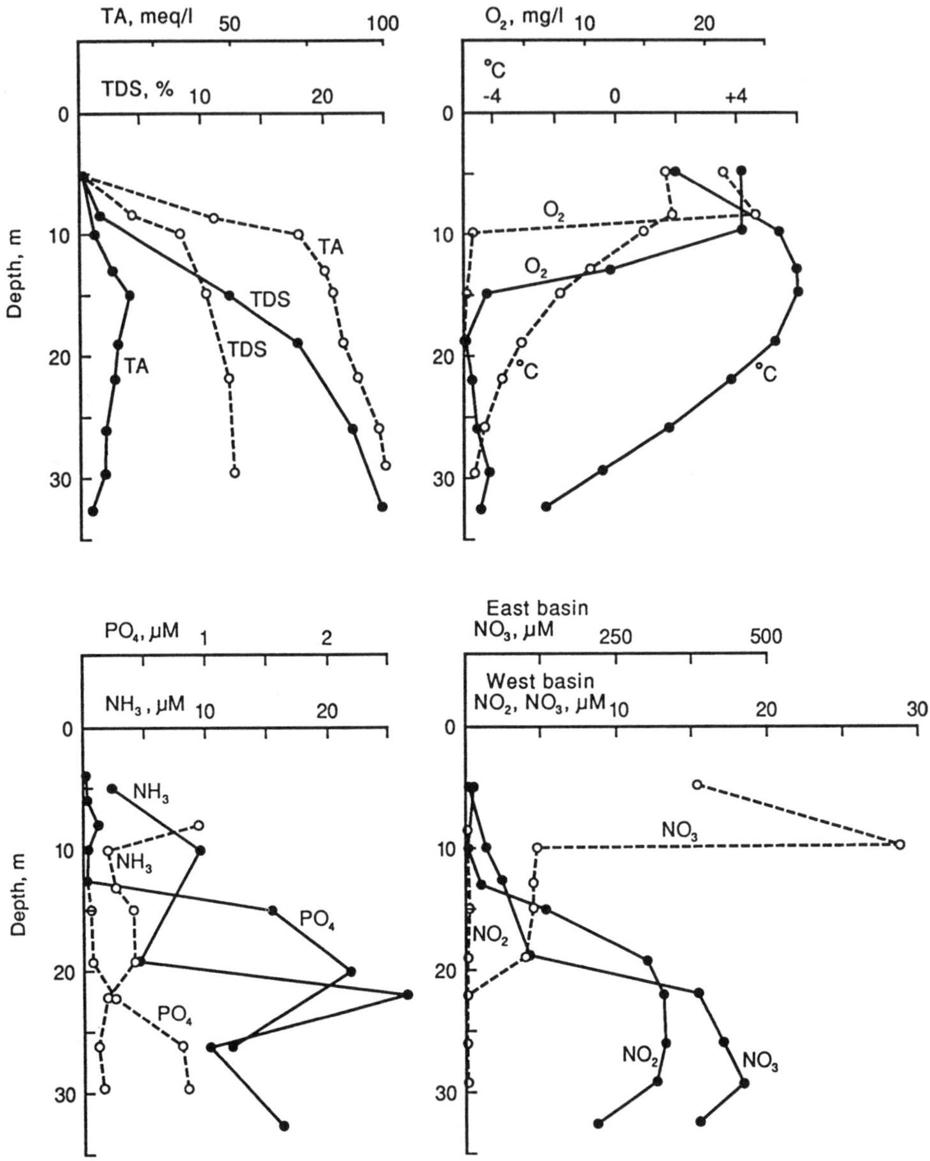

Figure 16.3 Some limnological features of Lake Bonney. Open symbols, west basin; closed symbols, east basin. Total dissolved solids, TDS; total alkalinity, TA; temperature, °C; oxygen, O_2; phosphate, PO_4; ammonia, NH_3; nitrate, NO_3; nitrite, NO_2. All data are from Torii et al. (1975) except phosphate in the east basin, which is from Fortner et al. (1976).

peak in alkalinity at the chemocline of the west basin (12–15 m depth). It is unknown whether the differences in the trends of alkalinity concentrations reflect differences in collection or analytical techniques, or whether significant differences occur intra-seasonally or inter-annually.

The east basin of Lake Bonney remains oxic, although O_2 values are low (1–2 $mg \cdot l^{-1}$) at the bottom (32.5 m). Like the west basin, it remains cold throughout the lake, with a temperature maximum at 15 m (6.0°C). The temperature maxima in both basins is due to solar heating (Hoare et al., 1964). The salinity profile of the east basin is quite different from the west basin. Between 10 and 15 m depth, salinity increases from 1.52 to 12.17%, and at the bottom it is 24.79%. The brines are moderately buffered, with a decrease in total alkalinity in the monimolimnion from 15 m (16.4 $meq \cdot l^{-1}$) to 32.5 m (3.70 $meq \cdot l^{-1}$). The major ions of both basins are Na^+, Mg^{2+}, and Cl^-. The high salinity is believed to have resulted from seawater evaporation or freeze-drying (Matsubaya et al., 1979).

Various nutrient concentrations have been reported, the differences apparently due to different techniques and dates of analysis (Fortner et al., 1976; Matsumoto et al., 1982). In analyses using improved techniques by Fortner et al. (1976), reactive phosphate in the 15–26 m-depth zone of Lake Bonney (probably the east basin, although the authors did not give the location) ranged from 112–205 $\mu g \cdot l^{-1}$, in comparison to concentrations of 0.34–1480 $\mu g \cdot l^{-1}$ they cited from the literature. Their nitrate values in those brines were 0.931–2.19 $mg \cdot l^{-1}$, while values in the literature they cited were 0–0.176 $mg \cdot l^{-1}$. At 26 m, they found 593 $\mu g \cdot l^{-1}$ nitrite, while earlier publications had reported values of 50–1100 $\mu g \cdot l^{-1}$ nitrite at that depth. The values for ammonia were similar to those for nitrate concentrations in the depth range of 12.5–26 m (1.20–6.0 $mg \cdot l^{-1}$) with total nitrogen in the range of 2.1 to 6.4 $mg \cdot l^{-1}$. In contrast, Hoehn et al. (1977) reported total N values up to 13.4 $mg \cdot l^{-1}$ and Weand et al. (1977) found total N values up to 21.6 $mg \cdot l^{-1}$. In both of those studies of the east basin nutrients, ammonia levels were greater than nitrate levels. The varied results of these studies emphasize the inherent difficulties of brine analysis. In the case of nutrients, errors in analyses can lead to incorrect interpretations concerning the potential fertility of a system and the extent and rates of such microbial processes as nitrification and denitrification.

The isotope fractionation of $\delta^{15}N$ in the east basin of Lake Bonney shows a trend for increasingly heavier values with depth between 20 and 30 m (+4.9‰ to +32.2‰) (Wada et al., 1984). The dissolved N_2 in the basin was also isotopically heavy (+1.5 to +2.5‰). The authors interpreted the accumulation of the heavy nitrogen to be a result of fractionation during nitrate reduction, suggesting there may be microsites of anaerobic conditions in an otherwise oxic monimolimnion. Alternatively, the nitrate trapped at the bottom during an earlier period in the basin's history could have been isotopically heavy. In contrast, the west basin of Lake Bonney, which has more than one order of magnitude less nitrate (see Figure 16.3), has relatively lighter nitrogen in the anaerobic monimolimnion. There was a marked decrease in nitrate values from 20 m (13.1 μM) to 30 m depth (1.7 μM), while total $\delta^{15}N$ decreased from +2.5‰ to +1.5‰ between 15 and 30 m depth. These findings were interpreted as evidence of denitrification.

Microbiology The reports of the distribution of microorganisms also show variations between investigators and year of the study. The ice-cover on Lake Bonney severely attenuates light in both basins. While 2.8–3.3% PAR was detected at 5–6 m, only 0.10–0.30% PAR was measured at 29 m (Parker et al., 1982). *Chroomonas lacustris* is probably the most abundant phytoplankter, but there are also significant numbers of green algae (*Chlamydomonas subcaudata, Chlorella vulgaris,* and an unidentified coccus), an unidentified flagellate, and possibly unidentified coccoid cyanobacteria (Goldman et al., 1967; Koob and Leister, 1972; Parker et al., 1982; Seaburg et al., 1983). In these studies, phytoplankton population maxima were found at different depths, nearly always in or above the chemocline. The report of a large accumulation of phytoflagellates at 30 m by Goldman et al. (1967) may be indicative of an inactive, sinking population. Parker et al. (1977) found that an accumulation of algae (1256 cells$\cdot l^{-1}$) at 20 m in the austral spring essentially remained unchanged one month later (1085 cells$\cdot l^{-1}$). Algal populations at different depths between 4 and 15 m all increased by factors of 2 to 100, with a maximum density of 5032 cells$\cdot l^{-1}$ at 4 m depth. These measurements again suggest that the hypersaline monimolimnion of Lake Bonney is a sink for algal cells rather than a zone of active photosynthesis.

Hoehn et al. (1977) detected no bacteria or yeasts below the chemocline at 15 m in the east basin of the lake, while Parker et al. (1977) reported relatively high algal counts and low bacterial counts below the chemocline. In contrast, Goldman et al. (1967) found an accumulation of bacteria at 20 m. Using the plate count method, Benoit et al. (1971) showed the greatest concentration of bacteria was near the ice interface, with total counts changing during the summer season. They used dilute and three-fourths-strength seawater media under aerobic conditions and detected low numbers of bacteria throughout the monimolimnion. In most cases, plate counts were lower for samples incubated at 0°C than at 15°C. In the austral spring (October), mostly yeasts (*Candida* and *Cryptococcus*) were observed in the chemocline at 10–15 m depth. They were absent from other depths and they could not be detected in subsequent months. The authors noted that the yeasts were present in the warmest zone of the lake (5°–8°C), which is about 10°–12°C below their optimum. A fungus (*Dendryphiella salina*) was isolated from the bottom water of Lake Bonney (Waguri, 1976), but it is unknown whether it is active at *in situ* salinities, particularly since the bottom of the lake is anaerobic.

Koob and Leister (1972) measured heterotrophic ^{14}C-acetate assimilation in the east basin of Lake Bonney. They found gross uptake decreased with depth from 5 to 30 m. Since they did not measure acetate concentrations in the water, these uptake values could not be translated into activity rates.

Benthic mats dominated by cyanobacteria were observed down to 14 m in the east basin (about 10% salinity), while diatom-dominated benthic mats were found at 14–16 m depth (about 10.5% salinity) (Wharton et al., 1983). The cyanobacterial mats contained *Lyngbya martensiana* and *Phormidium frigidum* along with lesser numbers of the green alga *C. subcaudata* and diatoms (*Navicula* spp.). The most salt-tolerant diatoms included *Navicula* spp., *Pinnularia cymatopleura,* and *Nitzschia angustata.* Parker and Simmons (1985) noted that phototrophic and heterotrophic bacteria were associated with anaerobic mats containing *Phormidium frigidum.*

Reliable determinations of rates of primary productivity in the hypersaline zones of Lake Bonney are lacking. Measurements using the ^{14}C method at different depths in such a density-stratified lake are complicated by uncertainties in calculating the appropriate carbonate alkalinity correction factor at each depth. Koob and Leister (1972) recognized this problem and presented their calculated data along with the raw counts per minute. In the study by Parker et al. (1977) of the productivity in the deep waters of Lake Bonney, the investigators rejected all data below 20 m depth because they were not convinced that either the ^{14}C method they employed or the conversion tables for total organic C values are applicable at high salinities. Productivity was measured from extracellular fixed ^{14}C in their study. They discussed the reasons why they and others have found high values of dark $^{14}CO_2$ fixation in the deep, hypersaline brines of Lake Bonney and Lake Vanda. They cite four possibilities: 1) the high content of DOC is stimulatory; 2) photorespiration activity is high; 3) injury of cells during the filtration of samples causes the release of ^{14}C products; and 4) high rates of abiotic $^{14}CO_2$ fixation have been found.

Parker et al. (1977) found carbon fixation rates in the mixolimnion were negligible for all dates between November 12 and January 17 except December 26, when 0–1938 mg $C \cdot m^{-3}h^{-1}$ was measured. Carbon fixation in the monimolimnion (12.5–20 m) for that date was 22–658 mg $C \cdot m^{-3}h^{-1}$, while even higher values (2–1677 mg $C \cdot m^{-3}h^{-1}$) were measured between November 12 and December 13. These productivity rates were based on calculations using the dissociation constants of the carbonate system in dilute lake water. Incorrect assumptions would result in estimates either too high or too low for productivity rates. Subsequent studies of quantum efficiencies and quantum yields of photosynthesis documented relatively high values at 9 and 12 m (Parker et al., 1982) and at 14 m depth (Seaburg et al., 1983). Without an empirical basis for calculating the activity and dissociation of bicarbonate and carbonate in solution in the brines of greater than seawater salinity, any estimates of primary productivity in and below the chemocline of Lake Bonney must be treated with caution.

There is no evidence of sulfate-reducing bacterial activity in either basin of Lake Bonney. Although the bottom waters of the west basin have a similar salinity as the bottom waters of Lake Vanda, the two lakes differ in the relative concentrations of Ca^{2+} and Mg^{2+}, and probably more significantly, in temperature. A comparison of the SO_4^{2-}/Ca^{2+} ratios with respect to depth of the three basins demonstrates that sulfate depletion occurs with depth in Lake Vanda, but the ratio remains roughly constant in the west basin of Lake Bonney (even in the anoxic regions of the lake) and actually increases in the oxic east basin. These observations suggest that as a group, sulfate-reducing bacteria may be constrained both by the combination of high salt and very low temperatures. In contrast, Ace Lake, a meromictic Antarctic lake of approximately seawater salinity, has an active sulfur cycle community in its cold (0.7°–3.3°C) monimolimnion (Burton and Barker, 1979; Hand, 1980). Comparative studies of rates of sulfate reduction in other very cold hypersaline lakes or even deep-sea hypersaline basins may elucidate the exact causes of the inhibition of this group of microorganisms in Lake Bonney.

Trace metals Trace metals dissolved in the east basin of Lake Bonney are present at higher concentrations in the monimolimnion than in the more dilute mixolimnion (Weand et al., 1976). The concentrations of Mn and Fe can be correlated with salinity, with Mn as high as 8.76 $mg \cdot l^{-1}$ and Fe as high as 1.48 $mg \cdot l^{-1}$ at 25 m. The concentrations of Cu (0.59 $mg \cdot l^{-1}$) and B (up to 94.2 $mg \cdot l^{-1}$) in the monimolimnion are possibly toxic to microorganisms, and the lesser concentrations of these trace elements in the mixolimnion are possibly inhibitory as well. Ni, Co, and Mo were measured in micromolar concentrations.

Organic geochemistry The two basins of Lake Bonney differ both in TOC and the relative amounts of different organic compounds. Matsumoto and Hanya (1977) found TOC at the bottom of both basins to be greater than values in shallower waters. TOC concentrations at the bottom of the west and east basins were 18.6 and 28.9 $mg \cdot l^{-1}$, respectively. These differences in TOC values are apparently largely due to concentration with respect to salinity. The ratio of TOC ($mg \cdot l^{-1}$)/TDS(%) for the west and east basins are 1.47 and 1.13, respectively, while that of Lake Vanda is significantly higher (about 5.60). The slightly larger ratio for the west basin may reflect relatively greater productivity in that basin since degradation is presumably low in both basins of the lake. The greater ratio in Lake Vanda may reflect overall greater organic productivity or a measurement during a part of the season when heterotrophic bacteria had not yet consumed much of the TOC. Parker et al. (1977) found somewhat higher TOC values (21–28 $mg \cdot l^{-1}$) at depths of $\geq$15 m in the west basin of Lake Bonney early in the austral summer, but by midsummer those values had decreased to 5–9 $mg \cdot l^{-1}$. These changes reflect increased microbial degradation of TOC accumulated during the prior fall, winter, and spring. They also determined that $\geq$90% of the organic matter is dissolved, not particulate, which they interpreted as products of the benthic microbial mats.

The EOC content of both basins of Lake Bonney is similar to that of Lake Vanda, with values ranging from 3.8 to 4.7 $mg \cdot l^{-1}$ (Matsumoto and Hanya, 1977). Fatty acids in the west basin of the lake (up to 7.8 $\mu g \cdot l^{-1}$) are also of comparable concentrations to those of Lake Vanda, while greater concentrations were recorded at depth in the east basin of the lake (up to 14 $\mu g \cdot l^{-1}$). The ratios of fatty acids:salinity are similar in the two basins, suggesting that fatty acids are conserved during brine concentration. The fatty acid profiles of both basins of Lake Bonney and of Lake Vanda are similar in carbon numbers, dominance of even over odd acids, and dominance of saturated acids. Methyl phenylacetate, a fungal metabolite, was found in low concentrations in the east basin of Lake Bonney. Matsumoto (1984a) also found significant quantities of phenolic acids in Lake Bonney, with values as high as 12 $\mu g \cdot l^{-1}$ at the bottom of the west basin. No hydrocarbons were detected in the brines of Lake Bonney, although they were found in the sediments.

16.4 Don Juan Pond

Don Juan Pond is an extremely hypersaline, ice-free body of water in the Wright Valley. It measures about 200 by about 700 m and has an average depth of 11 cm (Meyer et al., 1962). It is essentially a $CaCl_2$ brine. Meyer et al. (1962) reported it had a TDS content of 474 $g \cdot l^{-1}$ (presumably measured as the dried but relatively hydrated residue) while total ion content was 339 $g \cdot l^{-1}$. Harris et al. (1979) noted that there is a seasonal, dramatic variation in Don Juan Pond composition. The phase relations of the hydrated salts are under hydrological and temperature control in the summer, while temperature is the primary determinant of precipitation and dissolution in the winter. The peculiar chemical and physical nature of the Don Juan basin permits the seasonal precipitation of antarctite ($CaCl_2 \cdot 6\ H_2O$) and halite, but not hydrohalite. Individual trace components, such as sulfate and iron, also apparently undergo seasonal changes in concentrations. Sulfate ranges from <0.1 to 11 $mg \cdot l^{-1}$, and iron ranges from 0.8 to 23.7 $mg \cdot l^{-1}$ (Meyer et al., 1962; Cameron et al., 1972; Masuda et al., 1984).

The pH has been measured as high as 5.4 (Meyer et al., 1962) and as low as 3.8 (Cameron et al., 1972). While the mean annual temperature of the brine is $-18°C$ (Harris et al., 1979), temperatures as high as 8°C have been recorded (Cameron et al., 1972). The water activity was calculated to be 0.45 and the osmotic pressure was 75.6 atm.

Meyer et al. (1962) attributed the brown color of Don Juan Pond brine to a suspension of pyrite particles although Tomiyama and Kitano (1985) found no evidence for pyrite in the pond sediments. Variations in dissolved iron are probably due to precipitation, although there is no evidence of pyrite formation nor for any sulfide generation by sulfate-reducing bacteria. Other trace metals in solution in Don Juan Pond brines include Mn ($0.800\ mg \cdot l^{-1}$), Zn ($0.270\ mg \cdot l^{-1}$), and Co ($1.2\ \mu g \cdot l^{-1}$). These values are comparable to those found in the deep brines of Lake Vanda. Masuda et al. (1984) concluded that trace metals in trapped aerosols could be the sole source of Fe and Co, while Zn and Mn show some deviation from this pattern.

The origin of the salts in Don Juan Pond is intimately tied to the question of whether microorganisms can thrive in the brines. According to Tomiyama and Kitano (1985), $CaCl_2$-rich groundwater is an important source of brine. They reported $\delta^{34}S$ of the sulfate in Don Juan basin was +30.5‰. In a personal communication to those authors by Nakai, $\delta^{34}S$-SO_4^{2-} values were +31.8 to +37.5‰, an enrichment considered attributable to sulfate-reducing bacteria. Tomiyama and Kitano rejected the idea of sulfate-reducing bacteria because the pond lacks pyrite in its sediments. They also calculated that the amount of sulfate supplied through the groundwater represented only $\leq 0.5\%$ of the total sulfate in the pond basin. They concluded that the mechanism for $\delta^{34}S$ enrichment in Don Juan pond sulfate remains enigmatic since neither bacterial activity nor groundwater supply seemed probable. The beachline in the Don Juan basin shows that the pond was once 10 m deep (Priddle and Heywood, 1980). Sulfur enriched in the heavy isotope may be a relic of an earlier, more dilute stage of the Don Juan basin.

Microbiology The controversy of whether such dense $CaCl_2$ brines of low water activity (a_w) and low temperatures can support life remains unsettled. Meyer et al. (1962) found viable bacteria in the sediments but no growing algae. By incubating samples for 3 weeks at 20°C (considerably above ambient temperatures of the pond), they isolated several kinds of bacteria and one yeast. No anaerobes developed. They reported that the cultures grew at 0°–25°C in Don Juan pond water-based medium. They concluded that the bacteria were autochthonous in the pond since no airborne bacteria could be isolated in the vicinity.

Later reports by Benoit and Hall (1970), Cameron et al. (1972), and Horowitz et al. (1972) refuted the claims of life in Don Juan Pond. Benoit and Hall (1970) employed dilute media at 2°, 15°, and 20°C but detected no viable bacteria. They did not attempt plate counts with saline medium. Cameron et al. (1972) found significant counts of bacteria and fungi in air samples above the pond. They enriched for microbes at 20°C in dilute medium. Their only isolate was the bacterium *Achromobacter parvulus*, a strain that tolerated only 5% salt. Because no enrichments nor plate counts in very saline media were tried in these studies, the results cannot be regarded as absolute proof of the sterility of the pond brines. Horowitz et al. (1972) stated that the organisms from Don Juan Pond cannot grow in the $CaCl_2$ concentrations found in the pond. Cameron et al. (1972) made radiorespirometric tests of pond water with negative results, but they did not report any measurements in surficial sediments where bacterial activity would most likely be detected.

Siegel et al. (1979) presented evidence of active life in Don Juan Pond waters in their study of thin algal mats that cover a salt flat on one side of the pond. The mat was 2–5 mm thick. It contained cyanobacteria (*Oscillatoria*-like filaments), algae (non-motile *Chlorella*-like cells, motile *Dunaliella*-like cells, and other forms), occasional fungal hyphae, and motile and non-motile bacteria (red and colorless). The mat had areas of active bubble formation near it. Mats moistened with Don Juan brine contained ATP, proteins, polysaccharides, lipids, and demonstrated a variety of enzymatic activities. When the mat was suspended in pond water at 4°C, growth was slow but cells maintained osmotic integrity and pigmentation. Preliminary manometric experiments at 4°C showed that O_2 was produced in the light and CO_2 was released in the dark. Although this algal mat shows activity in Don Juan brine, it apparently formed during periods of higher (and probably more dilute) water in the Don Juan basin. No one has yet offered conclusive proof of *in situ* biological activity in Don Juan Pond brines.

In spite of lack of evidence for active life forms, the sediments of Don Juan pond contain organic matter, albeit very little (Matsumoto et al., 1984a). TOC was 0.18 $mg \cdot l^{-1}$ and total nitrogen was 0.021 $mg \cdot l^{-1}$. These values are about one-tenth the concentrations measured in Lake Vanda sediments. Very low concentrations of EOC, hydrocarbons, fatty acids, stenols, and stanols were also noted.

16.5 Deep Lake

Deep Lake, located in the Vestfold Hills, is a small (0.64 km^2), deep (36 m), ice-free, monomictic lake (Kerry et al., 1977) (Tables 1.1, 1.4, 2.1, and 2.2). It is hypersaline throughout (ca. 28% salinity), and the salinity is believed to be derived from evaporated seawater. Published brine analyses (Kerry et al., 1977; Campbell, 1978) have failed to note whether there is any dissolved sulfate in the lake water. Much or all of the SO_4^{2-} may have precipitated as mirabilite, as this salt and hydrohalite reach saturation in the extremely cold brine (Priddle and Heywood, 1980). Although the lake does not stratify chemically, it does undergo thermal stratification. Surface temperatures of the lake brine in the winter have been recorded as low as −20°C (Kerry et al., 1977), which is above the freezing point of −28°C (Priddle and Heywood, 1980). The surface warms to 10°C in the summer (Campbell, 1978). The lake remains oxic throughout the water column.

Nutrient concentrations were reported to be less than or equal to those of seawater (Kerry et al., 1977), with PO_4^{3-}-P values from a trace to 45.8 $\mu g \cdot l^{-1}$ and NO_3^--N values of 53–87 $\mu g \cdot l^{-1}$. However, Hand (1980) found very high concentrations of total nitrogen (about 7 $mg \cdot l^{-1}$). The pH is about 7.4 and the brine is poorly buffered (3.1–4.6 $meq \cdot l^{-1}$) (Kerry et al., 1977).

Planktonic life in Deep Lake is sparse. Campbell (1978) reported up to 10^5 $cells \cdot l^{-1}$ of two unidentified coccoid green algae. Wright and Burton (1981) have identified at least one of these as *Dunaliella* sp. Kerry et al. (1977) also noted centric and pennate diatoms, and rare occurrences of cyanobacteria and silicoflagellates, although they did not mention whether any of these cells were viable. The lake is devoid of invertebrates, amoebae, and ciliates. No $^{14}CO_2$ incorporation could be detected except a disturbed sediment once yielded an uptake rate of 0.18 mg $C \cdot m^{-3}$ h^{-1} (Campbell, 1978). Bacterial counts were also modest. About 10^5 $cells \cdot ml^{-1}$ were detected at all depths except at the beginning of holomixis (winter) when 10^6 $cells \cdot ml^{-1}$ were detected (Hand, 1980). The increase in bacterial counts was attributed to sediment resuspension since it occurred at the coldest and darkest time of the year.

Although bacteria were detected, there was no evidence of any bacterial activity in Deep Lake. They could not be cultured from the lake when the medium was greater than half the salinity of the lake brine. Hand (1980) interpreted the high concentrations of total nitrogen and TOC (50 $mg \cdot l^{-1}$) as evidence of the lack of bacterial degradation activity, and concluded that these nutrients were probably washed into the lake.

The sediments of Deep Lake were reported to show bands of mirabilite and dark sand, but no mention of reducing conditions or sulfide were made (Kerry et al., 1977). In comparison to the sediments of Lake Vanda and Lake Bonney, Deep Lake sediments have more TOC (4.6 $mg \cdot l^{-1}$), EOC (0.47 $mg \cdot l^{-1}$), and total nitrogen (0.99 $mg \cdot l^{-1}$) (Matsumoto et al., 1984a). Fatty acids, hydrocarbons, and stenols in the sediments are in comparable concentrations as in Lake Bonney sediments.

Deep Lake differs from Lake Bonney and Lake Vanda by its lack of a dilute mixolimnion. This factor, in combination with low nutrients, low temperatures, and limited light, apparently limit primary productivity and microbial heterotrophy.

16.6 Lakes Hunazoko and Suribati

Lakes Hunazoko and Suribati, located in the Syowa Oasis, have similar salinities (21.3 and 22.0%, respectively), high silica content (105–227 μM), pH (7.4–7.8 and 7.0–7.8, respectively), and alkalinities of about 5 mg $CO_2 \cdot l^{-1}$, but they are very different in other limnological respects (Tominaga and Fukui, 1981; Fukui et al., 1985) (Figure 16.4 and Tables 1.1, 1.4, 2.1, and 2.2). Lake Hunazoko is shallow (9.2 m) but it stratifies in the summer both thermally (10°C at the surface, −12°C at the bottom) and chemically (11.2% Cl^- at the surface, 15.8% at the bottom). During holomixis, the whole lake is −18.3°C. The entire lake remains oxic, with recorded oxygen concentrations of 2.17–

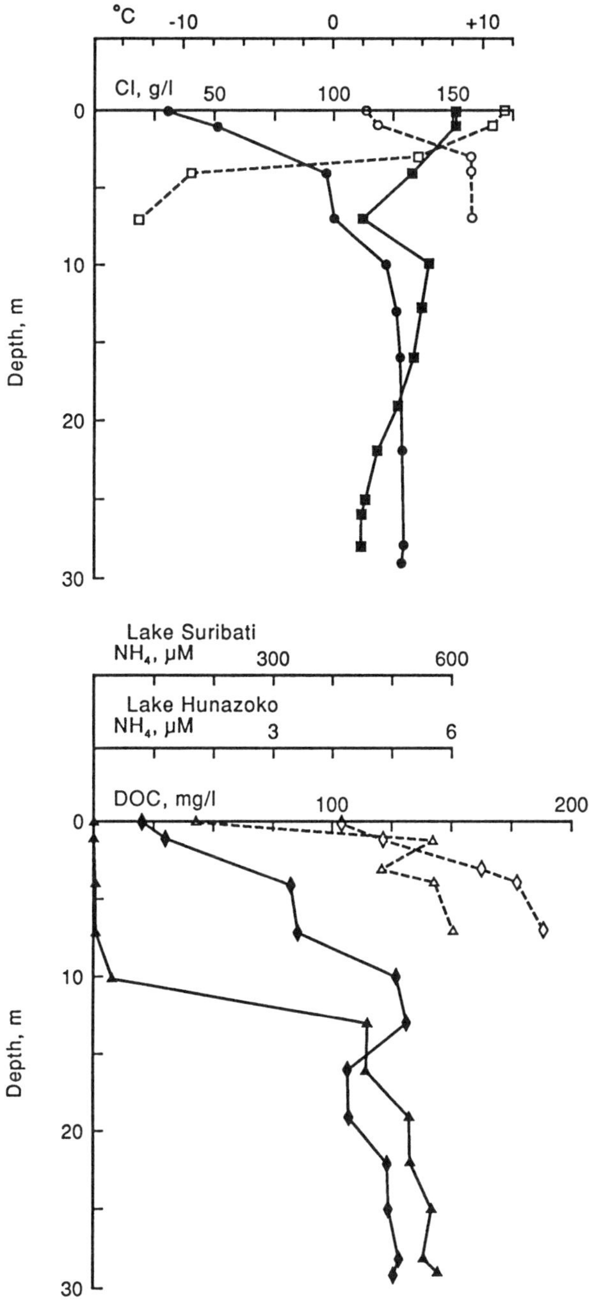

Figure 16.4 Some limnological features of Lakes Hunazoko and Suribati. Open symbols = Hunazoko, closed symbols = Suribati. Chloride (circles), temperature (squares), ammonia (triangles), DOC (diamonds). The data are from Fukui et al. (1985).

2.57 $mg \cdot l^{-1}$ (Fukui et al., 1985) and 2.6–2.8 $mg \cdot l^{-1}$ (Tominaga and Fukui, 1981). Inorganic nutrients were found in low concentrations: PO_4^{-3}-P = 0.25–0.45 μM, NO_2^--N = 0.1–0.4 μM, NO_3^--N = 0–1.9 μM, and NH_4^+-N = 1.7–6.0 μM (Fukui et al., 1985).

Lake Suribati, which is much deeper (31.2 m), also undergoes thermal stratification. While surface temperatures are the same as those for Lake Hunazoko, bottom temperatures go down to only 2°C. The lake is meromictic and remains anoxic below 10 m depth. The oxygen content in the mixolimnion during the summer was 5–6 $mg \cdot l^{-1}$ (Tominaga and Fukui, 1981). Although no nitrate or nitrite has been found at any depth in Lake Suribati, phosphate and ammonia concentrations correlate with the meromictic stratification. PO_4^{-3}-P in the mixolimnion was 0.05–0.40 μM, while in the monimolimnion concentrations were 54.0–66.7 μM. NH_4^+-N in the mixolimnion was 0–1.7 μM, with concentrations in the monimolimnion ranging from 458 to 575 μM (Fukui et al., 1985).

Dunaliella and the diatom *Tropidoneis* dominate the phytoplankton of both lakes (Tominaga and Fukui, 1981). In that study, planktonic density at the bottom of the photic zone in Lake Hunazoko (3 m depth) was rather high (5.5 mg chl $a \cdot m^{-3}$). Chlorophyll *a* concentrations in the mixolimnion of Lake Suribati were uniform (1 mg chl $a \cdot m^{-3}$). No phototrophic bacteria were detected. Experiments with isolates of the algae demonstrated that they were capable of low rates of photosynthesis under temperature and salinity conditions of the lakes, including measurable photosynthesis at temperatures below 0°C.

Both lakes have remarkably high concentrations of TOC. The concentration of TOC strongly correlates with salinity, which has been interpreted to be a result of passive concentration accompanying the concentration of seawater under freezing conditions (Tominaga and Fukui, 1981; Fukui et al., 1985). The high concentration of TOC and the green color of the sediments (due to chlorophyll) both indicate that bacterial degradation is probably rather slow (Tominaga and Fukui, 1981). No mention of sulfide in Lake Hunazoko sediments was made.

The bottom waters of Lake Suribati had 1 mg $S^{2-} \cdot l^{-1}$, indicating that sulfate-reducing bacteria are probably active during certain seasons. The bottom waters of Lake Suribati have significant dissolved SO_4^{2-} (8.1 $g \cdot l^{-1}$) as does the west basin of Lake Bonney (4.0–4.5 $g \cdot l^{-1}$). Even though Lake Suribati is more hypersaline than the west basin of Lake Bonney, sulfide has not been detected in Lake Bonney. The lack of activity of sulfate-reducing bacteria in that lake correlates with lower temperatures (−4.6°C vs. 2°C) and lower nutrient levels which further inhibit salinity-stressed populations.

Organic analyses of the presumably oxic sediments of Lake Hunazoko indicate they have moderately high concentrations of TOC (3.8 $mg \cdot l^{-1}$), EOC (1.1 $mg \cdot l^{-1}$) and, total nitrogen (0.75 $mg \cdot l^{-1}$) (Matsumoto et al., 1984a). Hydrocarbon concentrations are relatively low (0.38 $\mu g \cdot l^{-1}$), but fatty acids are rather abundant (120 $\mu g \cdot l^{-1}$). Those sediments have a higher concentration of phenolic acids than any of the other sediments from hypersaline Antarctic lakes measured (0.76 $\mu g \cdot l^{-1}$). Unfortunately, no organic geochemical analyses are available for Lake Suribati.

16.7 Conclusions

The hypersaline lakes and ponds of Antarctica share several common features: high total dissolved solids, extremely short growing season, and extremely low temperatures during at least part of the year. Apart from these similarities, the lakes and ponds are each unique. The published primary productivity rates are very low except in the more dilute parts of Lake Bonney and Lake Vanda. No primary productivity measurements have been published for Lakes Hunazoko and Suribati, but the chlorophyll *a* levels in those lakes and the measurable rates of photosynthesis by cultures of the phytoplankton indicate that appreciable photosynthetic rates may be possible under *in situ* conditions.

While certain groups of microorganisms can apparently grow in both temperate and Antarctic moderately hypersaline lakes (i.e., cyanobacteria, diatoms, and the green alga *Dunaliella*), other major groups have not been reported from the Antarctic hypersaline habitats: halobacteria, nitrogen-fixing bacteria, fermenting bacteria, and sulfur-oxidizing bacteria. The report of methane and sulfide near the benthos of the monimolimnion of Lake Vanda suggests that methanogens and sulfate-reducing bacteria can thrive at the moderate temperatures of the strong brines in that lake, but it is unknown whether they can survive the colder conditions of the other hypersaline lakes. Methanogenic, sulfate-reducing, phototrophic, and fermentative bacteria have been described from Antarctic environments at near-freezing temperatures in both freshwater environments (Yarrington and Wynn-Williams, 1985; Ellis-Evans, 1985) and saline Ace Lake (Burton and Barker, 1979; Hand, 1980). Ellis-Evans (1985) noted that sulfate-reducing bacteria isolated from Antarctic lakes were not psychrophilic, but had temperature optima of 32°C.

Nitrogen-fixing bacteria, which require anaerobic conditions to fix N_2, may be absent because the anoxic brines of the lakes are not generally nitrogen-limited. However, no extremely halophilic, nitrogen-fixing bacteria have been described. Likewise, no cold-tolerant halobacteria are known. The effects of low temperatures on the state of high salt concentrations inside the cells may prevent any members of this group from successful colonization of Antarctic environments. There is general agreement in the literature that the combination of low temperatures and high salt concentrations are major limiting factors of biological productivity in the hypersaline Antarctic lakes.

The apparent "absence" of certain groups of bacteria may also be due to the relative lack of effort to enrich for and identify them using appropriate media for halophiles. The report of red colonies of bacteria in the algal mats near Don Juan Pond and the identification of phototrophic bacteria in Lake Bonney mats both indicate potentially active phototrophic bacteria. Such microbial associations are common in hypersaline marine algal mats. Isolation of these and associated anaerobes may be successful under appropriate conditions of cultivation.

In addition to light limitations in some cases, algal productivity in hypersaline Antarctic lakes is apparently controlled by the same factors that limit bacteria. Attempts to enumerate viable phytoplankton by plate counts may have resulted in underestimates of populations because inappropriate (dilute) media were employed. Temperature tolerances of 128 strains of 35 algal taxa isolated from a variety of Antarctic lakes were measured by Seaburg et al. (1981), who found that 92% of the phytoplankton strains, and 74% of the benthic mat strains grew at 2°C. Some strains still grew at −1°C. Unfortunately, no media saltier than seawater were used. Measurements of the salt tolerances and the combined temperature-salt responses of these isolates would be invaluable for evaluating the ecology and productivity of hypersaline Antarctic lakes.

It is worthwhile comparing the microbiology of the hypersaline lakes with that of the Dry Valley saline soils, since they share some of the environmental factors that limit productivity. Cameron et al. (1970) listed the colonization order of microbes in such soils as: 1) non-pigmented, heterotrophic, aerobic bacteria; 2) pigmented, heterotrophic, microaerophilic bacteria; 3) actinomycetes; 4) cyanobacteria and coccoid green algae; 5) sulfate-reducing and N_2-fixing bacteria, fungi, and protozoa; 6) lichens; and 7) mosses and other algae. Primary productivity is absent in the harshest of the cold, saline desert soils, where the microflora must depend on "imported" organic carbon. Sulfate-reducing and N_2-fixing bacteria, important agents in nutrient cycles, are extremely sensitive to these conditions and are late-comers to well-established microbial floras. These observations parallel those made in the hypersaline Antarctic lakes. The authors noted that anaerobes, phototrophic bacteria, obligate psychrophiles, thermophiles, and obligate halophiles are generally not detected in the Dry Valley soils.

These observations are in agreement with other studies of Dry Valley soils (Benoit and Hall, 1970; Horowitz et al., 1972). Enrichment cultures of the soil around Don Juan Pond yielded actinomycetes and an abundant flora of non-halophilic, mesophilic bacteria (Benoit and Hall, 1970). The authors observed that many small ponds in the Dry Valley had large populations of phototrophic and sulfate-reducing bacteria. These bacteria, in addition to chemoautotrophic bacteria, were seldom detected in soils. Enrichments for bacteria from a saline soil in Taylor Valley demonstrated a strong temperature-salt correlation. Bacterial counts (per g soil) at 20°C in 0% NaCl were 7.2×10^4, while in 15% NaCl they were 1.4×10^3, and in 20% NaCl they were <40. At 2°C, viable bacteria were detected only in 0% and 5% NaCl media.

Although Antarctic hypersaline environments are rare examples of processes occurring in evaporitic habitats, they can be instructional for determining some of the factors that limit life, especially with respect to salinity, water activity, temperature, and light.

References

Angino, E.E. and **Armitage, K.B.** 1963. A geochemical study of Lakes Bonney and Vanda, Victoria Land, Antarctica. *Journal of Geology* 71: 89–95.

Barghoorn, E.S. and **Nichols, R.L.** 1961. Sulfate-reducing bacteria and pyritic sediments in Antarctica. *Science* 134: 190.

Benoit, R.E. and **Hall, Jr., C.L.** 1970. The microbiology of some Dry Valley soils of Victoria Land, pp. 697–701 in Holdgate, M.W. (editor), *Antarctic Ecology*, vol. 2, Academic Press, New York.

Benoit, R.E., Hatcher, R., and **Green W.** 1971. Biological profiles and some chemical characteristics of two permanently frozen Antarctic lakes, pp. 281–293 in Cairns, Jr., J. (editor), *The Structure and Function of Fresh-Water Microbial Communities*, Research Division Monograph 3, Virginia Polytechnic Institute and State University, Blacksburg.

Burton, H.R. 1981. Chemistry, physics and evolution of Antarctic saline lakes: a review. *Hydrobiologia* 82: 339–362.

Burton, H.R. and **Barker, R.J.** 1979. Sulfur chemistry and microbiological fractionation of sulfur isotopes in a saline Antarctic lake. *Geomicrobiology Journal* 1: 329–340.

Cameron, R.E., King, J., and **David, C.N.** 1970. Microbiology, ecology and microclimatology of soil sites in Dry Valleys of Southern Victoria Land, Antarctica, pp. 702–716 in Holdgate, M.W. (editor), *Antarctic Ecology*, vol. 2, Academic Press, New York.

Cameron, R.E., Morelli, F.A., and **Randall, L.P.** 1972. Aerial, aquatic and soil microbiology of Don Juan Pond, Antarctica. *Antarctic Journal of the United States* 7: 254–258.

Campbell, P.J. 1978. Primary productivity of a hypersaline Antarctic lake. *Australian Journal of Marine and Freshwater Research* 29: 717–724.

Canfield, D.E. and **Green W.J.** 1985. The cycling of nutrients in a closed-basin Antarctic lake: Lake Vanda. *Biogeochemistry Journal* 1: 233–256.

Ellis-Evans, J.C. 1985. Decomposition processes in maritime Antarctic lakes, pp. 253–260 in Siegfried, W.R., Condy, P.R. and Laws, R.M. (editors), *Antarctic Nutrient Cycles and Food Webs*, Springer-Verlag, New York.

Fortner, R.D., Weand, B.L., and **Hoehn, R.C.** 1976. Reevaluation of ortho-phosphorus and inorganic nitrogen levels in an Antarctic meromictic lake. *Hydrobiologia* 49: 229–232.

Fukui, F., Torii, T., and **Okabe, S.** 1985. Vertical distribution of nutrients and DOC in lake waters near Syowa Station, Antarctica. *Antarctic Record* 86: 28–35.

Goldman, C.R. 1964. Primary productivity studies in Antarctic lakes, pp. 291–299 in Carrick, R., Holdgate, M. and Prevost, J. (editors), *Biologie Antarctique*, Herman, Paris.

Goldman, C.R., Mason, D.T., and **Hobbie, J.E.** 1967. Two Antarctic desert lakes. *Limnology and Oceanography* 12: 295–310.

Hand, R.M. 1980. Bacterial populations in two saline Antarctic lakes, pp. 123–129 in Trudinger, P.A. and Walter, M.R. (editors), *Biogeochemistry of Ancient and Modern Environments*, Springer-Verlag, New York.

Harris, H.J.H., Cartwright, K., and **Torii, T.** 1979. Dynamic chemical equilibrium in a polar desert pond: a sensitive index of meteorological cycles. *Science* 204: 301–303.

Heywood, R.B. 1984. Antarctic inland waters, p. 279–344 in Laws, R.M. (editor), *Antarctic Ecology*, vol. 1, Academic Press, New York.

Hoare, R.A., Popplewell, K.B., House, D.A., Henderson, R.A., Prebble, W.M., and **Wilson, A.T.** 1964. Lake Bonney, Taylor Valley, Antarctica: a natural solar energy trap. *Nature (London)* 202: 886–888.

Hoehn, R.C., Parker, B.C., Fortner, R.D., Weand, B.L., Craft, J.A., Lane, L.S., Stavros, R.W., Sugg, Jr., H.G., and **Whitehurst, J.T.** 1977. Nitrogen and phosphorus availability to plankton and benthic communities in Lake Bonney, Southern Victoria Land, Antarctica, pp. 859–872 in Llano, G.A. (editor), *Adaptations Within Antarctic Ecosystems*, Smithsonian Institution, Washington, D.C.

Horowitz, N.H., Cameron, R.E., and **Hubbard, J.S.** 1972. Microbiology of the Dry Valleys of Antarctica. *Science* 176: 242–245.

Jones, L.M. and **Faure, G.** 1967. Origin of the salts in Lake Vanda, Wright Valley, Southern Victoria Land, Antarctica. *Earth and Planetary Science Letters* 31: 101–106.

Kerry, K.R., Grace, D.R., Williams, R., and **Burton, H.R.** 1977. Studies on some saline lakes of the Vestfold Hills, Antarctica, pp. 839–858 in Llano, G.A. (editor), *Adaptations Within Antarctic Ecosystems*, Smithsonian Institution, Washington, D.C.

Koob, D.D. and **Leister, G.L.** 1972. Primary productivity and associated physical, chemical, and biological characteristics of Lake Bonney: a perennially ice-covered lake in Antarctica, p. 51–68 in Llano,G.A. (editor), *Antarctic Terrestrial Biology*, Antarctic Research Series, vol. 20, American Geophysical Union, Washington, D.C.

Love, F.G., Simmons, Jr., G.M., Parker, B.C., Wharton, Jr., R.A., and **Seaburg, K.G.** 1983. Modern Conophyton-like microbial mats discovered in Lake Vanda, Antarctica. *Geomicrobiology Journal* 3: 33–48.

Masuda, N., Nishimura, M., and **Torii, T.** 1982. Distribution and origin of some trace metals in Lake Vanda, Antarctica. *Antarctic Record* 75: 25–36.

Masuda, N, Nakaya, S., and **Torii, T.** 1984. Origin of salt in Antarctic saline lake waters through trace element analysis, pp. 194–203 in Nagata, T. (editor), *Proceedings of the Fourth Symposium on Antarctic Geosciences*, Memoirs of National Institute of Polar Research, Special Issue 33, Tokyo.

Matsubaya, O., Sakai, H., Torii, T., Burton, H., and **Kerry, K.** 1979. Antarctic saline lakes—stable isotopic ratios, chemical composition and evolution. *Geochimica et Cosmochimica Acta* 43: 7–25.

Matsumoto, G. and **Hanya, T.** 1977. Organic carbon and fatty acids in Antarctic saline lakes. *Antarctic Record* 58: 81–88.

Matsumoto, G., Tanaka, Y., and **Torii, T.** 1982. Nutrient matters in saline lakes of McMurdo Oasis in 1976–77 summer season. *Antarctic Record* 74: 109–118.

Matsumoto, G.I., Torii, T., and **Hanya, T.** 1984a. A review of organic geochemistry in Antarctica, pp. 204–217 in Nagata, T. (editor), *Proceedings of the Fourth Symposium on Antarctic Geosciences*, Memoirs of National Institute of Polar Research, Special Issue 33, Tokyo.

Matsumoto, G., Torii, T., and **Hanya, T.** 1984b. Vertical distribution of organic constituents in an Antarctic lake: Lake Vanda. *Hydrobiologia* 111: 119–126.

Matsumoto, G.I., Torii, T., and **Hanya, T.** 1985. Sterols and fatty acids in foams from Antarctic lakes of the Dry Valleys in South Victoria Land. *Geochemical Journal* 19: 91–96.

Meyer, G.H., Morrow, M.B., Wyss, O., Berg, T.W., and **Littlepage, J.L.** 1962. Antarctica: the microbiology of an unfrozen saline pond. *Science* 138: 1103–1104.

Morikawa, H., Minato, I., Ossaka, J., and **Hayashi, T.** 1975. The distribution of secondary minerals and evaporites at Lake Vanda, Victoria Land, Antarctica, p. 45–59 in Torii, T. (editor), *Geochemical and Geophysical Studies of Dry Valleys, Victoria*

Land in Antarctica, Memoirs of National Institute of Polar Research, Special Issue 4, Tokyo.

Nakai, N., Kiyosu, Y., Wada, H., and **Takimoto, M.** 1975a. Stable isotope studies of salts and water from Dry Valleys, Antarctica. I. Origin of salts and water, and the geologic history of Lake Vanda, pp. 30–44 in Torii, T. (editor), *Geochemical and Geophysical Studies of Dry Valleys, Victoria Land in Antarctica,* Memoirs of National Institute of Polar Research, Special Issue 4, Tokyo.

Nakai, N., Wada, H., Kiyosu, Y., and **Takimoto, M.** 1975b. Stable isotope studies on the origin and geological history of water and salts in the Lake Vanda area, Antarctica. *Geochemical Journal* 9: 7–24.

Nakai, N. and **Mizutani, Y.** 1977. Geologic history of the Dry Valleys, Antarctica, based on the stable isotope studies. *Antarctic Record* 58: 244–253.

Nelson, C.S. and **Wilson, A.T.** 1972. Bathymetry and bottom sediments of Lake Vanda, Antarctica. *Antarctic Journal of the United States* 7: 97–99.

Osnitskaya, L.K. and **Chudina, V.I.** 1978. Photosynthetic bacteria from the Vanda Lake in the Antarctic. *Mikrobiologiya* 46: 131–137.

Parker, B.C., Hoehn, R.C., Paterson, R.A., Craft, J.A., Lane, L.S., Stavros, R.W., Sugg, Jr., H.G., Whitehurst, J.T., Fortner, R.D., and **Weand, B.L.** 1977. Changes in dissolved organic matter, photosynthetic production, and microbial community composition in Lake Bonney, Southern Victoria Land, Antarctica, pp. 873–890 in Llano, G.A. (editor), *Adaptations Within Antarctic Ecosystems,* Smithsonian Institution, Washington, D.C.

Parker, B.C., Simmons, Jr., G.M., Seaburg, K.G., Cathey, D.D., and **Allnutt, F.C.T.** 1982. Comparative ecology of plankton communities in seven Antarctic oasis lakes. *Journal of Plankton Research* 4: 271–286.

Parker, B.C. and **Simmons, Jr., G.M.** 1985. Paucity of nutrient cycling and absence of food chains in the unique lakes of southern Victoria Land, p. 238–244 in Siegfried, W.R., Condy, P.R., and Laws, R.M. (editors), *Antarctic Nutrient Cycles and Food Webs.* Springer-Verlag, New York.

Priddle, J. and **Heywood, R.B.** 1980. Evolution of Antarctic lake ecosystems, p. 51–66 in Bonner, W.N. and Berry, R.J. (editors), *Ecology in the Antarctic,* Academic Press, New York.

Seaburg, K.G., Parker, B.C., Wharton, Jr., R.A., and **Simmons, Jr., G.M.** 1981. Temperature-growth responses of algal isolates from Antarctic oases. *Journal of Phycology* 17: 353–360.

Seaburg, K.G., Kaspar, M., and **Parker, B.C.** 1983. Photosynthetic quantum efficiences of phytoplankton from perennially ice-covered Antarctic lakes. *Journal of Phycology* 19: 446–452.

Siegel, B.Z., McMurty, G., Siegel, S.M., Chen, J., and **LaRock, P.** 1979. Life in the calcium chloride environment of Don Juan Pond, Antarctica. *Nature (London)* 280: 828–829.

Sugiyama, J., Sugiyama, Y., Iizuka, H., and **Torii, T.** 1967. Report of the Japanese summer parties in Dry Valleys, Victoria Land, 1963–1965. IV. Mycological studies of the Antarctic fungi. Part 2. Mycoflora of Lake Vanda, an ice-free lake. *Antarctic Record* 28: 23–32.

Tominaga, H. and **Fukui, F.** 1981. Saline lakes at Syowa Oasis, Antarctica. *Hydrobiologia* 82: 375–389.

Tomiyama, C. and **Kitano, Y.** 1985. Salt origin in the Wright Valley, Antarctica. *Antarctic Record* 86: 17–27.

Torii, T., Yamagata, N., Nakaya, S., Murata, S., Hashimoto, T., Matsubaya, O., and **Sakai, H.** 1975. Geochemical aspects of the McMurdo saline lakes with special emphasis on the distribution of nutrient matters, p. 5–29 in Torii, T. (editor), *Geo-*

chemical and Geophysical Studies of Dry Valleys, Victoria Land in Antarctica, Memoirs of National Institute of Polar Research, Special Issue 4, Tokyo.

Torii, T. and **Yamagata, N.** 1981. Limnological studies of saline lakes in the Dry Valleys, p. 141–159 in McGinnis, L.D. (editor), *Dry Valley Drilling Project,* American Geophysical Union, Washington, D.C.

Vincent, W.F., Downes, M.T., and **Vincent, C.L.** 1981. Nitrous oxide cycling in Lake Vanda, Antarctica. *Nature (London)* 292: 618–620.

Wada, E., Imaizumi, R., Nakaya, S., and **Torii, T.** 1984. ^{15}N abundance in the Dry Valley area, South Victoria Land, Antarctica: eco-physiological implications of microorganisms, pp. 130–139 in Hoshiai, T. and Fukuchi, M. (editors), *Proceedings of the Sixth Symposium on Polar Biology,* Memoirs of National Institute of Polar Research, Special Issue 32, Tokyo.

Waguri, O. 1976. Isolation of microorganisms from salt lakes in the Dry Valley, Antarctica, and their living environment. *Antarctic Record* 57: 80–96.

Weand, B.L., Hoehn, R.C., and **Parker, B.C.** 1976. Trace element distributions in an Antarctic meromictic lake. *Hydrobiological Bulletin* 10: 104–114.

Weand, B.L., Hoehn R.C., and **Parker, B.C.** 1977. Nutrient fluxes in Lake Bonney—a meromictic Antarctic lake. *Archiv für Hydrobiologie* 80: 519–530.

Wharton, Jr., R.A., Parker, B.C., and **Simmons, Jr., G.M.** 1983. Distribution, species composition and morphology of algal mats in Antarctic Dry Valley lakes. *Phycologia* 22: 355–365.

Wilson, A.T. 1970. The McMurdo Dry Valleys, pp. 21–30, in Holdgate, M.W. (editor), *Antarctic Ecology,* vol. 1, Academic Press, New York.

Wilson, A.T. 1979. Geochemical problems of the Antarctic dry areas. *Nature (London)* 280: 205–208.

Wilson, A.T. and **Wellman, H.W.** 1962. Lake Vanda: an Antarctic lake. *Nature (London)* 196: 1171–1173.

Wilson, A.T., Holdsworth, R., and **Hendy, C.H.** 1974. Lake Vanda: source of heating. *Antarctic Journal of the United States* 9: 137–138.

Wright, S.W. and **Burton, H.R.** 1981. The biology of Antarctic saline lakes. *Hydrobiologia* 82: 319–338.

Yarrington, M.R. and **Wynn-Williams, D.D.** 1985. Methanogenesis and the anaerobic micro-biology of a wet community at Signy Island, pp. 229–233 in Sigfried, W.R., Condy, P.R. and Laws, R.M. (editors), *Antarctic Nutrient Cycles and Food Webs,* Springer-Verlag, New York.

17

Lake Eyre and Other Temperate Lakes

Unlike hypersaline environments of marine origin, terminal continental lakes (which have no outlet to the ocean) may have distinct chemical compositions dependent on the weathering of local rocks, salt transport, and evaporation. Australian Lake Eyre, the lakes of Saskatchewan, and other smaller, temperate lakes have not received the scientific attention that the Great Salt Lake, the Dead Sea, or the alkaline lakes of Africa and North America (discussed in Chapters 18–20) have. However, the few microbiological studies on the lesser-known lakes described here indicate that microbial activity can be found in extreme salinities and in brines of unusual composition in temperate climates.

17.1 Lake Eyre

Lake Eyre, South Australia, is an extremely large, saline, playa basin. The highest point on the shoreline lies about 5 m below sea level and the lowest point about 10 m below sea level in this terminal lake. The whole lake occupies about 9500 km^2. Madigan Gulf, which measures about 32 $\times$ 40 km and is about 2 m deep, occupies the deepest part of Lake Eyre North. The lake is mostly dry, but periodic floods fill the lake with varying amounts of water. Upon evaporation, the deeper parts of the lake contain practically all the lake salts (Bonython, 1956). The dominant evaporite minerals in the lake are gypsum and halite. The salt composition is different from the ocean and from inflowing water and probably reflects a complex history of contributions from the underlying basin materials and windborne salts.

Nodules of native sulfur were found at the lee shore on the surface and at a sediment depth of over 30 cm. The identification of various groups of microbes in the sediments and a description of their activities in different salt concentrations led Baas-Becking and Kaplan (1956) to conclude that the sulfur nodules had been formed from gypsum as a result of microbiological activities. The following description of the biogeochemistry of Madigan Gulf, Lake Eyre, was taken from the works of Bonython (1956) and Baas-Becking and Kaplan (1956). Other general descriptions of Lake Eyre were referenced by Bonython (1956). At the time of these studies, the center of the Madigan Gulf basin had a halite crust of about 28 cm thickness, below which was a 1-cm thick layer of black mud. From about 30 to 260 cm depth the sediment was mostly gypsum with a thin layer of halite at about 90 cm depth. Below 260 cm, the sediments were largely clay and dolomite.

Bonython described the composition of the salt crust, including the colors at different levels (Table 17.1). The rusty color near the surface was mainly due to iron oxide, with a small contribution by halobacteria and *Dunaliella*. The purple color at 13–15 cm depth was found to contain small, amorphous, mucoid masses which could have been of bacterial or of algal origin, but they were not described. The purple layer clearly coincides with a dilution event in the lake and with $CaSO_4$ precipitation (compare SO_4^{2-} and Ca^{2+} content of the salt at 13 cm vs. 13–15 cm). Because evaporites can precipitate below the sediment-brine interface, it is impossible to tell if the purple layer formed after the precipitation of nearly pure halite, or if halite crystallized below the organic sediment after the cells were deposited. It is unknown whether the organic matter represents planktonic material or the remains of an active microbial mat. The brownish layer at 12 cm depth probably represents dust that had settled during a freshening event.

Baas-Becking and Kaplan (1956) described the organic and microbiological composition of sulfur nodules which were found in mud along with wind-blown gypsum and organic detritus. The nodules contained abundant organic

Table 17.1 Chemical and color description of the salt crust and of the brine from borehole 93/54, Madigan Gulf, Lake Eyre[a]

Depth, cm	NaCl, %	SO_4^{2-}, %	Ca, %	Mg, %	Color
0–3	99.0	0.46	0.12	0.05	Clear to rusty
4–8	97.5	0.90	0.32	0.02	Pink
8–11	96.8	1.37	0.53	0.01	Pink
12					Brownish
13	94.5	2.61	0.91	0.05	White
13–15	99.1	0.43	0.11	0.04	Purple
18–20	97.6	0.66	0.21	0.03	White
20–23	94.3	3.31	0.64	0.07	Pink
24–28	97.5	1.51	0.42	0.02	White; grey at bottom
	NaCl	SO_4^{2-}	Ca	Mg	
Ground brine[b]	308.8	17.9	0.28	6.56	

[a] From Bonython (1956).
[b] In $g \cdot l^{-1}$

matter. They observed or obtained enrichment cultures for a variety of microbes from brine, salt, mud, and sulfur samples (Table 17.2). The sulfur nodules proved to be a rich source of bacteria involved in the sulfur cycle. The muds and brines were of near-neutral pH (6.72–7.20) and the redox potentials were −60 to −110 mV in the zones of the most active sulfate reduction.

The mechanism for elemental sulfur formation was described as an integrated process involving the following steps and their respective agents:

$$\text{Anaerobic fermentation of organic detritus} \rightarrow H_2$$

$$\text{Gypsum} + H_2 + \text{sulfate-reducing bacteria} \rightarrow S^{2-}, \text{FeS}$$

$$S^{2-}, \text{FeS} + O_2 \rightarrow S^0 \text{ (abiotic or thiobacteria)}$$

$$S^0 + \text{thiobacteria} \rightarrow H_2SO_4$$

$$H_2SO_4 + CaCO_3 \rightarrow Ca^{2+} + CO_3^{2-}$$

$$Ca^{2+} + \text{brine } SO_4^{2-} \rightarrow \text{secondary gypsum crust around } S^0 \text{ nodules}$$

Enrichments of brines, salts, and muds for various groups of bacteria in 5–25% NaCl demonstrated that certain metabolic processes or microorganisms responsible for those processes were inhibited by high salt concentrations and that each of these environments had different microbiological potential (Tables 17.3 and 17.4). Fermentation of glucose and cellulose were the least

Table 17.2 Microorganisms in Lake Eyre[a]

Type	Identification	Source
Chlorophytes	*Dunaliella minuta, D. parva, D. euchlorld,* unnamed, spindle-shaped alga, 15 μm long	Brine, salt
Diatoms	*Pleurosigma* sp.	Diluted brine
	Amphora coffeaeformis	Brine
Protozoans	Colorless ciliates, flagellates	Brine
Other eukaryotes	Chytrid fungi	With *Lyngbya*
	Parartemia	Brine (one locality)
Cyanobacteria	*Lyngbya* sp.	Various materials
Bacteria[b]	Spirilla, long rods, halobacteria	Brine
	Sulfate reducers on H_2	Gypsum, mud, sulfur
	Sulfate reducers on lactate	Gypsum, mud, sulfur
	Sulfate reducers on pyruvate	Surface mud only
	Sulfur oxidizers on thiosulfate	Gypsum, deep mud, sulfur
	Sulfur oxidizers on sulfur	Gypsum, mud, sulfur
	Glucose fermenters	Brine, salt, surface mud
	Denitrifiers	Brine, surface mud
	Aerobic cellulose decomposers	Surface mud, sulfur
	Anaerobic cellulose decomposers	Brine, salt, mud, gypsum
	Methanogens	Brine, salt surface mud

[a] From Baas-Becking and Kaplan (1956).
[b] Identified from enrichment cultures.

Table 17.3 Effects of NaCl concentration on anaerobic activity in brine, salt crust, and surface mud of Lake Eyre[a]

	Source		
% NaCl	Surface brine	Pink salt	Surface mud
Glucose fermentation, ml H_2 formed after 14 days, 30°C			
5	0.70	0.25	3.15
10	3.15	0.05	6.25
15	0	4.50	11.25
20	5.65	5.65	9.40
25	0.75	2.90	10.00
Denitrification, ml gas formed after 28 days, 30°C			
5	1.25	0.02	1.25
10	1.88	0	0.83
15	0.25	0	0.13
20	0.07	0	0.06
25	0.02	0	0
Methanogenesis, ml gas × 10 formed after 14 days, 30°C			
5	0	0.16	0.25
10	0.25	0.16	0.25
15	0.25	0.10	0
20	0	0	0
25	0	0	0

[a] From Baas-Becking and Kaplan (1956).

Table 17.4 Anaerobic cellulose decomposition in brine, salt, and sediment from Lake Eyre incubated in different concentrations of NaCl[a]

	Source					
% NaCl	Surface brine	Pink salt	Surface mud	Deep mud	Gypsum	Sulfur mud
pH after 29 days, 30°C						
5	5.60	—	3.80	2.80	3.42	
10	2.87	4.43	2.72	2.66	2.93	
15	3.53	3.92	2.80	4.33	2.85	
20	4.03	4.55	2.57	2.49	2.70	
25	3.62	3.70	2.47	2.44	2.70	
Sulfate reduction in the presence of H_2[b]						
5	+++		0	++		0
10	+		++++++	+		++++++
15	+		++	0		0
20	+		++++++	++++		0
25	+		++	+		0

[a] From Baas-Becking and Kaplan (1956).
[b] 0, no sulfate reduction; +, sulfate reduction (extent indicated by the number of + symbols).

affected by high salt. The highest rates of glucose fermentation were observed in 20% salt. Aerobic decomposition was reported only in surface mud and mud associated with sulfur nodules between 6.25 and 18.75% NaCl. Denitrification and especially methanogenesis were retarded or completely inhibited at high salinities. The details for those enrichments were not given. It is unknown whether nitrate was added to the enrichments for denitrifiers, whether the authors assayed for nitrite, or whether the kind of gas formed was determined. Although the development of sulfate-reducing bacteria was only subjectively quantified, it is clear that sulfate reduction can occur in 20–25% salt as long as there are appropriate carbon and energy sources. The authors were unable to isolate green and purple phototrophic bacteria except for one instance from a surface mud sample using 20% NaCl.

Lake Eyre probably resembles the Great Salt Lake of North America during its early, playa stages (Spencer et al., 1984). Although it lacks a permanent water column, the periodic wetting and drying cycles result in the accumulation of organic matter that is subject to microbial degradation, even under very saline conditions. Enrichment culture studies indicate that anaerobic degradation of at least some forms of organic carbon can be potentially quite rapid, even at brine concentrations near halite saturation.

17.2 Other Australian lakes

A number of hypersaline lakes have been described from Australia. Microbial mats have been documented from several of these lakes in West Australia (Government House Lake, 16.4% TDS; Lake Cowan, 29.8% TDS), South Australia (Deep Lake, 12.5% TDS; Coorong lakes, 17.5% TDS), and Victoria (West Basin Lake, 9–13% TDS) (Bauld, 1981). Although microbial mats as well as lake salinities were noted, rates of microbial activity were not reported.

Pink Lake, Victoria, is a shallow (up to 70 cm deep) hypersaline lake with a salinity that fluctuates between 9% and 24% (Marchant and Williams, 1977) and a conductivity that fluctuates between 180 and 410 milliSiemens (mS)$\cdot$cm^{-1} (Hammer, 1981). The pH also fluctuates seasonally between 7.6 and 8.6. The lake is characterized by negligible phosphate and no nitrate (ammonium was not measured). *Dunaliella salina* was always present in Pink Lake except during periods of the highest salinities. The photosynthetic rates measured in Pink Lake were maximal in July (461 mg C$\cdot$m$^{-3}\cdot$d^{-1}) and undetectable in February and March. These rates are rather modest, which is probably attributable to the low nutrients and high salinity. The photosynthetic rates were often greater at 0.5 m depth, suggesting photoinhibition or a greater accumulation of cells with depth.

The sediments of Pink Lake are loose and black, indicating sulfate reduction. In the top 3 cm, the sediments had 9.7–11.8% organic matter by weight (Marchant and Williams, 1977). About 2% of the total organic material was water-soluble. Much of the organic matter was believed to be fecal pellets of *Parartemia zietziana*, a benthic-grazing brine shrimp endemic to Australia. Organic N was 1.8 $mg \cdot g^{-1}$ sediment. In another study by Aasen et al. (1969), crystalline salts from the lake were found to contain about 0.75% organic matter. About 14% of the organic matter was β-carotene from *Dunaliella*. Total lipid-soluble material constituted nearly 50% of the organic matter in the sediment.

17.3 Saskatchewan, Canada

The saline lakes of the plains region of Saskatchewan, Canada, are dominated by Mg^{2+}, Na^+, and SO_4^{2-}. The composition of eight extremely hypersaline lakes of Saskatchewan is given in Table 17.5 (see also Table 1.1). The very saline lakes are somewhat alkaline (pH 7.9–9.15), although some of the less saline lakes of the region have a pH as high as 9.8 (Hammer, 1978). NH_4^+ is quite high and NO_3^- is scarce (Table 17.6). PO_4^{3-} is also high in spite of the presence of high concentrations of Mg^{2+}. The N:P ratios of Little Manitou and Patience Lakes are close to the Redfield ratio of 16 (the average N:P ratio of living organisms), but that of west Chaplin Lake is very low.

The lakes undergo extreme temperature fluctuations between summer and winter (30° to −3°C) (Hammer and Haynes, 1978). The permanent lakes are typically ≤10 m deep. The deeper lakes stratify and higher salinities tend to prolong the period of stratification. In lakes >3 m deep, the sediments tend to be fine-grained, organic-rich, and well laminated (Last and Schweyen, 1983).

In the permanent lakes, the precipitation of calcite occurs due to the effects of biological activities, and calcite, gypsum, and mirabilite precipitate as a result of temperature fluctuations, freezing-out, and brine mixing (Last and Schweyen, 1983). The authors noted that very thin calcite layers (varves)

Table 17.5 The composition of eight hypersaline lakes of Saskatchewan, Canada[a]

Lake	Salinity, %	pH	Mg^{2+}	Na^+	SO_4^{2-}	Cl^-	HCO_3^- + CO_3^{2-}
Little Manitou	8.14	8.8	9.52	12.3	39.6	18.0	0.985
Aroma	8.52	9.1	9.12	12.5	58.9	3.3	0.715
Chaplin (east)	10.8	9.15	1.55	29.6	65.0	3.1	2.44
Bitter	19.2	7.9	29.7	18.2	95.5	44.5	1.84
Whiteshore	20.1	8.7	7.38	50.4	129	10.0	0.841
Patience	20.7	8.8	2.63	55.2	17.2	105	0.126
Chaplin (west)	21.4	8.8	5.05	59.6	132	11.2	2.76
Muskiki	34.2	7.9	29.0	62.2	237	11.4	1.47

[a]From Hammer (1978). Values given in $g \cdot l^{-1}$ except salinity (%).

Table 17.6 Nutrient composition of three hypersaline lakes of Saskatchewan, Canada[a]

Lake	NH_4^+-N	NO_3^--N	o-PO_4^{3-}-P	Total P	N:P[b]	Fe	SiO_2
Little Manitou	60.0	0.31	0.41	2.00	15.2	0.54	0.65
Patience	26.5	0.05	0.35	2.03	13.1	7.50	2.0
Chaplin (west)	13.1	0.20	2.81	7.00	1.9	2.85	1.55

[a] From Hammer (1978). Concentrations in μM.
[b] (Nitrate-N + Ammonium-N)/Phosphate-P.

formed in the sediments of some of the saline lakes, including the hypersaline Little Manitou Lake.

Gypsum in Waldsea Lake, where the salinity reaches 4.2%, is found in different forms and textures. The physical, chemical, and biological processes that affect the crystal morphology have not been defined. Both prismatic and lenticular crystals are found, while rounded and pitted grains suggest dissolution or reworking processes. In the playa lakes of the region, evaporite precipitation results in the formation of efflorescent crusts and intrasedimentary crystals.

Planktonic primary productivity of several of the saline lakes of the region has been measured, including Little Manitou Lake (8.14% salinity, 5.2 m deep) (Haynes and Hammer, 1978). A photosynthetic rate of 1.19 g $C\cdot m^{-2}\cdot d^{-1}$ was measured, a rate comparable to other hard-water, meso-eutrophic, temperate lakes (Table 2.2). During spring, a single recording of a rate of 0.475 g $C\cdot m^{-3}\cdot d^{-1}$ was made. During the periods of the highest productivity, 5.8–14.1 $mg\cdot m^{-3}$ chlorophyll *a* was detected along with 20.5–46.3 $mg\cdot m^{-3}$ pheopigments in the euphotic zone. During periods of lower productivity, only 0.6–3.9 $mg\cdot m^{-3}$ chlorophyll *a* was found. The phytoplankton of the lake consisted primarily of the diatom *Chaetoceros elmorei* and the green alga *Rhizoclonium hieroglyphicum.*

The distribution and abundance of algae in the saline lakes of Saskatchewan have been documented in detail by Hammer et al. (1983). Many of these taxa are given in Table 2.1. The most saline lakes are dominated by the green alga *Dunaliella salina, Rhizoclonium hieroglyphicum,* and *Ctenocladus circinnatus.* A variety of diatoms are also found in these sulfate-rich lakes, including *Nitzschia* spp., *Hantzschia amphioxys,* and the cosmopolitan *Amphora coffeaeformis.* Various cyanobacteria (both filamentous and coccoid) are abundant or common, including the nitrogen-fixing species *Nodularia spumigena* (in $\geq$20% TDS) and *Anabaena wisconsinense* (17.8–42.8% TDS). No measurements of N_2 fixation in these hypersaline lakes of Saskatchewan have been reported. The authors did not mention whether algal or cyanobacterial mats develop in those extremely saline lakes. The high concentrations of phosphate may enhance the development of phytoplankton and effectively shade the benthos.

17.4 Hot Lake, Washington (U.S.A.) and other small lakes

Some other small, neutral pH, hypersaline lakes that have been briefly described include Lake Assal (French Somaliland, about 40% salinity) (Brisou et al., 1974), Zuni Salt Lake (New Mexico, U.S.A., 20.6% salinity) (Bradbury, 1971), and several lakes in the Soviet Union (Kurochkin, 1960). The Russian lakes, with salinities up to NaCl saturation, were reported to be pink. Extremely halophilic bacteria were isolated from those lakes.

Hot Lake, a $MgSO_4$ (Epsom salt) lake, was studied by Anderson (1958). The lake occupies a former Epsom salt pit. The basin has been subsequently altered and the limnological characteristics have changed. However, a brief description of the former status of Hot Lake demonstrates that $MgSO_4$ brines can be highly productive. The lake was 3.5 m deep and density-stratified (Table 17.7). A chemocline below 1 m depth caused solar heating of the monimolimnion. During July, 1955, the temperature in 0–1 m depth was 27°–28°C and at 2.0 m depth was 50.5°C. At 3.0 m depth it was 36.2°C. The monimolimnion was described as stagnant, dark brown, and full of H_2S. A layer of phototrophic bacteria (probably *Chlorobium*) was detected at 2.0 m depth where the salinity was near 39.2%. Such halophilic or halotolerant green phototrophic bacteria have not been described elsewhere. The surface waters were oxygenated in the spring, but anoxic by autumn. *Artemia salina* inhabited the surface waters along with rotifers, occasional copepods, and an ostracod. At $\geq$1 m depth, thick cyanobacterial mats were found consisting of *Plectonema nostocorum*, *Oscillatoria chlorina*, *Anacystis thermalis*, and *Gomphosphaeria aponina*. Because none of these species is reported in other hypersaline systems, the taxonomy of the mat-forming species should be reevaluated. Anderson did note that the mat extended into the lower regions of the hypolimnion and suggested the cyanobacteria were capable of H_2S metabolism. H_2S metabolism in halophilic cyanobacteria was much later demonstrated in *Oscillatoria limnetica* of Solar Lake, Sinai (Cohen et al., 1975a, 1975b).

Table 17.7 Chemical constituents of Hot Lake, Washington, during stratification in August, 1955[a]

Constituent	Surface	3 m	Ratio, 3 m:surface
Total dissolved solids, %	16.1	39.2	2.43
pH	8.2	7.8	–
Mg^{2+}	22.8	56.6	2.48
Na^+	0.89	1.56	1.75
Ca^{2+}	0.64	0.72	1.13
SO_4^{2-}	104	244	2.35
HCO_3^-	3.15	3.06	0.97
Cl^-	1.67	1.88	1.13

[a] From Anderson (1958). Constituents in $g \cdot l^{-1}$

References

Aasen, A.J., **Eimhjellen, K.E.**, and **Liaaen-Jensen, S.** 1969. An extreme source of β-carotene. *Acta Chemica Scandinavica* 23: 2544–2545.

Anderson, G.C. 1958. Some limnological features of a shallow saline meromictic lake. *Limnology and Oceanography* 3: 259–270.

Baas-Becking, L.G.M. and **Kaplan, I.R.** 1956. The microbiological origin of the sulphur nodules of Lake Eyre. *Transactions of the Royal Society of South Australia* 79: 52–65.

Bauld, J. 1981. Occurrence of microbial mats in saline lakes. *Hydrobiologia* 81: 87–111.

Bonython, C.W. 1956. The salt of Lake Eyre—its occurrence in Madigan Gulf and its possible origin. *Transactions of the Royal Society of South Australia* 79: 66–91.

Bradbury, J.P. 1971. Limnology of Zuni Salt Lake, New Mexico. *Geological Society of America Bulletin* 82: 379–398.

Brisou, J., **Courtois, D.**, and **Denis, F.** 1974. Microbiological study of a hypersaline lake in French Somaliland. *Applied Microbiology* 27: 819–822.

Cohen, Y., **Padan, E.**, and **Shilo, M.** 1975a. Facultative anoxygenic photosynthesis in the cyanobacterium *Oscillatoria limnetica*. *Journal of Bacteriology* 123: 855–861.

Cohen, Y., **Jørgensen, B.B.**, **Padan, E.**, and **Shilo, M.** 1975b. Sulfide dependent anoxygenic photosynthesis in the cyanobacterium *Oscillatoria limnetica*. *Nature (London)* 257: 489–492.

Hammer, U.T. 1978. The saline lakes of Saskatchewan III. Chemical characterization. *Internationale Revue der gesamten Hydrobiologie* 63: 311–335.

Hammer, U.T. 1981. A comparative study of primary production and related factors in four saline lakes in Victoria, Australia. *Internationale Revue der gesamten Hydrobiologie* 66: 701–743.

Hammer, U.T., and Haynes, R.C., 1978. The saline lakes of Saskatchewan II. Locale, hydrography and other physical aspects. *Internationale Revue der gesamten Hydrobiologie* 63: 179–203.

Hammer, U.T. Shamess, J., and **Haynes, R.C.** 1983. The distribution and abundance of algae in saline lakes of Saskatchewan, Canada. *Hydrobiologia* 105: 1–26.

Haynes, R.C. and **Hammer, U.T.** 1978. The saline lakes of Saskatchewan IV. Primary production by phytoplankton in selected saline ecosystems. *Internationale Revue der gesamten Hydrobiologie* 63: 337–351.

Kurochkin, B.I. 1962. Relationship of some halophiles to various concentrations of sodium chloride. *Microbiology* (English translation) 31: 56–59.

Last, W.M. and **Schweyen, T.H.** 1983. Sedimentology and geochemistry of saline lakes of the Great Plains. *Hydrobiologia* 105: 245–263.

Marchant, R. and **Williams, W.D.** 1977. Organic content of some saline lake sediments in Western Victoria. *Australian Journal of Marine and Freshwater Research* 28: 269–275.

Spencer, R.J., **Baedecker, M.J.**, **Eugster, H.P.**, **Forester, R.M.**, **Goldhaber, M.B.**, **Jones, B.F.**, **Kelts, K.**, **Mckenzie, J.**, **Madsen, D.B.**, **Rettig, S.L.**, **Rubin, M.**, and **Bowser, C.J.** 1984. Great Salt Lake, and precursors, Utah: the last 30,000 years. *Contributions to Mineralogy and Petrology* 86: 321–334.

18

Great Salt Lake

The Great Salt Lake, Utah (U.S.A.), a remnant of glacial Lake Bonneville, is an extremely saline (up to 33.0% salinity) and extremely large (about 3900 km^2), terminal lake in the Great Basin of North America (Figure 18.1). Before the construction of a causeway across the lake, the whole lake had a salinity of about 20.0%. After the completion of the causeway in 1959, the South Arm (Figure 18.2a), which receives most of the inflow from the surrounding mountains, had an average salinity of about 12.0%, while the North Arm (Figure 18.2b) became extremely hypersaline (33.0% salinity) (Post, 1977). Great Salt Lake has shown dramatic changes in water level during historic times, depending primarily on the extent of winter snowfall in the mountains to the east. From relatively low water levels in the early 1970s, the lake rose to an historic high level in the mid-1980s. The causeway which divides the lake has been breached intentionally within recent years to prevent the shores of the South Arm from drowning due to excessive inflow. Since human activity has altered the salinity regime of the lake, and therefore biological activity, older (pre-1959) reports may not represent the ecology of the lake as it is now.

The following discussion will center largely on the descriptive microbiology of the lake, since there are few published reports describing *in situ* microbial processes. The hydrologic and chemical history of the lake during the Holocene has also been documented in great detail (Spencer et al., 1984, 1985a, 1985b). While data describing the organic geochemistry and *in situ* degradation in the sediments are incomplete, evidence gleaned so far from sediment and porewaters accumulated during the last 30,000 years offer a unique opportunity to evaluate and interpret biological processes and remains in this hypersaline basin.

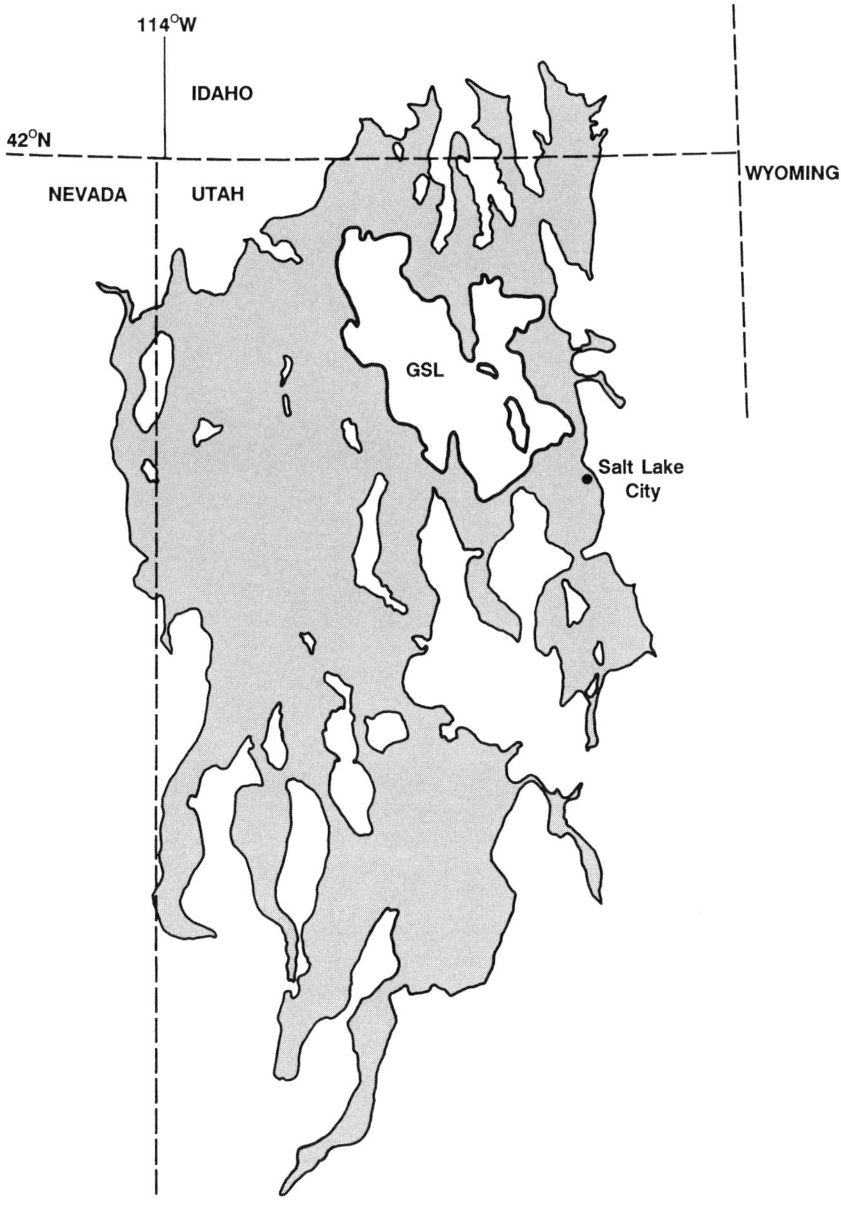

Figure 18.1 The recent Great Salt Lake (GSL) and its dilute precursor, Lake Bonneville, 17,000 years B.P. Lake Bonneville is indicated by the shaded area. Adapted from Morrison (1966).

a)

b)

Figure 18.2 Great Salt Lake. a) A view of the South Arm. Photograph taken from the top of Stansbury Island, looking east. b) A salt spring and salt flats in the Northern Arm. Photograph taken in the late 1960's during a low-water period of the lake.

18.1 Chemistry and physics

In a well-documented review, Post (1977) described the general limnology of the Great Salt Lake and compared it to the Dead Sea. The Great Salt Lake is much larger in area than the Dead Sea (3900 vs. 800 km^2), shallower (10 vs. 340 m deep), and subject to greater temperature changes ($-5°$ to 35°C vs. 21°–36°C). The ion composition and pH of the two lakes differ as well (Table 1.1). NaCl constitutes 86% (w/w) of the salts in the Great Salt Lake and only 78% of the salts in the Dead Sea. Sulfate concentrations of 281 mM were measured in the North Arm of the Great Salt Lake while they were only 4 mM in the Dead Sea. The high sulfate concentration not only affects the dissolved concentrations of Ca^{2+} (7.5 mM in the North Arm, 430 mM in the Dead Sea), but it also causes a yearly cycle of mirabilite precipitation in the winter and dissolution in the spring in the North Arm. Both lakes have reported sulfide accumulations at depth, but neither sulfate-reduction nor the isolation of sulfate-reducing bacteria has been confirmed (discussed below).

Post (1977, 1981) published nutrient analyses of the North Arm brines for 1973–1976. Some of these measurements are presented in Table 1.4. Orthophosphate values ranged from 3 to about 17 μM, and total P ranged from 630 to 1100 $\mu g \cdot l^{-1}$, except near the sediments where even higher concentrations were measured. Nitrate and nitrite were absent. Ammonia ranged from 0 to 60 μM but soluble nitrogen was always present in $mg \cdot l^{-1}$ quantities. Degens et al. (1964) reported the Great Salt Lake had 490 $\mu g \cdot l^{-1}$ dissolved amino acids, with the relative proportions of basic, neutral, and acidic amino acids as basic > neutral > acidic. No N_2 fixation (acetylene reduction technique) has been detected in the lake (Post, 1977).

18.2 Bacteria

An early report of bacterial populations in the Great Salt Lake gave viable counts (plate count technique) of 200–625 $cells \cdot ml^{-1}$ (Daniels, 1917). Five different species were isolated and briefly described. Among these, three produced color (yellow, orange, and violet). The later studies of Smith and ZoBell demonstrated unequivocally that an indigenous bacterial flora existed in the Great Salt Lake and that freshwater and terrestrial bacteria introduced into the brine could not survive. Smith and ZoBell (1937a) set up enrichment cultures using hypersaline lake water (33.6% total dissolved solids), and estimated that 50% of the Great Salt Lake bacteria required >7% salt to grow and that 96% of the bacterial population could not grow without salt. A similar percentage could not grow in seawater-based medium (ZoBell et al., 1937). By immersing sterile glass slides in lake water, Smith and ZoBell (1937b, 1937c) showed that bacteria would attach to the slides and microcolonies would develop (Figure 18.3). Even after 6 hr submergence, an appreciable

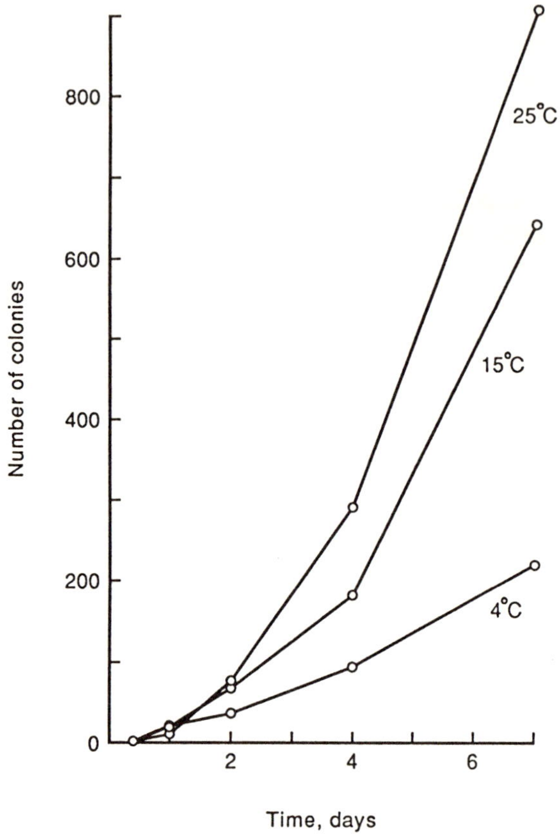

Figure 18.3 Rate of development of colonies of bacteria attached to sterile glass slides immersed in the brines of Great Salt Lake (from Smith and ZoBell, 1937b).

number of bacteria were attached in 4°C brine. Controls showed that dead bacteria or bacteria from other sources inoculated in Great Salt Lake brine did not attach. Different sites in the lake yielded different numbers of attached bacteria after 24 hr submergence (40–1100 microcolonies·cm^{-2}), indicating a heterogeneity of sizes of active bacterial populations.

Smith and ZoBell (1937b) determined that the bacteria were mostly small bacilli, but vibrio-like and coccoid forms were also observed. They counted at least nine morphological varieties of bacteria, seven of which occurred in microcolonies. These included ovoid rods, slender rods, spindle rods, spirochaetes, sheathed forms, alga-like forms, and "bizarre" forms. About one-third of the bacteria had capsules and some had structures suggestive of spores. Less than 10% stained Gram positive.

ZoBell et al. (1937) also inoculated full-strength and diluted Great Salt

Lake water with sewage, seawater, and soil at 25°C, and mouth washings at 37°C, and measured viability after incubation. None of the sewage or oral bacteria grew in 50% or 75% lake water (14–21% salinity). Less than 1% of the soil and marine bacteria survived 50% or 75% lake water. In 10–25% lake water, survival rates of 4–18% of the sewage and oral bacteria, 13–30% of the soil bacteria, and 10–64% of the marine bacteria were recorded. Full-strength lake water killed over 90% of the sewage bacteria during a one-minute exposure, but only 60% of the soil bacteria were killed by this exposure. For *Escherichia coli* and *Staphylococcus albus*, 25% lake water in peptone medium was bactericidal. The results of Burdyl and Post (1979) are similar to those of ZoBell et al. (1937).

Although relatively low bacterial plate counts (using 25% NaCl, no temperature given) were determined by Daniels (1917), Post (1977) found high numbers of bacteria by plate counts (in 22% NaCl medium at 35°C) in the North Arm ($5.4 \times 10^4 \cdot ml^{-1}$ in the winter, $5.5 \times 10^6 \cdot ml^{-1}$ in the summer). Plate counts in the South Arm were not reported. Direct counts, which typically yield larger numbers, were made in both arms of the lake. Bacterial counts in the North Arm between 1974 and 1976 were $4.0–10 \times 10^7 \cdot ml^{-1}$ at all times of the year except after an algal bloom or in samples taken near the sediment surface. Those counts were $1.0–2.4 \times 10^8 \cdot ml^{-1}$. Post noted that the bacteria appeared to be distributed uniformly in the North Arm, including the anoxic hypolimnion. These observations contrast those of Smith and ZoBell (1937b), who found a great heterogeneity in microcolony-forming bacteria. Post (1977) also noted that the bacterial counts were less in the more dilute South Arm ($<10^6 \cdot ml^{-1}$), but no details of their distribution were described.

Isolated bacteria from the North Arm include strains of *Halobacterium* and *Halococcus* (Post, 1977). Some isolates can hydrolyze chitin, and some can digest the cell wall of *E. coli*. Some of the isolates were reported to produce acid from carbohydrates. It is uncertain which, if any, of these strains are the same as those observed in the earlier studies. Seven different halophages specific for halobacteria have been isolated from the lake, but their role in bacterial ecology has yet to be elucidated.

The presence of 3–6 mg $H_2S \cdot l^{-1}$ in the brine of the South Arm (Stephens and Gillespie, 1976), suggests anaerobic bacterial activity. Two obligately anaerobic bacteria have been isolated from the Great Salt Lake (Chapters 5 and 6). *Haloanaerobium praevalens*, a eubacterium found in both arms of the lake, transforms methionine to methylmercaptan. This is significant in the Great Salt Lake ecosystem since methane is derived principally from methylmercaptan and methanol in the lake, not from the more typical precursors, acetate and CO_2-H_2. Paterek and Smith (1985, 1988) also described a moderately halophilic methanogen, *Methanohalophilus mahii*, from sediments of the South Arm. It does not produce methane from CO_2-H_2, formate, or acetate, but rather from methanol and methylamines. No sulfate-reducing bacteria have been reported from Great Salt Lake sediments.

Direct microscopic observations of North Arm brines were reported to show at least two types of phototrophic bacteria: *Ectothiorhodospira*-types and amoebobacter-types (Brock, 1979). *Ectothiorhodospira* was isolated but no details were given. Denitrifying bacteria were also isolated, although neither nitrate nor nitrite are detectable in the lake.

Ward and Brock (1978) measured bacterial degradation of glutamate, mineral oil, and hexadecane in both arms of the Great Salt Lake and associated salt evaporation pans. There was an inverse relationship between salinity and degradation rates for these compounds. The rate limitations were apparently not due to low oxygen concentrations nor the availability of organic nutrients. Using radiolabeled substrates, the investigators found low rates of glutamic acid oxidation at high salinities (Figure 18.4) but no hexadecane degradation could be detected at these salinities. Enrichment cultures on mineral oil were only positive in ≤20% salinity. These results confirm the popular (but rarely empirically demonstrated) notion that rates of biodegradation decrease with increasing salinity and that the degradation of pure hydrocarbons is extremely retarded in strong brines.

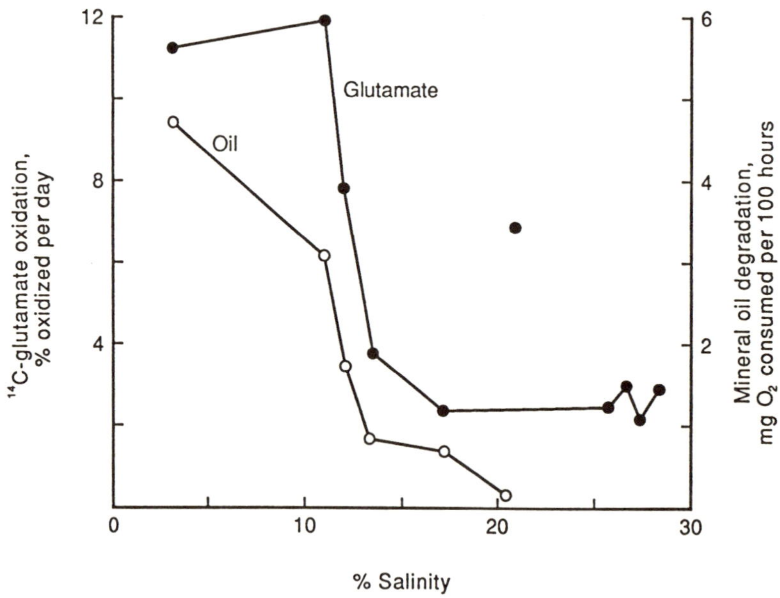

Figure 18.4 Glutamate and mineral oil oxidation rates in brines from salt evaporation ponds and Great Salt Lake. Glutamate oxidation (closed circles) was calculated from tabulated data and mineral oil oxidation (open circles) was estimated from graphic data of Ward and Brock (1978).

18.3 Algae, cyanobacteria, and primary productivity

While the early report of Daniels (1917) listed a few taxa of algae, the large monograph by Felix and Rushforth (1979) documents the occurrences and general salinity tolerances of four cyanobacteria, seven chlorophytes, one dinoflagellate, and 17 diatoms. A total of 77 species have been documented from the Great Salt Lake, but only 29 species were growing in the plankton and periphyton at the time of their study. The most salt-tolerant species are listed in Table 2.1. Table 18.1 lists the highest salinities recorded by Felix and Rushforth for taxa capable of living in >10% salinity. These salinities can be compared to maximum salinities tolerated by other taxa in Table 2.3. The illustrations and photomicrographs of the identified algae in the publication of Felix and Rushforth should serve as a useful guide for identifying algae from other hypersaline environments.

Brock (1976) characterized the cyanobacterium *Aphanothece halophytica* (*Coccochloris elabens*) isolated from the North Arm from a mixed algal-phototrophic bacterial mat. The green layer of the mat contained mostly the palmelloid stage of *Dunaliella, A. halophytica,* and a filamentous cyanobacterium (probably *Phormidium*). The purple layer contained several unidentified phototrophic bacteria. Enrichments for *A. halophytica* showed that this organism grew most rapidly in 16–23% NaCl. In 30% NaCl, it grew very slowly. *A. halophytica* was also found in the South Arm where it formed mats.

Table 18.1 Location and prevalence of some cyanobacteria and algae in the Great Salt Lake[a]

Organism	Site of collection[b]
Cyanobacteria	
Coccochloris elabens	South Arm (common)
Microcoleus lyngbyaceus	South Arm (rare)
Chlorophytes	
Dunaliella salina	North Arm (common)
D. viridis	South Arm (common), North Arm (rare)
Diatoms	
Biddulphia levis	South Arm (common)
Navicula sp.	South Arm (common)
N. graciloides	South Arm (abundant)
N. tripunctata	South Arm (abundant)
Entomoneis pulchra	South Arm (common)
Amphora coffeaeformis	South Arm (the most common diatom)
A. delicatissima	South Arm (common)
Rhopalodia musculus	South Arm (abundant)
Nitzschia epithemioides	South Arm (common)
N. palea	South Arm (common)

[a] From Felix and Rushforth (1976).
[b] South Arm maximum measured salinity was 12.9%; North Arm maximum measured salinity was 34.2%.

In the South Arm, the dominant phytoplankter is *Dunaliella viridis*. The standing crop is about 10^6 cells$\cdot l^{-1}$ during most of the year. In April of both 1971 and 1973, short blooms producing 2.5×10^8 cells$\cdot l^{-1}$ were recorded (Stephens and Gillespie, 1976). Reeflike benthic bioherms (microbial mats) composed of cyanobacteria (*Coccochloris elabens* and *Oscillatoria* sp.) occur in shallow water of the South Arm. The distribution of diatoms was noted by Felix and Rushforth (1976) but their contribution to the algal biomass was not calculated.

Stephens and Gillespie (1976) calculated primary productivity rates in the South Arm of the Great Salt Lake. Recognizing the limitations of the $^{14}CO_2$ method due to the lack of reliable dissociation constants for the carbonate species in the brines, corrections for the carbonate alkalinity to total CO_2 were not attempted. The uncorrected estimates averaged 145 g $C\cdot m^{-2}\cdot yr^{-1}$ for two stations in 1973. The greatest production occurred during a *Dunaliella* bloom in March and April when daily carbon fixation rates averaged 2.13 g $C\cdot m^{-2}$. A small bloom in August contributed 5% of the annual primary productivity. The highest daily rate measured was 7.66 g $C\cdot m^{-2}$ in August. Factors that limited total primary productivity included low winter temperatures, self-shading (in the April blooms), available nitrogen, and grazing by the brine shrimp *Artemia salina*. Rates of productivity and biomass in Great Salt lake are compared to other hypersaline systems in Table 2.2.

In the North Arm the plankton is devoid of most of the species found in the South Arm (Post, 1977). Halobacteria dominate the bacterial flora and *Dunaliella salina* is by far dominant over *D. viridis*. No other algal species have been noted in the plankton. *Dunaliella*-bacterial crusts have been observed on the shores of the North Arm (Post, 1980), with both cyanobacteria and phototrophic bacteria present (Brock, 1976). Cultures of *Dunaliella* isolated from 25% salt grew better in 16% (w/v) NaCl than in higher salinities (Brock, 1975). $^{14}CO_2$ uptake of the natural population concentrated by centrifugation had a very low salinity optimum (3.5% salt) which could not be explained. Because there was no apparent correction for the changes in carbonate alkalinity at the different salinities used in the incubations, the results of that experiment may reflect lower concentrations of CO_2 (less dilution of the radiolabel) in lower salt.

The estimated biomass contributed by brine shrimp, *Dunaliella* spp., and bacteria (viable counts and direct counts) in the North Arm plankton was (in $g\cdot m^{-3}$): 0.1, 25.4, 22, and 300, respectively (Post, 1977). Depending on the method used to enumerate the bacteria, their biomass is either approximately equal to or 10-fold greater than that of the primary producers. This is rather unusual for an unpolluted aquatic system. The rather high contribution by bacteria suggests that *Dunaliella* and *Artemia* are excreting large amounts of reduced carbon and/or the bacteria are living at the expense of DOC and POC brought in from the South Arm and concentrated by evaporation. The flow of carbon to the bacteria in this system is poorly known.

18.4 Protozoa and fungi

Protozoa were recognized in Great Salt Lake water by Vorhies (1917), who found two or three varieties including the most common, *Amoeba limax*. Pack (1919) found two ciliates in the lake, *Uroleptus packii* and *Prorodon utahensis*. The effects of salinity on these protozoans are described in Chapter 10. A microcosm of North Arm brine and sediments maintained unfed in an aquarium in the laboratory developed large populations of four types of protozoans: a wedge-shaped and a long, thin flagellate, amoebae of two different sizes, and a *Tetramitus*-like organism (Post, 1977). It is possible that some of these forms represent alternate morphologies of the same species. A ciliate population developed in a similar microcosm that was lighted and fed, but these microorganisms were not observed in the lake proper by Post (1977). Another heterotrophic microorganism, the fungus *Cladosporium*, was found on a submerged piece of wood in the lake (Cronin and Post, 1977). No other published accounts of the mycology of this environment are available.

Although this volume is not devoted to the macroorganisms in hypersaline environments, the grazing activities, fecal matter accumulation, and the contributions to productivity and nutrient cycles by brine shrimp and brine flies in the Great Salt Lake are considerable (Eardley, 1938; Post, 1977).

18.5 Biogeochemistry

The nitrogen cycle has been fairly well documented (Post, 1977), but the carbon and sulfur cycles are not as well known. With the exception of the isolation of several anaerobic bacteria, there is little published information on the anaerobic processes in the lake. The North Arm was anoxic below 6 m in August, 1975, and anoxic at the bottom in February, 1976 (Post, 1977). Stephens and Gillespie (1976) stated that South Arm was meromictic, with the brine below 8 m anaerobic and slightly denser than the surface waters. These data indicate that the whole lake bottom is anaerobic but the types and rates of transformations of organic matter at the bottom are poorly known.

In a study of the evolution of the lake, Spencer et al. (1985b) determined that NaCl-rich hydrothermal springs and $CaCO_3$-dominated river waters were the source of ions for the lake. A mass balance of Na^+ and Cl^- in all the sinks (lake brines, pore fluids, salts, sediments, and overflow losses) accounted for 97–98% of the total inputs calculated from river and spring influxes. A similar calculation of the sinks of sulfate ions accounted for only 87% of the inputs. The authors concluded that a combination of sulfate reduction, evaporite precipitation, and mass transport of sulfate to pore fluids has resulted in sulfate depletion in the brine. No estimate was given of the importance of bacterial sulfate reduction.

Analyses of sediment cores up to 6.5 m in length from the South Arm of the Great Salt Lake show that five distinct periods of lake history can be discerned based on sedimentological criteria (Spencer et al., 1984). These sediments comprise over 30,000 years of Lake Bonneville history. Prior to 32,000 yr before the present, the basin was an ephemeral saline lake. Units I (the youngest), II, IV, and V have brine shrimp egg capsules and pellets, indicating hypersaline periods. Only the top 10–20 cm of Unit III (total length = 160 cm) contains remains of brine shrimp. The rest of Unit III contains abundant ostracod valves from a period when the lake was one of fresher water.

It is interesting to interpret the biological activities of the freshwater period with those of the more saline periods. Unit III has generally less organic carbon than the other layers (0–2% vs. up to 4%), some of the lowest $\delta^{13}C$ values (−28 to −30‰ vs. −20 to −29‰), the highest sulfur/carbon ratios (3.0 vs. an average of 0.1), and the greatest $\delta^{34}S$ values (0‰ vs. −18 to −26‰). In one core sample, >70% of the sulfide was in the form of pyrite in Unit III while in Unit I, 50% of the sulfide was in the form of pyrite and 50% was organically bound. The high sulfur/carbon values and pyrite concentrations in Unit III may be related to an increased abundance of Fe or a more complete oxidation of organic matter. The high $\delta^{34}S$ values may indicate sulfate depletion and the bacterial reduction of the very heavy residual sulfate. The authors found the $\delta^{34}S$ fractionation of the sulfate at the modern lake bottom was +1.6‰, indicating that sulfate-reducing bacteria were probably active.

The investigators interpreted the relatively light carbon in Unit II as being derived from a different source or from bacterial recycling. Methanogenesis and bacterial oxidation of the methane would result in light organic carbon. The high sulfur/carbon ratios may partially reflect the lower organic fallout into the sediments, and the more complete degradation of organic matter under less saline conditions. The interpretations of the data of Spencer et al. (1984) support the popular theories regarding reduced rates of degradation under extremely hypersaline conditions. These interpretations are supported by the *in situ* degradation studies of Ward and Brock (1978).

References

Brock, T.D. 1975. Salinity and the ecology of *Dunaliella* from Great Salt Lake. *Journal of General Microbiology* 89: 285–292.

Brock, T.D. 1976. Halophilic blue-green algae. *Archives of Microbiology* 107: 109–111.

Brock, T.D. 1979. Ecology of saline lakes, pp. 29–47 in Shilo, M. (editor), *Strategies of Microbial Life in Extreme Environments*, Dahlem Konferenzen, Berlin.

Burdyl, P. and **Post, F.J.** 1979. Survival of *Escherichia coli* in Great Salt Lake water. *Water, Air, and Soil Pollution* 12: 237–246.

Cronin, E.A. and **Post, F.J.** 1977. Report of a dematiaceous hyphomycete from the Great Salt Lake, Utah. *Mycologia* 69: 846–847.

Daniels, L.L. 1917. On the flora of Great Salt Lake. *American Naturalist* 51: 499–506.

Eardley, A.J. 1938. The sediments of the Great Salt Lake, Utah. *Bulletin of the American Association of Petroleum Geologists* 22: 1305–1411.

Felix, E.A. and **Rushforth, S.R.** 1979. The algal flora of the Great Salt Lake, Utah, U.S.A. *Nova Hedwigia* 31: 163–194.

Morrison, R.B. 1966. Predecessors of Great Salt Lake, pp. 77–104 in Stokes, W.L., ed., *The Great Salt Lake, Guidebook to the Geology of Utah 20,* Utah Geological Society, Salt Lake City.

Pack, D.A. 1919. Two ciliata of Great Salt Lake. *Biological Bulletin* 36: 273–282.

Paterek, J.R. and **Smith, P.H.** 1985. Isolation and characterization of a halophilic methanogen from Great Salt Lake. *Applied and Environmental Microbiology* 50: 877–881.

Paterek, J.R. and **Smith, P.H.** 1988. *Methanohalophilus mahii* gen. nov., sp.nov., a methylotrophic halophilic methanogen. *International Journal of Systematic Bacteriology* 38: 122–123.

Post, F.J. 1977. The microbial ecology of the Great Salt Lake. *Microbial Ecology* 3: 143–165.

Post, F.J. 1980. Oxygen-rich gas domes of microbial origin in the salt crust of the Great Salt Lake, Utah. *Geomicrobiology Journal* 2: 127–139.

Post, F.J. 1981. Microbiology of the Great Salt Lake North Arm. *Hydrobiologia* 81: 59–69.

Smith, W.W. and **ZoBell, C.E.** 1937a. An autochthonous bacterial flora in Great Salt Lake. *Journal of Bacteriology* 33: 118.

Smith, W.W. and **ZoBell, C.E.** 1937b. Direct microscopic evidence of an autochthonous bacterial flora in Great Salt Lake. *Ecology* 18: 453–458.

Smith, W.W. and **ZoBell, C.E.** 1937c. Direct microscopic evidence of an indigenous bacterial flora in Great Salt Lake. *Journal of Bacteriology* 33: 87.

Spencer, R.J., Baedecker, M.J., Eugster, H.P., Forester, R.M., Goldhaber, M.B., Jones, B.F., Kelts, K., Mckenzie, J., Madsen, D.B., Rettig, S.L., Rubin, M., and **Bowser, C.J.** 1984. Great Salt Lake, and precursors, Utah: the last 30,000 years. *Contributions to Mineralogy and Petrology* 86: 321–334.

Spencer, R.J., Eugster, H.P., Jones, B.F., and **Rettig, S.L.** 1985a. Geochemistry of Great Salt Lake, Utah I: Hydrochemistry since 1850. *Geochimica et Cosmochimica Acta* 49: 727–737.

Spencer, R.J., Eugster, H.P., and **Jones, B.F.** 1985b. Geochemistry of Great Salt Lake, Utah II: Pleistocene-Holocene evolution. *Geochimica et Cosmochimica Acta* 49: 739–747.

Stephens, D.W. and **Gillespie, D.M.** 1976. Phytoplankton production in the Great Salt Lake, Utah, and a laboratory study of algal response to enrichment. *Limnology and Oceanography* 21: 74–87.

Vorhies, C.T. 1917. Notes on the fauna of Great Salt Lake. *American Naturalist* 51: 494–499.

Ward, D.M. and **Brock, T.D.** 1978. Hydrocarbon biodegradation in hypersaline environments *Applied and Environmental Microbiology* 35: 353–359.

Zeikus, J.G., Hegge, P.W., Thompson, T.E., Phelps, T.J., and **Langworthy, T.A.** 1983. Isolation and description of *Haloanaerobium praevalens* gen. nov. and sp. nov., an obligately anaerobic halophile common to Great Salt Lake sediments. *Current Microbiology* 9: 225–234.

ZoBell, C.E., Anderson, D.Q., and **Smith, W.W.** 1937. The bacteriostatic and bactericidal action of Great Salt Lake water. *Journal of Bacteriology* 33: 253–262.

19

Dead Sea

19.1 General chemistry

The Dead Sea (see map, Figure 14.1), like the Great Salt Lake, has undergone recent physical and chemical changes as a result of human influence. Before 1979, this very deep lake was characterized by two basins: a small, shallow (up to about 8 m deep) southern basin (Figure 19.1a), and a larger, deeper (340 m deep), density-stratified, meromictic northern basin (Figure 19.1b). Details of the Dead Sea have been presented by Neev and Emery (1967), Nissenbaum (1975), and Steinhorn (1985). The Jordan River supplies the major source of water for this terminal lake. After about 1964, the use of water for irrigation from the Jordan River and its watershed reduced the inflow so much that evaporation in the Dead Sea exceeded inflow. Between 1930 and 1979, the water in the lake dropped 10 m and the southern basin was completely exposed. By 1975, the mixolimnion had increased in salinity to the point that cooler winter temperatures in the surface layers reduced the density differences between the upper and lower layers of the lake. This resulted in increased mixing, deepening of the mixolimnion, and eventually complete overturn in February, 1979 (Steinhorn et al., 1979). The changes in the structure of the Dead Sea before overturn were summarized by Beyth (1980) and Steinhorn and Assaf (1980) while those during the 1975–1979 period were reported by Steinhorn et al. (1979) and Steinhorn (1985). Table 19.1 lists some of the chemical constituents of the lake before, during, and after the 1979 overturn. The differences in the composition of the deep lake layers between 1975 and 1978 are most likely due to differences in analytical techniques or errors.

The Dead Sea is essentially a Mg-Na-Cl lake. At the time of overturn,

a)

b)

Figure 19.1 Dead Sea. a) View of the southern basin, showing the extensive salt deposits which have developed at the periphery of the lake. b) View of the northern basin. In the foreground is the effluent of a small hot spring which has its source nearby in the Dead Sea basin.

the lake was saturated and even slightly supersaturated with respect to gypsum and halite (Katz et al., 1981; Steinhorn, 1983). The lake lies in a deep depression in a northern extension of the African Rift Valley system. As of 1978, the surface was 402 m below sea level (Steinhorn, 1985). Large, saline lakes are known to have occupied this rift zone since the Oligocene. The

Table 19.1 Composition of the major chemical components of the Dead Sea before and after overturn (in $g \cdot l^{-1}$)

	Depth			
	Before overturn (1975)		Before overturn (1978)	After overturn (1979)
	0–40 m	40–330 m	Below 200 m	Below 200 m
Total dissolved solids, %	30.0	33.2	34.0	34.0
pH	6.3	5.9	6.1	6.2
Mg	36.2	42.4	44.2	44.1
Na	38.5	39.7	39.5	40.5
Ca	16.4	17.2	17.2	17.1
K	6.5	7.6	7.6	7.8
Cl	196.9	219.3	224.9	226.0
Br	4.6	5.3	5.4	5.3
SO_4^{2-}	0.6	0.4	0.456	0.444
HCO_3^-	0.2	0.2	0.246	0.259
References[a]	1, 2	1, 2	3	3

[a] (1) Neev and Emery (1967); (2) Nissenbaum (1975); (3) calculated from Steinhorn (1985) based on density at 25°C.

large salt deposits of the Sdom Formation that border the Dead Sea in the south were apparently precipitated in the Upper Tertiary from an earlier saline lake (Bentor, 1961). The current lake is thought to be 12,000–70,000 yr old. The Dead Sea is not a relict body of seawater, but is a true terminal lake whose salts are derived from the Jordan River and highly saline springs (Bentor, 1961) or from the residual brines of an earlier saline lake (Neev and Emery, 1967).

Trace metal concentrations in Dead Sea brines and sediments were reported by Nissenbaum (1974, 1975). Schonberger et al. (1985) tested several methods of heavy metal determination designed to overcome interferences with chemical assays of the heavy brines of the Dead Sea. Their values for Cu and Pb were one to more than two orders of magnitude smaller than the concentrations published by Nissenbaum, while Cd values were zero to two orders of magnitude smaller than those presented by Nissenbaum (Table 19.2). These large differences are not due to analysis before and after lake overturn. Analyses of other heavy metals are presumably subject to chemical interferences and the concentrations listed by Nissenbaum (1975) should be

Table 19.2 Some trace metals of the Dead Sea (in $\mu g \cdot l^{-1}$).

Depth, m	Cu	Pb	Cd	Zn	Mn	Fe	Reference[a]
0–10	3.43–38.01	0.02–26.4	0.03–10.28	nd	nd	nd	2
0–40	300	300	10	500	4000	15	1
40–330	500	120	8	500	7100	10	1

[a] (1) Nissenbaum (1975); (2) Schonberger et al. (1985).
nd, not determined.

regarded as maximum values. It is noteworthy that heavy metal concentrations as high as 750 $\mu g \cdot l^{-1}$ Zn, 750 $\mu g \cdot l^{-1}$ Cu, and 450 $\mu g \cdot l^{-1}$ Pb in excess of the ambient concentrations of these metals did not inhibit enrichment cultures of algae and bacteria from the Dead Sea (Oren, 1983d).

19.2 Microbiology

The study of the microbiology of the Dead Sea during the last half century can be roughly divided into three epochs of intensive investigations: the microbiological surveys by Elazari-Volcani published in the 1940's, the microbiological and biogeochemical studies by Nissenbaum, Kaplan, and coworkers in the 1970's, and investigations by Oren after the overturn of the lake. The following discussion is an introduction to the descriptive biology of the Dead Sea during the last 50 years followed by an account of *in situ* microbial activities and their biogeochemical traces.

Many kinds of enrichments for bacteria from the Dead Sea have been tried with mixed success (Table 19.3). Some isolates of both eubacterial and archaebacterial halophiles have been further characterized. Details of these strains are given in Chapters 5 and 6. A detailed review of the microorganisms isolated from the Dead Sea was also given by Oren (1988).

Isolation of bacteria from some positive enrichments were not pursued (i.e., sulfur-oxidizers). Other bacteria, particularly the sulfate-reducers, could not be sub-cultured after their initial enrichments (Kaplan and Friedman, 1970). Nissenbaum (1975) reported the isolation of a *Clostridium*-like sulfate-reducer. Anaerobic enrichments by Oren for sulfate-reducers were negative but they resulted in the isolation of obligately anaerobic, fermenting bacteria, *Sporohalobacter lortetii, S. marismortui,* and *Halobacteroides halobius* (Oren, 1983b; Oren et al., 1984, 1987). The positive demonstration of both sulfur-oxidizers and reducers would confirm the existence of an active sulfur cycle in the Dead Sea.

Both the high salt and the peculiar composition of the Dead Sea may exclude major groups of bacteria, including the phototrophic bacteria and some members of the nitrogen cycle (e.g., N_2-fixers and nitrifiers). The absence of these same nitrogen bacteria was noted in the Great Salt Lake ecosystem as well (Post, 1977).

The dominant or only alga of the Dead Sea is *Dunaliella parva* (Oren and Shilo, 1982) although it was called *D. viridis* in earlier reports (Elazari-Volcani, 1940b; Kaplan and Friedman, 1970). Other algae have been reported, but they were apparently washed into the lake (Elazari-Volcani, 1940a, 1943a).

Elazari-Volcani (1943a) reported the isolation of an amoeba from Dead Sea mud samples. It grew optimally in 15–18% salt. It survived in saturated salt and grew slowly in 6% salt. It had an alternate, flagellated morphology. A ciliate which grew in 21% salt was also isolated from shallow water (Elazari-

Table 19.3 Bacterial enrichment studies in the Dead Sea

Kind of bacteria	Results of enrichments	Identification	Reference[a]
Coliforms	−		1
Urea decomposers	−		1
Nitrogen fixers	−		1
Nitrifiers	−		1
Denitrifiers	+	*"Halobacterium maris-mortui"* (red) *Micrococcus morrhuae* (red) *"Chromobacterium maris-mortui"* (white but produced a soluble blue pigment) *"Pseudomonas halestorgus"* (grey-white to brown) *Flavobacterium halmephilum* (yellow)	1, 2
	−		3
Thiosulfate oxidizers	+		1, 2
	−		3
Cellulose decomposers (aerobic)	+		1, 2, 3
Fibrinolytic bacteria	+		1
Lactate utilizers (aerobic)	+	*Sarcina* sp.	1
Phototrophic bacteria	−		3
Luminescent bacteria	−		3
Lignin decomposers	−		3
Chitin decomposers	+ (faint)		3
Petroleum oxidizers	+		2
Glucose fermenters	+		1, 2
Protein or yeast extract decomposers:			
Aerobic	+	*Haloferax volcanii*	4
		Halobacterium sodemense	7
			1, 3
Anaerobic	+	*Halobacteroides halobius*	8
		Sporohalobacter marismortui	9
		S. lortetii	6
Sulfate reducers	+	Could not be sub-cultured	3, 5

[a] (1) Elazari-Volcani (1940b); (2) Elazari-Volcani (1943b); (3) Kaplan and Friedman (1970); (4) Mullakhanbhai and Larsen (1975); (5) Nissenbaum (1975); (6) Oren (1983b); (7) Oren (1983c); (8) Oren et al., (1984); (9) Oren et al. (1987).
−, negative results, +, successful enrichment cultures.

Volcani, 1940b). Neither these organisms nor any other protozoans or higher organisms have been found in more recent investigations.

Both the halobacteria and *Dunaliella* live in the Dead Sea near their upper salt limits (Oren, 1981; 1988). At the time of overturn, the algae were limited to the upper 10 m of the water column and the bacteria were mostly in the upper 10–25 m. A slight freshening of the surface layers of the Dead Sea resulted in a bloom of both algae and bacteria in the summer of 1980 (Oren, 1981, 1983d). Up to 1.0×10^7 bacteria·ml^{-1} were found during the bloom.

The population then decreased to 5×10^6 cells·ml^{-1} for over one year. Before the overturn, 2–9 $\times 10^6$ bacteria·ml^{-1} were found near the surface (Kaplan and Friedman, 1970). The bloom was primarily composed of *H. volcanii* and *H. sodomense* with no mention of "*H. marismortui*" nor halotolerant eubacteria.

The bacterial population had measurable concentrations of bacteriorhodopsin (Oren and Shilo, 1981). In the autumn of 1981, a low rate of light-stimulated, bacteriorhodopsin-mediated, CO_2 fixation (0.023 mg C·m^{-3}·d^{-1}) was measured near the lake surface (Oren, 1983a). Carbonate values in all the CO_2 uptake studies by Oren were based on the calculations of Sass and Ben-Yaakov (1977). Controls that chemically blocked CO_2 fixation by the few *Dunaliella* cells (4–6 cells·ml^{-1}) demonstrated that light-stimulated CO_2 uptake was possible by a natural population of halobacteria. While it was suggested that the halobacteria may contribute significantly to light-dependent CO_2 assimilation in the Dead Sea, there is no evidence that CO_2 is reduced during the process and therefore cannot be termed primary productivity.

The summer 1980 bloom of bacteria in the Dead Sea accompanied a bloom of *Dunaliella parva*, with population densities of algae as high as 8.8 $\times 10^3$ cells·ml^{-1} (Oren and Shilo, 1982). A variety of closely monitored environmental data and algal activities according to depth and salinity were presented, including distribution of cells and CO_2 assimilation, penetration of solar radiation, and growth rates of cultured cells. On August 7, 1980, the maximal rate of photoassimilation of CO_2 by *Dunaliella* was 6.9 mg C·m^{-3}·h^{-1} (Oren, 1981). On December 14, 1980, and January 1, 1981, CO_2 uptake rates (which were expressed as a per cent of the rates at the surface) were maximal at the surface and negligible at 10 m depth.

Experiments showed that dilution of Dead Sea water and the addition of phosphate are prerequisites for a *Dunaliella* bloom, while dilution, phosphate, and a suitable carbon source (a *Dunaliella* bloom or glycerol) are required for a bloom of halobacteria (Oren and Shilo, 1982, 1985; Oren, 1983d, 1985). The slow growth rates reported in Dead Sea enrichment studies (Nissenbaum, 1975) may reflect limited growth activity in the strong brine. Phosphate is present in low or negligible quantities in the Dead Sea. Salt interference prevents the accurate measurement of this nutrient. Values of 30–40 μg P·l^{-1} were reported by Oren (1981).

Combined nitrogen is plentiful in the Dead Sea. Total nitrogen, which has been reported to be 2–8 mg·l^{-1}, is predominantly ammonia and dissolved organic nitrogen (Neev and Emery, 1967). Dissolved organic carbon (DOC) in the lake before it turned over was 4–8 mg·l^{-1}, but the constituents in the DOC pool were not described.

Before the lake turned over, the brine below 50 m had a negative redox potential (Neev and Emergy, 1967) and contained 0.23–0.56 mg·l^{-1} H_2S (Nissenbaum and Kaplan, 1976). Sulfate depletion with depth was found in the water column. Sulfate concentrations of 529 and 402 mg·l^{-1} were measured at 0 and 300 m depth, respectively (Nissenbaum and Kaplan, 1976). Gypsum precipitated from the upper water column but disappeared with depth (Neev and Emery, 1967). Both the sulfate depletion and gypsum disappearance suggested bacterial sulfate reduction.

19.3 Stable isotopes

In a study of the stable isotopes of sulfur and carbon in the Dead Sea, Nissenbaum and Kaplan (1976) found evidence of intensive sulfate reduction. At the Ein Gedi station, $\delta^{34}S$ values for sulfate of the water column, porewater (at 330 m depth), and sediments were +13.6 to +15.9‰, +20.6‰, and +12.7‰, respectively. All inflowing sources of sulfate, except one small spring, had positive $\delta^{34}S$ values. Reduced sulfur in the porewaters and sediments all showed evidence of isotope fractionation: $\delta^{34}S$ of both H_2S and FeS were −16.3‰ and organic-bound S was −19.6‰. The $\delta^{34}S$ value of H_2S in the lake was −19.6 to −21.7‰. Although small amounts of FeS were detected in the sediments, no pyrite or elemental sulfur was found at depth in the lake. Because sulfate reduction has not been actually measured in the sediments or water column, and positive identification of cultures of sulfate reducers have not been made, it is premature to interpret these isotope fractionations to be a result of on-going biological processes, the remains of biological activity in the brine of a previous saline lake in the same basin, or a combination of both.

The $\delta^{13}C$ values of carbonate in the water column and sediments are all negative (−0.3 to −15.4‰). Nissenbaum and Kaplan (1976) suggested the dissolved CO_3^{2-} represents a partial equilibration of the inorganic carbon introduced from the atmosphere and the Jordan River ($\delta^{13}C$ = −13‰). The oxidation of dissolved organic matter ($\delta^{13}C$ = −24‰) had little influence on the inorganic carbonate pool except near the oxic-anoxic boundary before the lake turned over. Friedman (1965) suggested that bacterial oxidation of organic matter led to the precipitation of calcite. Nissenbaum and Kaplan (1976) refuted this theory noting that Friedman's samples were collected near the lake edge and hence are probably not representative of processes in the bulk of the lake. In addition, the $\delta^{13}C$ and $\delta^{18}O$ fractionations of carbonates in the lake were highly dependent upon the size of the analyzed grains. The larger grains had more negative isotopic fractionation values, which probably reflects the influence of trapped detritus processed with the carbonates during analysis. They concluded that carbonate precipitation in the lake is basically a non-biological process. However, it may be triggered by increased concentrations of calcium at depth (Neev and Emery, 1967) that result from gypsum dissolution, which in turn is presumably driven by bacterial sulfate reduction.

19.4 Organic geochemistry

A variety of organic compounds have been found in Dead Sea sediments. Kaplan and Baedecker (1970) measured 520 $\mu g \cdot kg^{-1}$ phytanic acid, 800 $\mu g \cdot kg^{-1}$ dihydrogenphytol, and 17 $\mu g \cdot kg^{-1}$ phytane from the reducing sediments at 300 m depth. They suggested that the phytane was a hydrolysis product of halobacterial lipids. These and other lipids were further characterized by Anderson et al. (1977). Total sediment lipids were 24.5–39.4 mg extractable lipids per 100 g dry sediment, with phytanol at 2.0–2.5 mg·100 mg $sediment^{-1}$ and phytanyl diether at 1.5–2.3 mg·100 mg $sediment^{-1}$. Fatty acids constituted only 4.3% of the extractable neutral lipids of one sample. C_{16} + C_{18} fatty acids constituted 70% of the fatty acids, with saturated acids predominating. C_{12}–C_{15} fatty acids constituted the rest. Hydrocarbons accounted for only 1.1% of the neutral lipids.

In samples washed with water and acid to remove inorganic salts, Nissenbaum (1972) found the total organic content of Dead Sea sediments was 0.23–0.40% while Anderson et al. (1977) reported values of 0.38–0.87%. It is unknown how much organic matter was removed by washing the sediments before analysis, since most cells would lyse under those preparatory conditions. Nissenbaum et al. (1972) also found chlorophyll *a* but not bacterioruberin in the sediments. Reducing sediments had higher concentrations of amino acids than oxidizing sediments (747–794 $\mu g \cdot g$ dry $sediment^{-1}$ vs. 62.4–123.2 $\mu g \cdot g$ dry $sediment^{-1}$) as well as more humic and fulvic acids.

19.5 Conclusion

While the Dead Sea is not a lifeless lake, it challenges the limit to microbial life. As has been noted in other hypersaline habitats, the presence of reducing conditions and sulfide could not be positively correlated with the unequivocal demonstration of halophilic or halotolerant, sulfate-reducing bacteria. The report of positive enrichments for sulfate reducers and the cultivation of *Clostridium*-like anaerobes that reduced sulfate suggest that sulfide generation may at least in part be the result of fermentative rather than respiratory metabolism. With the advent of overturn and the diminished role of algae and bacteria in the surface of the lake, many of the clues to past and present microorganisms and their processes probably reside only in the sediments.

References

Anderson, R., Kates, M., Baedecker, M.J., Kaplan, I.R., and **Ackman, R.G.** 1977. The stereoisomeric composition of phytanyl chains in lipids of Dead Sea sediments. *Geochimica et Cosmochimica Acta* 41: 1381–1390.

Bentor, Y.K. 1961. Some geochemical aspects of the Dead Sea and the question of its age. *Geochimica et Cosmochimica Acta* 25: 239–260.

Beyth, M. 1980. Recent evolution and present stage of Dead Sea brines, pp. 155–165 in Nissenbaum, A. (editor), *Hypersaline Brines and Evaporitic Environments*, Developments in Sedimentology 28, Elsevier Scientific, New York.

Elazari-Volcani, B. 1940a. Algae in the bed of the Dead Sea. *Nature (London)* 145: 975–977.

Elazari-Volcani, B. 1940b. *Studies on the Microflora of the Dead Sea*. Ph.D. Thesis. Hebrew University of Jerusalem. 119 pp.

Elazari-Volcani, B. 1943a. A dimastigamoeba in the bed of the Dead Sea. *Nature (London)* 152: 275–277.

Elazari-Volcani, B. 1943b. Bacteria in the bottom of the Dead Sea. *Nature (London)* 152: 274–275.

Friedman, G.M. 1965. On the origin of aragonite in the Dead Sea. *Israel Journal of Earth-Sciences* 14: 79–85.

Kaplan, I.R. and **Baedecker, M.J.** 1970. Biological productivity in the Dead Sea. Part II. Evidence for phosphatidyl glycerophosphate lipid in sediment. *Israel Journal of Chemistry* 8: 529–533.

Kaplan, I.R. and **Friedman, A.** 1970. Biological productivity of the Dead Sea. Part I. Microorganisms in the water column. *Israel Journal of Chemistry* 8: 513–528.

Katz, A., Starinsky, A., Taitel-Goldman, N., and **Beyth, M.** 1981. Solubilities of gypsum and halite in the Dead Sea and in its mixtures with seawater. *Limnology and Oceanography* 26: 709–716.

Mullakhanbhai, M.F. and **Larsen, H.** 1975. *Halobacterium volcanii* spec. nov., a Dead Sea halobacterium with a moderate salt requirement. *Archives of Microbiology* 104: 207–214.

Neev, D. and **Emery, K.O.** 1967. The Dead Sea. Depositional processes and environment of evaporites. *Geological Survey of Israel Bulletin* 41: 1–147.

Nissenbaum, A. 1974. Trace elements in Dead Sea sediments. *Israel Journal of Earth-Sciences* 23:111-116.

Nissenbaum, A. 1975. The microbiology and biogeochemistry of the Dead Sea. *Microbial Ecology* 2: 139–161.

Nissenbaum, A. and **Kaplan, I.R.** 1976. Sulfur and carbon isotopic evidence for biogeochemical processes in the Dead Sea ecosystem, pp. 309–325 in Nriagu, J.O. (editor), *Environmental Biogeochemistry*, Vol. 1, *Carbon, Nitrogen, Phosphorus, Sulfur and Selenium Cycles*, Ann Arbor Science, Ann Arbor.

Oren, A. 1981. Approaches to the microbial ecology of the Dead Sea. *Kieler Meeresforschungen* 5: 416–424.

Oren, A. 1983a. Bacteriorhodopsin-mediated CO_2 photoassimilation in the Dead Sea. *Limnology and Oceanography* 28: 33–41.

Oren, A. 1983b. *Clostridium lortetii* sp. nov., a halophilic obligatory anaerobic bacterium producing endospores with attached vacuoles. *Archives of Microbiology*. 136: 42–48.

Oren, A. 1983c. *Halobacterium sodomense* sp. nov., a Dead Sea halobacterium with an extremely high magnesium requirement. *International Journal of Systematic Bacteriology* 33: 381–386.

Oren A. 1983d. Population dynamics of halobacteria in the Dead Sea water column. *Limnology and Oceanography* 28: 1094–1103.

Oren, A. 1985. The rise and decline of a bloom of halobacteria in the Dead Sea. *Limnology and Oceanography* 30: 911–915.

Oren, A. 1987. A procedure for the selective enrichment of *Halobacteroides halobius* and related bacteria from anaerobic hypersaline sediments. *FEMS Microbiology Letters* 42: 201–204.

Oren, A. 1988. The microbial ecology of the Dead Sea. *Advances in Microbial Ecology* 10: 193–229.

Oren, A. and **Shilo, M.** 1981. Bacteriorhodopsin in a bloom of halobacteria in the Dead Sea. *Archives of Microbiology* 130: 185–187.

Oren, A. and **Shilo, M.** 1982. Population dynamics of *Dunaliella parva* in the Dead Sea. *Limnology and Oceanography* 27: 201–211.

Oren, A. and **Shilo, M.** 1985. Factors determining the development of algal and bacterial blooms in the Dead Sea: a study of simulation experiments in outdoor ponds. *FEMS Microbiology Ecology* 31: 229–237.

Oren, A., Weisburg, W.G., Kessel, M., and **Woese, C.R.** 1984. *Halobacteroides halobius* gen. nov., sp. nov., a moderately halophilic anaerobic bacterium from the bottom sediments of the Dead Sea. *Systematic and Applied Microbiology* 5: 58–70.

Oren, A., Pohla, H., and **Stackebrandt, E.** 1987. Transfer of *Clostridium lortetii* to a new genus *Sporohalobacter* gen. nov. as *Sporohalobacter lortetii* comb. nov., and description of *Sporohalobacter marismortui* sp. nov. *Systematic and Applied Microbiology* 9: 239–246.

Post, F.J. 1977. The microbial ecology of the Great Salt Lake. *Microbial Ecology* 3: 143–165.

Sass, E. and **Ben-Yaakov, S.** 1977. The carbonate system in hypersaline solutions: Dead Sea brines. *Marine Chemistry* 5: 183–199.

Schonberger, E., Kassovicz, J., and **Shenhar, A.** 1985. Determination of some heavy metals in the Dead Sea. *International Journal of Environmental Analytical Chemistry* 21: 283–297.

Steinhorn, I. 1983. In situ salt precipitation at the Dead Sea. *Limnology and Oceanography* 28: 580–585.

Steinhorn, I. 1985. The disappearance of the long term meromictic stratification in the Dead Sea. *Limnology and Oceanography* 30: 451–472.

Steinhorn, I., Assaf, G., Gat, J.R., Nishry, A., Nissenbaum, A., Stiller, M., Beyth, M., Neev, D., Garber, R., Friedman, G.M., and **Weiss, W.** 1979. The Dead Sea: deepening of the mixolimnion signifies the overture to overturn of the water column. *Science* 206: 55–57.

Steinhorn, I. and **Assaf, G.** 1980. The physical structure of the Dead Sea water column, 1975–1977, pp. 145–153 in Nissenbaum, A. (editor), *Hypersaline Brines and Evaporitic Environments,* Developments in Sedimentology 28, Elsevier Scientific, New York.

20

Hypersaline, Alkaline Lakes

20.1 Introduction

Hypersaline, alkaline (>pH 9) lakes constitute a special class of extremely saline lakes in closed basins. The major ions are typically Na^+, Cl^-, HCO_3^- and CO_3^{2-}. Sulfate is proportionately low. The high concentrations of HCO_3^- and CO_3^{2-} are largely responsible for buffering these lakes at such high pH. From both biological and geochemical viewpoints, hypersaline, alkaline lakes of the African Rift Valley are the best known. Unfortunately, much of the literature describing *in situ* biological activity in those lakes and their sediments concerns the less saline lakes of the region (Talling et al., 1973; Hammer, 1981; Melack, 1981). However, the reports of the activity and isolation of bacteria involved in the reductive part of the sulfur cycle in the more extremely hypersaline, alkaline lakes of Africa may support the theory that these microorganisms can play a major role in the development of alkaline conditions in such closed basins.

The biogeochemistry of Lake Magadi (a Rift Valley lake in Kenya) and the Wadi Natrun lakes (sub-sea level depressions west of the Nile Delta basin, Egypt) are reviewed here. Several hypersaline, alkaline lakes from the Great Basin of western North America have also been investigated and the main points regarding their productivity and biogeochemistry are discussed. Because important geological deposits (e.g., the oil-bearing, Eocene Green River Formation of western North America) at least partially accumulated under conditions similar to those of modern, alkaline lakes (Eugster and Hardie, 1975), these systems deserve to be regarded as more than mere geochemical curiosities.

20.2 Lake Magadi, Kenya

Lake Magadi (Figure 20.1) is an irregularly-shaped basin about 35 km long, that is situated at an elevation of 580 m (Eugster, 1970). Trona ($NaHCO_3 \cdot Na_2CO_3 \cdot 2H_2O$) deposits cover about 75 km^2 of the basin and they are up to 40 m thick locally. The basin has been saline for at least 10,000–20,000 years. The lake is usually flooded during March and April and it dries again by June or July.

Ephemeral runoff, groundwater, and saline hot springs feed the lake, with the springs being the chief inflow into the lake. The composition of the inflows and lake brines, and the interpretation of geochemical events involved in the concentration, precipitation, and dissolution of brines and salts have been described in detail (Eugster, 1970, 1980; Jones et al., 1977). A comparison of some of the constituents is presented in Table 20.1. Analyses of conservative ions (Na^+, Cl^-, and Br^-) in the springs indicate they are derived by evaporative concentration of the dilute rivers and groundwater. Variations in concentrations of some of the non-conservative ions in the springs are attributed to precipitation (Mg^{2+}, Ca^{2+}, SiO_2, HCO_3^-, and CO_3^{2-}), degassing (CO_2), mixing with dilute groundwater (CO_3^{2-}/HCO_3^- ratios), ion exchange (K^+), and sorption with subsequent biological reduction (SO_4^{2-}). In stronger brines, the precipitation of trona leads to HCO_3^- depletion and Cl^- enrichment. Other sodium salts precipitate in strongly concentrated brines, including mirabilite, villiaumite (NaF), and thermonatrite ($Na_2CO_3 \cdot H_2O$).

The data indicate that 96–98% of the SO_4^{2-} of the dilute inflow is lost by the time the brines are concentrated in the hot springs and in the lake. Jones et al. (1977) suggested that sulfate can salt-out with SiO_2, and both Eugster (1970) and Jones et al. (1977) recognized the potential role of sulfate-reducing bacteria in the black muds associated with trona. Sediment sulfide analyses have not been published. The activitiy of sulfate-reducing bacteria remains to be measured in order to evaluate their role in sulfate removal.

Microbiology The high concentrations of phosphate in all the brines of Lake Magadi and the relatively high concentrations of dissolved organic carbon (DOC) in the interstitial brines suggest that the potential for biological productivity is high. Jones et al. (1977) found the DOC in borehole brines was 79–320 $mg \cdot kg^{-1}$. Eugster (1980) indicated on the basis of their red color that the biological activity of the concentrated brines was high. The red color could possibly be attributed to phototrophic bacteria and/or alkalophilic halobacteria, since both types of microorganisms have been isolated from Lake Magadi (Imhoff, 1988; Tindall, 1988). Imhoff et al. (1981) described a new purple sulfur phototrophic bacterium, *Ectothiorhodospira vacuolata*, a strict anaerobe that grows optimally at pH 6.5–10.0 in up to 10% NaCl.

Tindall et al. (1980, 1984) isolated red-colored rods and cocci from the trona crusts. These bacteria belong to the genera *Natronobacterium* and *Na-*

Figure 20.1 Lake Magadi and other closed basin lakes (shaded areas) of the East Rift of Ethiopia, Kenya, and Tanzania. Adapted from Eugster and Hardie (1978).

tronococcus (see Chapter 6). They grow optimally in medium with 4.0 M NaCl at pH 9. Microscopic examination of the orange-pink crusts of trona revealed many rod-shaped bacteria. These bacteria are strict heterotrophs and depend on the DOC in the brines for carbon sources. It is most likely that these

Table 20.1 A comparison of some of the constituents of Lake Magadi inflows and brines[a]

Source of water sample	Temperature, °C	pH	TDS, $g \cdot kg^{-1}$	Cl^-, $g \cdot kg^{-1}$	SO_4^{2-}, $g \cdot kg^{-1}$	SO_4^{2-}/Cl^- ratio, molar average	PO_4^{3-}, $mg \cdot kg^{-1}$
Dilute inflow	15°–24°	6.4–8.2	0.067–0.384	0.0037–0.042	0.0024–0.026	3.03	0–0.3
Hot springs	35°–86°	8.82–9.81	9.6–45.4	1.55–9.15	0.073–0.321	0.0980	1.1–18
Lake brines	26°–52°	9.50–11.22	11.7–324	23.9–106	0.598–2.61	0.0688	23–97
Interstitial brines	—	9.65–10.85	82.7–305	24.4–98.7	0.236–1.85	0.0292–0.0672[b]	50–150

[a] Data compiled from Eugster (1970) and Jones et al. (1977).
[b] Range of values from four different boreholes.
TDS, total dissolved solids.

alkalophilic halobacteria cause the red color of the Lake Magadi brines. Because they contain diether-linked polar lipids (see Chapters 4 and 6), lipid analyses of the brine and salts would indicate the relative contribution of these bacteria to the microbial biomass in Lake Magadi.

Tindall (1988) also noted that the top 10 cm of the crystalline trona deposits in Lake Magadi had a laminated microbial community of alkalophilic halobacteria and phototrophs in four distinct zones. The uppermost red-orange zone consisted of the heterotrophic alkalophilic halobacteria. This layer was underlain by a green zone of various cyanobacteria that were not further described. Below the cyanobacteria was a violet layer of *Ectothiorhodospira* which was underlain by a second green layer of presumably another phototrophic bacterium. Black sulfidic brine percolated below the four colored zones. Such a stratified community is similar to that found in microbial mats of other hypersaline environments except that the trona salt community has a colored zone of heterotrophs overlying the phototrophic microorganisms.

No descriptions have been made of the sources of organic matter and the rates of primary productivity in the lake and its associated solar salterns. Because the most important source of brines to the lake comes from spring discharge, and the springs may be recharged from a hot groundwater reservoir (Eugster, 1970), the springs may bring in decomposition products forming in the sediments and include perhaps organic matter produced during earlier stages of the lake history. Because oxygenic phototrophs are a minor component of Lake Magadi saltern brines, it is possible that the only *in situ* primary productivity in the salterns (that is, reduction of CO_2 into organic matter) may be due to the activity of anoxygenic phototrophic bacteria.

The nature of the dissolved organic matter in the lake is unknown. Jones et al. (1977) found that clarified interstitial brines were pale yellow to deep amber or coffee-colored, although it cannot be assumed that all the coloration was due to dissolved humic matter. Although only five data points are available, it appears that interstitial brine DOC is largely conservative with brine concentration, with maximum values of 320 mg DOC per kg brine at 28.7% salinity. The subjective interpretation of high biological productivity coupled with the observation that DOC content is conservative with brine concentration poses the problem of determining the nature of microbial decomposition in such hypersaline, alkaline systems.

This question is important because bacterial decay has been invoked as a mechanism to recharge the brines with CO_2 and cause the precipitation of interstital trona (Bradley and Eugster, 1969; Eugster and Hardie, 1978). Because trona precipitation depletes the brine of CO_2, further precipitation of surficial trona crusts can continue only if CO_2 is added from atmospheric or biological sources. Eugster and Hardie (1978) suggested that biological CO_2 recharge is the more rapid and effective process, even at the sediment surface.

20.3 The Wadi Natrun

The Wadi Natrun is a low-lying depression of the Libyan desert in north-central Egypt. Several eutrophic, extremely hypersaline, alkaline lakes have been described from the area (Jannasch, 1957; Abd-el-Malek and Rizk, 1963; Imhoff et al., 1979). The lakes are fed by underground seepages from the Nile River that pass through burdi (grass) swamps. Once the rather dilute swamp water (0.1–0.8% TDS) reaches the lakes, it is subjected to evaporative concentration. Some chemical analyses of Imhoff et al. (1979) are presented in Figure 20.2.

Abd-el-Malek and Rizk (1963) demonstrated that the activity of sulfate-reducing bacteria was responsible for the increase in pH between the swamps (pH 6.8–7.2) and the lakes (pH 9.2–11.5). In two of the same lakes studied by Imhoff et al. (1979) (Hamra and Zugm), Abd-el-Malek and Rizk found the ratio of the concentrations of sulfate to bicarbonate (reported in meq, w/w) in porewaters feeding the lakes dramatically decreased from 23.3–45.8 at 100 m distance from the lakes to 0.06–0.1 at 2 m distance from the lakes. A similar decrease was noted near Lake Om-Risha (2.8 to 0.48) but not near Lake Rizunia (1.24 to 0.48) where very high counts of sulfate-reducing bacteria were found in sediments at 100 m distance from the lake (Table 20.2). These

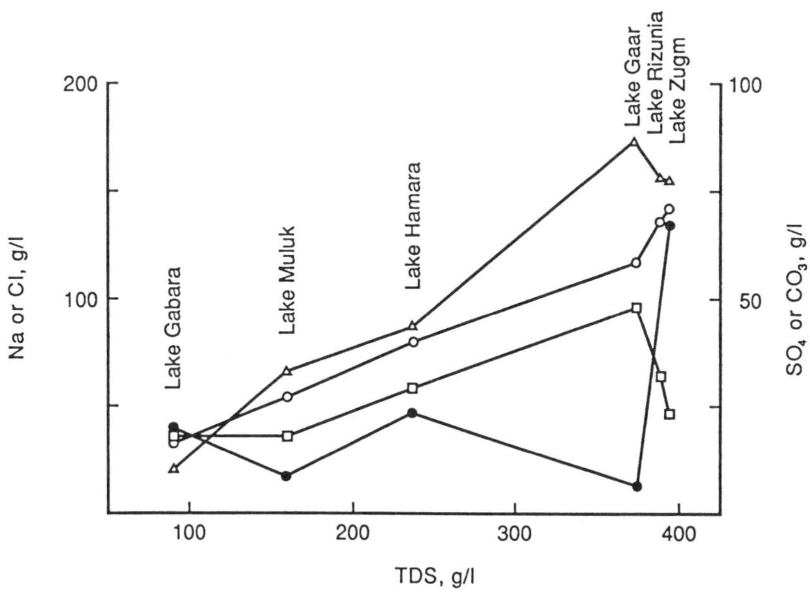

Figure 20.2 Major ion concentrations vs. TDS in six Wadi Natrun lakes (from Imhoff et al., 1979).

Table 20.2 Viable counts of sulfate-reducing bacteria in sediments near Wadi Natrun lakes[a]

	Distance from lakes		
	2 m	50 m	100 m
Lake Om-Risha			
Salinity, %	8.6	4.3	4.0
Viable count:			
Porewater, per ml	5×10^5	3×10^4	5×10^2
Soil, per g	3×10^4	1×10^2	2×10^1
Lake Hamra			
Salinity, %	8.3	4.0	3.8
Viable count:			
Porewater, per ml	3×10^6	3×10^4	3×10^2
Soil, per g	8×10^4	5×10^3	7×10^1
Lake Zugm			
Salinity, %	6.3	4.7	4.6
Viable count:			
Porewater, per ml	4×10^5	2×10^4	1×10^2
Soil, per g	2×10^4	5×10^3	2×10^1
Lake Rizunia			
Salinity, %	7.2	—	3.1
Viable count:			
Porewater, per ml	8×10^5	—	5×10^4
Soil, per g	4×10^5	—	4×10^4
Lake Rizunia			
Salinity, %	6.7	—	3.2
Viable count:			
Porewater, per ml	2×10^5	—	5×10^4
Soil, per g	7×10^5	—	5×10^4

[a] From Abd-el-Malek and Rizk (1963). Two transects of Lake Rizunia were performed.

ratios are indicative of the activity of sulfate-reducing bacteria but the actual ion concentrations measured in the lake waters by Abd-el-Malek and Rizk are quite different from those reported by Imhoff et al. (1979). In the earlier study, Cl^- was not noted as the dominant anion in any lake and both SO_4^{2-} and $HCO_3^- + CO_3^{2-}$ concentrations were extremely different in some cases (see Fig. 20.1).

Viable counts of sulfate-reducing bacteria also increased with proximity to the lakes (Table 20.2), although no sulfate-reducing bacteria were detected in the lake waters. Lake muds apparently were not tested. The source of organic matter for the sulfate-reducing bacteria is the burdi grass in the swamps. Abd-el-Malek and Rizk (1963) noted that the swamps have a characteristic smell of sulfide and the black color of iron sulfide. The lakes may also contain some sulfide. Concentrations of 0–43 μM were measured by Imhoff et al. (1979). Abd-el-Malek and Rizk (1963) measured the viable counts of sulfate-reducing bacteria in Lake Rizunia only (389 $g \cdot l^{-1}$ TDS according to Imhoff et al. [1979]). In soil samples taken 2 m from the lake, 4–7 $\times 10^5$ bacteria$\cdot g^{-1}$ were found while 5×10^2 to 3×10^4 bacteria$\cdot g^{-1}$ were found in sediments at the lake shore.

Imhoff et al. (1979) described six Wadi Natrun lakes, including several of the same lakes studied by Abd-el-Malek and Rizk (1963). Physical features of the lakes (e.g., area, depth, temperature, and stratification) were not given in either report. Assuming that the lakes represent stages of evaporation of a common source, Na^+, Cl^-, and SO_4^{2-} appear to be fairly conservative until both chloride and sulfate apparently precipitate in salinities greater than about 375 $g \cdot l^{-1}$ TDS (see Figure 20.2). Upon loss of these ions, CO_3^{2-} content increases. In Lake Gabara, the most dilute lake (91.9 $g \cdot l^{-1}$ TDS), the data indicate relative SO_4^{2-} enrichment. This enrichment may be derived from sulfate-enriched inflows, since Imhoff et al. (1979) indicated that sulfur was leaving the system (the lake surface had a low redox potential and smelled of H_2S). Sulfide content was 43 μM in Lake Gabara, suggesting an active community of sulfate-reducing bacteria that could deplete the lake of sulfate. In addition, dark metabolism of phototrophic sulfur bacteria leads to sulfide formation.

Nutrient and organic carbon concentrations in the six lakes were high. The ranges of concentrations measured by Imhoff et al. (1979) were: 116–6830 μM phosphate, 53–237 μM nitrate, 2–461 μM ammonia, and 136–1552 $mg \cdot l^{-1}$ organic carbon. The correlation coefficients between the nutrients, major ions, and total dissolved solids of the six lakes demonstrate several interesting phenomena (Table 20.3). TDS, Na^+, and Cl^- strongly correlate. Nutrients are strongly correlated with CO_3^{2-} content, with PO_4^{3-} and NO_3^- showing positive correlation, and NH_4^+ showing negative correlation. PO_4^{3-} strongly correlates with NO_3^- (positive) and NH_4^+ (negative), although the correlation between nitrate and ammonia is rather weak. The high degree of correlation between CO_3^{2-} and both PO_4^{3-} and NO_3^- suggests that the lack of divalent cations in the nutrient-rich watershed helps to maintain these ions in solution.

Organic carbon shows no correlation with salinity or nutrient levels. The lack of correlation between these two factors sharply contrasts with the data

Table 20.3 Correlation coefficients between dissolved substances in six Wadi Natrun lakes[a]

	PO_4^{3-}	NO_3^-	NH_4^+	TDS	Na^+	CO_3^{2-}	Cl^-	SO_4^{2-}	Org C
PO_4^{3-}	—	+0.813	−0.854	+0.556	+0.542	+0.937	+0.415	−0.170	−0.398
NO_3^-	+0.813	—	−0.457	+0.519	+0.530	+0.904	+0.403	−0.146	−0.254
NH_4^+	−0.854	−0.457	—	−0.676	−0.647	−0.722	−0.580	−0.085	+0.135
TDS	+0.556	+0.519	−0.676	—	+0.998	+0.566	+0.982	+0.647	+0.448
Na^+	+0.542	+0.530	−0.647	+0.998	—	+0.551	+0.987	+0.659	+0.441
CO_3^{2-}	+0.937	+0.904	−0.722	+0.566	+0.551	—	+0.416	−0.202	−0.225
Cl^-	+0.415	+0.403	−0.580	+0.982	+0.987	+0.416	—	+0.737	+0.525
SO_4^{2-}	−0.170	−0.146	−0.085	+0.647	+0.659	−0.202	+0.737	—	+0.734
Org C	−0.398	−0.254	+0.135	+0.448	+0.441	−0.225	+0.525	+0.734	—

[a] Calculated from the data of Imhoff et al. (1979).
TDS, total dissolved solids.

from Lake Magadi. However, Lake Magadi is a single basin of concentration whereas the Wadi Natrun lakes reflect the variations in microbial geography, composition, and activity in individual lakes. There is a modest correlation between DOC and SO_4^{2-} content. Sulfate also shows some correlation with Cl^- content. Because sulfate reduction is only one of many processes affecting the CO_3^{2-} and SO_4^{2-} content in such lakes, it can be expected to find little correlation between the two ions.

The microbial composition of these eutrophic lakes was studied by Imhoff et al. (1979) and Imhoff (1988). Lake Gabara (9.19% salinity) had a bloom of cyanobacteria (including *Spirulina*) and an apparently very active community of sulfate reducers that caused the whole water column to remain anaerobic. Lake Muluk (15.9% salinity) had cyanobacterial mats (*Phormidium* and *Synechococcus*) and halophilic phototrophic bacteria (*Ectothiorhodospira halochloris* and *E. halophila*). The water of Lake Hamara (23.8% salinity) was mostly clear and saturated with O_2. The authors stated that alkalophilic halobacteria tinted the water light red and phototrophic bacteria were observed in the muds. Lake Gaar (37.4% salinity) was unique with a population of the chlorophycean alga *Dunaliella salina*. These algae were apparently responsible for depleting the lake of CO_2 and maintaining supersaturated levels of O_2. Lake Gaar, along with Lake Rizunia (38.9% salinity) and Lake Zugm (39.4% salinity), also harbored populations of alkalophilic halobacteria, phototrophic bacteria, and a few cyanobacteria. Black mud and sulfide were noted in the most saline lakes, suggesting that sulfate reduction occurs in salinities of nearly 40% in alkaline environments. Bacteria isolated from the lakes include phototrophic bacteria (Imhoff and Trüper, 1977, 1981; Imhoff et al., 1978; Imhoff, 1988), a halophilic bacillus (Weisser and Trüper, 1985), and alkalophilic halobacteria (Soliman and Trüper, 1982; Tindall, 1985, 1988).

20.4 Big Soda Lake

Several hypersaline, alkaline lakes are found in the western part of the Great Basin of North America. This region, also known as the basin and range province, consists of a series of north-south trending ranges and valleys. Some of the valley basins are virtually dry salt lakes, while others have terminal lakes of various salinities and alkalinities. Four lakes are discussed here: Mono, Owens, Searles, and Big Soda. Other lakes of the region that have been investigated include Red and Green Ponds (Arizona) (Cole et al., 1967) and Lake Abert (Oregon) (Jones et al., 1969).

No one extremely saline, alkaline lake of the region has been studied intensively from the microbiological point of view. The meromictic Big Soda Lake (Nevada, U.S.A.) has a rather dilute mixolimnion down to 34.5 m depth (2.6% salinity, pH 9.7) underlain by a moderately hypersaline monimolimnion (8.8% salinity, pH 9.7) down to 65 m depth. Meromixis was initiated when irrigation caused the local water table to rise (Kimmel et al., 1978). The microbiology and productivity of the mixolimnion have been described in detail (Axler et al., 1978; Priscu et al., 1982; Cloern et al., 1983a, 1983b). Comparative studies between the mixolimnion and the anoxic monimolimnion were conducted by Zehr et al. (1987), Smith and Oremland (1987), and Iversen et al. (1987).

Table 20.4 shows the composition of the monimolimnion. The very high concentration of sulfide (12 mM) and the non-conservative increase of SO_4^{2-} from the concentration measured in the mixolimnion suggest biological reduction. If the monimolimnion represents mixolimnion water concentrated 4.15 × (based on Cl^- concentrations), and the loss of Na^+, K^+, Mg^{2+}, and Ca^{2+} were all due to precipitation with SO_4^{2-}, there would still be a deficit of about 50 $meq \cdot l^{-1}$ sulfate in the monimolimnion. If Na^+ and K^+ are conservative ions, then the loss of SO_4^{2-} is even greater.

The six-fold increase in alkalinity between the mixolimnion and monimolimnion also suggests that the extremely alkaline condition of the lake may in part be due to sulfate reduction *in situ*. The very high concentration of sulfide in the monimolimnion reflects the probable lack of iron in the sediments, the high pH (which retards the escape of sulfide as volatile H_2S), and the lack of mixing due to density stratification. Smith and Oremland (1987) studied bacterial sulfate reduction rates in the water column. In May, the average rate of sulfate reduction in the monimolimnion was 3000 nmol $SO_4^{2-} \cdot l^{-1} \cdot d^{-1}$ while it was only 900 nmol $SO_4^{2-} \cdot l^{-1} \cdot d^{-1}$ in October. Much lower rates were measured in the anoxic zone of the mixolimnion (Table 20.5). Stimulation by Fe^{2+} demonstrated that sulfate reduction in the monimolimnion was iron-limited. Estimates of the amount of organic carbon min-

Table 20.4 Chemical constituents of Mono Lake and the monimolimnion of Big Soda Lake[a]

	Mono Lake $mg \cdot l^{-1}$	Big Soda Lake, monimolimnion $mg \cdot l^{-1}$	Big Soda Lake, ratio monimolimnion/mixolimnion	Big Soda Lake, monimolimnion minus mixolimnion $meq \cdot l^{-1}$
TDS	90,000	88,000	3.38	—
Na^+	29,500	28,000	3.50	+870
K^+	1500	1100	3.55	+20.2
Mg^{2+}	33.4	6	0.04	−11.5
Ca^{2+}	4.1	0.8	0.16	−0.21
Cl^-	17,600	27,000	4.15	+578
HCO_3^-	30,100[b]	24,000	6.00	+328
SO_4^{2-}	10,300	6700	1.20	+22.9
NH_3	1.0	45	>450	ca. +2.65 (as NH_4^+)
H_2S	—	410	>2050	ca. +12 (as HS^-)
DOC	62	60	3.00	—

[a] Mono Lake data from Mason (1967) and Winkler (1977); Big Soda Lake data calculated from Cloern et al. (1983b).
[b] $HCO_3^- + CO_3^{2-}$

Table 20.5 Some bacterial processes in Big Soda Lake[a]

Process	Aerobic mixolimnion	Anaerobic mixolimnion	Monimolimnion
Sulfate reduction[b]	—	25–600	900–3000
Methanogenesis[b]	—	0.1–1.0	1.6–12
Methane oxidation[b]		2.0–6.0	49–85
Glutamate turnover[c]	0.01–0.03	0.01–0.03	0–0.02
Thymidine assimilation[d]	2.7–14.2	4.4–36.1	1.0–2.1
Bacterial abundance (viable count × 10^6 per ml)	0.16–7.60	0.41–10.9	0.62–25.3

[a] From Zehr et al. (1987), Smith and Oremland (1987), and Iversen et al. (1987).
[b] $nmol \cdot l^{-1} \cdot d^{-1}$.
[c] Percent per hour of added radiotracer assimilated.
[d] $pmol \cdot l^{-1} \cdot h^{-1}$.

eralized by sulfate reduction were much greater than the measured fluxes of particulate organic carbon sinking through the lake. The authors suggested that productivity by benthic microbial mats and macrophytes in the mixolimnion was the most probable source of the "missing" organic carbon.

The sediments of Big Soda Lake produced significant amounts of methane (Oremland et al., 1982). Experiments demonstrated that methanogenesis was stimulated by methanol, trimethylamine, and methionine, but not by H_2, acetate, or formate. Enrichment cultures with methanol produced CH_4 with a $\delta^{13}C$ content of $-72‰$ to $-77‰$ lighter than the added methanol. Enrichment cultures led to the isolation of a methanogenic coccus that grew optimally at pH 9.7. Smith and Oremland (1983) also reported the anaerobic degradation of oxalate in Big Soda Lake sediments. Nearly all the methane oxidation occurred in the anoxic zones of the lake, with the greatest rates recorded in the monimolimnion (Table 20.5). Measured rates of CH_4 oxidation were greater than measured rates of methanogenesis (Iversen et al., 1987). The accumulation of organic-rich, laminated sediments (Oremland et al., 1982; Cloern et al., 1983b) indicate that not only is anaerobic degradation incomplete, but that there are seasonal cycles of sedimentation. The high DOC content of the monimolimnion ($60\ mg \cdot l^{-1}$) (Kimmel et al., 1978) also indicates incomplete degradation.

The highly productive mixolimnion is characterized by a seasonal succession of blooms of diatoms, phototrophic bacteria, and chemoautotrophic bacteria (nitrifiers and sulfur-oxidizers) with a total annual productivity estimated to be around 500 $g\ C \cdot m^{-2}$ (Cloern et al., 1983a). The greatest accumulation of bacteria occurred in the chemocline (anoxic mixolimnion) although relatively high concentrations of bacteria were also counted in the monimolimnion (Table 20.5). Measurements of rates of glutamate turnover and thymidine incorporation showed the cells in the monimolimnion were relatively inactive while the greatest rates were found at or just below the layer of phototrophic bacteria in the anoxic mixolimnion (Zehr et al., 1987).

20.5 Mono, Owens, and Searles Lakes

Three other lakes situated at the eastern base of the Sierra Nevada mountains are in the same general area as Big Soda Lake. These three California lakes are of interest not only scientifically but because of the impact that human activity has had on their properties.

Mono Lake This moderate size lake (about 10 km × 20 km) (Figure 20.3) is a terminal lake fed by two major streams and numerous springs. Since 1941, the city of Los Angeles has diverted water that would normally run into the lake. During the next two decades, the lake level dropped 14 m and the water increased in salinity from 4.8% to 9.1% (Melack, 1983). The pH is 9.7–10 (Mason, 1967; Winkler, 1977).

The lake is usually monomictic and stratified from spring to autumn. The hypolimnion is anoxic from June to September (Melack, 1983). After two seasons of heavy snowfall (1982–3), the city of Los Angeles allowed more water to enter the lake and a meromictic condition followed for several seasons until the fresher water largely mixed with the deeper water. Periodic meromixis may also be a natural feature of such lakes.

The major chemical constituents of Mono Lake reported by Winkler (1977)

Figure 20.3 Mono Lake, California. Tufa ($CaCO_3$) towers, formed from Ca^{2+}-rich springwater discharged into the lake when the lake level was higher, are now exposed at the southwestern end of this hypersaline, alkaline lake.

are largely similar to those of the monimolimnion of Big Soda Lake (Table 20.4). There is a dynamic productivity cycle in Mono Lake (Mason, 1967; Winkler, 1977; Melack, 1983). The largest standing crop of phytoplankton occurs in the winter when Secchi disc readings are as low as 0.7 m (Melack, 1983). Summer measurements of Secchi disc depth were 10 m. At 2 m depth, up to 65 $\mu g \cdot l^{-1}$ chlorophyll *a* was recorded in the winter while only 1 $\mu g \cdot l^{-1}$ was found in the summer at that depth. At 20 m depth, relatively high concentrations of chlorophyll *a* were found throughout the year (25–55 $\mu g \cdot l^{-1}$).

The phytoplankton is dominated by several diatoms, cyanobacteria, and species of *Nannochloris* and *Chlamydomonas* (green algae) (Winkler, 1977). References to *Coccomyxa* (Melack, 1983) and *Palmellococcus* (Mason, 1967) may be synonymous with *Nannochloris*, a 3-μm cell. Identification of 10–15 μm-long species of *Dunaliella* by Mason (1967) was not confirmed by Winkler (1977), who recognized the dominance of the diatom *Nitzschia communis*. Mason (1967) noted that a small *Nitzschia* (20–35 μm long) dominated in the fall and that a larger *Nitzschia* (60 μm long) dominated in the winter. Several other species of diatoms were recorded, including the cosmopolitan hypersaline species *Amphora coffeaeformis* (Winkler, 1977). The presence of numerous fungal hyphae, gemmae, and arthrospores in the summer plankton and the growth of *Chytridomycetes* in diatom frustules was also reported.

The clarity of the upper levels of the lake in the summer is attributed to grazing by the dense populations of brine shrimp, *Artemia monica* (Winkler, 1977). The zooplankton also includes several protozoans and rotifers. Extremely large populations of migratory birds remain at the lake for part of the year to feed on both brine shrimp and brine fly larvae (*Ephydra hians*). Their effect on nutrient cycling in the lake should not be ignored, although there are no published estimates of their actual role in nutrient regeneration.

Primary productivity rates at a shallow water station (10 m depth) and a deep water station (22 m depth) were measured by Winkler (1977). At the shallow water station, rates decreased from 142 to 81.5 mg $C \cdot m^{-2} \cdot h^{-1}$ through the course of the summer. The highest value was 160 mg $C \cdot m^{-2} \cdot h^{-1}$. In the 0–10 m depth range of the deep water station, photosynthetic rates generally increased through the course of the summer, from 29.1 to 118 mg $C \cdot m^{-2} \cdot h^{-1}$. The highest value was 142 mg $C \cdot m^{-2} \cdot h^{-1}$. In the lower 12 m, primary productivity decreased from 209 to 11.5 mg $C \cdot m^{-2} \cdot h^{-1}$ through the course of the summer. On one occasion, no carbon fixation at all could be detected. The cells in the hypolimnion were found to be active but were not receiving enough light for photosynthesis. Because a large proportion of the chlorophyll-containing cells remain in the dark for much of the year, dark metabolism and possible heterotrophy in these algae may be important for maintaining a potentially active population. In addition, the mechanisms of osmotic regulation and buoyancy control for the phytoplankton has not been established.

Nitrogen limitation may prevent the phytoplankton from reaching even greater densities. While very high phosphate concentrations (74.6 $mg \cdot l^{-1}$) were found (Winkler, 1977), NH_4^+ was relatively scarce (1 $mg \cdot l^{-1}$) and NO_3^- was low (5.2–66 $mg \cdot l^{-1}$) (Winkler, 1977) or absent (Melack, pers. comm.) Regeneration of ammonia in the sediments and hypolimnion as well as from animal excretion should be considered. The very high concentrations of phosphate found in this alkaline lake are at least in part due to the lack of Ca^{2+}, Mg^{2+}, and Fe^{3+} ions that typically precipitate PO_4^{3-} in hard water lakes. A similar situation was noted in the Wadi Natrun lakes (Imhoff et al., 1979).

Bacterial populations in the plankton were measured by Mason (1967). In the winter, a high density of bacteria ($8.0 \times 10^7 \cdot ml^{-1}$) accompanied the dense population of phytoplankton. Mason noted that the bacterial population decreased through the spring and summer, and increased again with the fall bloom of phytoplankton. In experiments in the summer measuring the biological oxygen demand (BOD) of pelagic water samples from which metazoans were excluded, Mason found the O_2 consumption was two or more times that of the controls when the samples were amended with organic nitrogen (casein, chitin, uric acid, glycine, or n-acetyl glucosamine). Sucrose, acetate, and dextrose showed little or no stimulation of the BOD. These experiments suggest that nitrogen limitation is indeed an important environmental factor in Mono Lake.

Methanogenesis has been found in Mono Lake sediments (R. Oremland, pers. comm.). Mason (1967) noted that both nitrate and sulfate reduction might be occurring in the hypolimnion. An analysis made in early autumn showed a nitrate depletion of 1.4 $mg \cdot l^{-1}$ in the hypolimnion. In one measurement, sulfate in the hypolimnion was depleted by 0.34 $g \cdot l^{-1}$. In a July measurement, when O_2 was detected at the bottom, the sulfate concentration was found to be slightly higher than in the epilimnion. Mason suggested sulfide oxidation was responsible for regenerating SO_4^{2-}. No sulfate reduction was detected.

Interstitial sulfide concentrations of 15–40 mM were detected in shallow Mono Lake sediments (Javor, unpubl. data). The highest concentrations were associated with delta sediments near the inflow of a freshwater stream although the porewater was as saline and alkaline as the rest of the lake. The ability to retain high concentrations of sulfide in the sediments, a phenomenon also found in Big Soda Lake, is probably related to the lack of iron and the high pH.

DOC in Mono Lake (62 $mg \cdot l^{-1}$) (Mason, 1967) was similar in concentration to that of the equally saline hypolimnion of Big Soda Lake (Table 20.4). Mason noted that the DOC was responsible for lowering the surface tension of Mono Lake water. The DOC was intimately tied to the formation of stable foams.

Owens Lake Owens Lake, which lies south of Mono Lake, had a body of water until the inflow was diverted to the Owens Valley aqueduct in 1913

(Friedman et al., 1976). By 1921, the lake was nearly dry. About 1.5–3 m of salts had deposited during those 8 years. During the winters of 1938–9 and 1969–70, excessive run-off from the Sierra Nevada mountains flooded the lake. During the latter flood, the lake was covered to 2.4 m depth. The compositional and isotopic changes in the brines in the period following that flood were studied by Friedman et al. (1976). Their observations on the color changes in the brines (Table 20.6) indicate that green phytoplankton can survive a salinity of at least 13.6%. By the end of 1970, brine salinities approached 30% and the brine color was dark yellow green. It is not known whether it supported an active, chlorophyll-containing phytoplanktonic population. The decrease in brine visibility during that period may have been due to passive concentration of cells with evaporation.

During January, 1971, freshwater decreased the salinity to at least 24%, and the presence of a light olive-grey color was noted. Cultures of the phototrophic bacteria *Ectothiorhodospira halochloris* and *E. abdelmalekii* are of this hue (Imhoff, 1988). It can be speculated that one or both of these bacteria bloomed in the lake at that time. Further increases in salinities led to a brown color.

Phosphate levels in Owens Lake, as in other alkaline, hypersaline lakes, are very high. During the phytoplankton bloom of early 1970, 1–2 mM PO_4^{3-} was measured. During the subsequent evaporation, PO_4^{3-} levels increased to nearly 14 mM. Friedman et al. noted that Li^+ remained conservative during the evaporation sequence. The PO_4^{3-}/Li^+ ratios increased during the evaporation sequence, perhaps as a result of decomposition of the phytoplankton. In fact, the PO_4^{3-}/Li^+ ratio increased rather linearly with respect to TDS. For all the PO_4^{3-}/Li^+ and TDS values listed in Table 20.6, the correlation coefficient is 0.61. If the phosphate concentration during the phytoplankton bloom is excluded, the correlation coefficient is 0.84. If the possibly deviant value of PO_4^{3-} in the 38.8% salinity brine is also excluded, the r value of PO_4^{3-}/

Table 20.6 Salinity, pH, phosphate, and color of Owens Lake, California[a]

Date	TDS, %	pH	PO_4^{3-}, mM	PO_4^{3-}/Li^+	PO_4^{3-}/TDS $\times 10^3$ w/w	Color and visibility
26-II-70	13.6	9.6	0.97	40.4	0.675	Light green, 10 cm
12-VIII-70	25.0	9.6	1.89	37.9	0.719	Light green, 5 cm
XI-70	29.7	9.62	2.23	37.9	0.714	Dark yellow-green, 5 cm
19-I-71	24.0	9.31	3.00	34.6	1.19	Light olive grey
19-I-71	25.3	9.35	3.13	35.2	1.183	Light olive grey
13-IV-71	38.8	9.5	4.16	35.9	1.02	Light olive brown
7-VII-71	45.0	10.5	8.82	46.6	1.86	Light brown, >40 cm
20-VII-71	47.0	11.0	13.6	46.1	2.74	Light brown
1985–6[b]	—	—	—	—	—	Pink

[a] From Friedman et al. (1976).
[b] Javor, personal observations.
TDS, total dissolved solids.

Li^+ versus TDS is 0.98 for the remaining six determinations made over an 18-month period. Because phosphate is present in more than minute trace concentrations in many alkaline, extremely hypersaline systems, it might be a useful indicator of the degree of brine concentration in such evaporite deposits. It remains to be determined whether the activities of microorganisms perturb the large concentrations of phosphate in the brines of relatively lower salinities.

Searles Lake Searles Lake is a largely dry, evaporite deposit in eastern California. The lake is fed by the water leaving the Owens Valley and has a variety of minerals, although it is relatively devoid of sulfates (Smith and Haines, 1964). The authors noted that the brines had a high organic content. Gaylussite ($Na_2CO_3 \cdot CaCO_3 \cdot 5\ H_2O$), a mineral commonly associated with mud layers in the lake sediments, sometimes had inclusions that were interpreted to be mud or red microorganisms. Bien and Schwartz (1965) observed a variety of preserved microorganisms in subrecent salt samples.

Smith and Haines (1964) noted that the occasional occurrence of elemental sulfur in the evaporites was probably associated with bacterial sulfate reduction and sulfide oxidation even though there was a relative lack of sulfate. Another possible source of elemental sulfur could have been the extracellular sulfur produced by *Ectothiorhodospira* in the photic zone of the lake at an earlier stage in its history. Holser and Kaplan (1966) measured the ^{34}S distribution of sulfur minerals in Searles Lake. The average $\delta^{34}S$ values for all the salts and brines were +14.6‰ and +15.2‰, respectively. The high concentration of dissolved sulfide (107 $mg \cdot l^{-1}$) was light ($\delta^{34}S = -24.7‰$). The sulfate in those same interstitial brines (9100 $mg \cdot l^{-1}$) was correspondingly heavy ($\delta^{34}S = +16.5‰$). The freshwater source for the lake (Owens River) had relatively light SO_4^{2-} ($\delta^{34}S = +8.5‰$), which led to the suggestion that most of the biological reduction took place in Owens Lake before the brine entered the Searles Lake basin.

Longinelli and Craig (1967) measured the $\delta^{18}O$ variations in H_2O and SO_4^{2-} of Mono, Owen, and Searles lakes and compared them to mean ocean seawater and to hypersaline systems (Table 20.7). The differences do not correlate with temperature or salinity. While biological sulfate reduction could account for the increasingly heavier values of sulfate from the Mono-Owens-Searles Lake chain, possible atmospheric input of ^{18}O in these high surface-to-volume dry lakes cannot be ignored. The stable isotope chemistry of freshwater-derived hypersaline lakes should be compared with the dilute inflows.

Microbiology of Owen and Searles Lakes Both Owens Lake and Searles Lake are now essentially playa systems that are exploited commercially for their soda salts. Shallow residual brines in Owens Lake are red in color. Tew (1980) noted the presence of the phototrophic bacterium *Ectothiorhodospira* in Owens Lake while Tindall (1985) and Morth and Tindall (1985) reported

Table 20.7 The $\delta^{18}O$ values of H_2O and SO_4^{2-} in extremely hypersaline lakes[a]

Source	$\delta^{18}O$-H_2O	Dissolved $\delta^{18}O$-SO_4^{2-}	Sediment $\delta^{18}O$-SO_4^{2-}
Oceanic water	0	+9.5	—
Mono Lake, Calif.	−0.46	+17.8	+7.9
Owens Lake, Calif.	+3.92	+14.4	+23.4
Searles Lake, Calif.	+4.22	+23.2	+48.7
Great Salt Lake, Utah	−3.2	+13.3	+16.4
Dead Sea	+4.72	+12.97	+0.9
Dead Sea	+4.36	+13.34	+0.54
Red Sea hot brines	+1.21	+7.21–7.51	—

[a] From Longinelli and Craig (1967). All values in ‰. The $\delta^{18}O$ of atmospheric O_2 was +23‰.

the isolation of red, alkalophilic, halophilic archaebacteria. Tew (1980) noted that *Ectothiorhodospira* survived prolonged entrapment both in mirabilite and in the residual brine, but not in anhydrous Na_2SO_4 (thenardite). The presence of water, although chemically-bound, is apparently important for the entrapped cells to remain osmotically stable.

Nehrkorn and Schwartz (1961) tried a variety of enrichments for microorganisms from Owens Lake and Searles Lake. They used both natural and artificial seawater supplemented with NaCl (pH 7.5–8), and synthetic brine that included borax (32% TDS, pH 10). Various carbon and nitrogen sources were tried at 20° and 40°C in the light and dark, and aerobically and anaerobically. At 20°C in the light and dark, very few enrichments were positive. Notably, both the natural and artifical seawater media with various C and N sources gave positive results from enrichments of pH 9, red surface brines of Owens Lake.

Aerobic enrichment cultures in seawater and artifical brine at 40°C in the dark gave generally positive results. Red or turbid cultures developed from these enrichments. The artificial alkaline brine medium was particularly successful, yielding positive enrichments from all the surface salts and brines from both lakes, but not from borehole brines. Enrichments for sulfate-reducing bacteria in the artificial brine yielded positive results (black color) in five of the ten inocula from the two lakes. The sulfate-reducing bacteria were vibrios and short, spiral-forming chains. Pure cultures were obtained in seawater medium with 25% NaCl, at pH 8. This is the highest salt tolerance ever reported for pure cultures of sulfate-reducing bacteria. The authors did not establish whether they were alkalophiles. In the nearby Saline Valley dry lake (pH 7.8–9), NaCl-saturated mud below the halite crust was noted to be black, presumably from sulfate reduction (Hardie, 1968).

Nehrkorn and Schwartz (1961) isolated four strains of aerobic bacteria from Owens Lake and seven strains from Searles Lake. Several of their isolates were extreme halophiles and at least one from Owens Lake was an alkalophile as well. From Searles Lake they isolated *Sarcina littoralis*, a red extreme halophile that forms packets of cells. It grew in 10–30% NaCl at pH 6–10. They also isolated a pseudomonad (probably a halobacterium or natronobacterium) with similar growth requirements as *S. littoralis*. From both lakes they isolated *Flavobacterium* (orange to ochre in color) that grew optimally in 10–20% salt, but tolerated 5–30% NaCl. Its optimal pH was not given. Another strain they called *Flavobacterium*, isolated from Owens Lake, grew optimally in 10–20% NaCl at pH 10. It tolerated pH 8–12. Curiously, this strain grew slightly better with 20% Na_2SO_4 than with 20% NaCl. Even better growth was obtained in media with 10 or 15% Na_2SO_4 plus 10 or 5% NaCl, respectively. K^+ could not substitute for Na^+. This is the only reported case of a halophilic bacterium with a preference for sulfate. This strain is probably the first alkalophilic, extremely halophilic bacterium ever reported but its uniqueness was not recognized at the time.

In the second part of the study of halophilic microorganisms, enrichments were made from evaporating ponds and boreholes of the extremely alkaline (pH 9.4–9.9), hypersaline salt works of Sosa Texcoco (Mexico) (Jaschof and Schwartz, 1961). Some of the taxonomic assignments given by the authors are probably wrong, including their descriptions of the chlorophycean algae *Chlorella* and *Chlamydomonas* in 22.3–23.6% salinity brines. However, they made a variety of observations on the effects of salinity on aerobic and anaerobic enrichments, and the ability of certain microorganisms (i.e., thiobacilli) to tolerate a variety of salt concentrations. All of their isolates grew optimally in pH 8–9 and tolerated up to pH 12. In the most concentrated brines they recognized both rods and coccoid bacteria. Again, the uniqueness of these bacteria was not recognized until Tindall and co-workers established the separate classification of alkalophilic halophiles among the archaebacteria (Tindall et al., 1984).

20.6 Conclusion

Hypersaline, alkaline lakes tend to be rich in phosphate, a factor that promotes organic productivity. The alkalinity itself may be caused by the degradation of organic matter by sulfate-reducing bacteria. Extremely high concentrations of sulfide can remain in porewater solution due to the high pH combined with the lack of iron. Although a variety of microorganisms have been isolated from these environments, little is known about their rates of activities in nature or the organic geochemistry of their natural milieux. It has been established that hypersaline, alkaline basins have produced the organic matter associated with some petroleum deposits. Because of the distinctive biochemical composition of the variety of microorganisms that inhabit these environments (i.e., pigments and polar lipids), the problems of tracing the microbial chain of events in petroleum genesis and alteration should provide a challenge for biogeochemists.

References

Abd-el-Malek, Y. and **Rizk, S.G.** 1963. Bacterial sulphate reduction and the development of alkalinity. III. Experiments under natural conditions in the Wadi Natrun. *Journal of Applied Bacteriology* 26: 20–26.

Axler, R.P., Gersberg, R.M., and **Paulson, L.J.** 1978. Primary productivity in meromictic Big Soda Lake, Nevada. *Great Basin Naturalist* 38: 187–192.

Bien, E. and **Schwartz, W.** 1965. Geomikrobiologische Untersuchungen. VI. Über das Vorkommen konservierter toter und lebender Bakterienzellen in Salzgesteinen. *Zeitschrift für Allgemeine Mikrobiologie* 5: 185–205.

Bradley, W.H. and **Eugster, H.P.** 1969. Geochemistry and paleolimnology of the trona deposits and associated authigenic minerals of the Green River Formation of Wyoming. *United States Geological Survey Professional Paper* 496-B, 71 pp.

Cloern, J.E., Cole, B.E., and **Oremland, R.S.** 1983a. Autotrophic processes in meromictic Big Soda Lake, Nevada. *Limnology and Oceanography* 28: 1049–1061.

Cloern, J.E., Cole, B.E., and **Oremland, R.S.** 1983b. Seasonal changes in the chemistry and biology of a meromictic lake (Big Soda Lake, Nevada, U.S.A.). *Hydrobiologia* 105: 195–206.

Cole, G.A., Whiteside, M.C., and **Brown, R.J.** 1967. Unusual monomixis in two saline Arizona ponds. *Limnology and Oceanography* 12: 584–591.

Eugster, H.P. 1970. Chemistry and origin of the brines of Lake Magadi, Kenya. *Mineralogical Society of America Special Paper* 3: 213–235.

Eugster, H.P. 1980. Lake Magadi, Kenya, and its precursors, pp. 195–232 in Nissenbaum, A. (editor), *Hypersaline Brines and Evaporitic Environments, Developments in Sedimentology* 28, Elsevier Scientific, New York.

Eugster, H.P. and **Hardie, L.A.** 1975. Sedimentation in an ancient playa-lake complex: the Wilkins Peak Member of the Green River Formation of Wyoming. *Geological Society of America Bulletin* 86: 319–334.

Eugster, H.P. and **Hardie, L.A.** 1978. Saline lakes, pp. 237–293 in Lerman, A. (editor), *Lakes. Chemistry Geology Physics*, Springer-Verlag, New York.

Friedman, I., Smith, G.I., and **Hardcastle, K.G.** 1976. Studies of Quaternary saline lakes—II. Isotopic and compositional changes during desiccation of the brines in Owens Lake, California, 1969–1971. *Geochimica et Cosmochimica Acta* 40: 501–511.

Hammer, U.T. 1981. Primary productivity in saline lakes: a review. *Hydrobiologia* 81: 47–57.

Hardie, L.A. 1968. The origin of the Recent non-marine evaporite deposit of Saline Valley, Inyo County, California. *Geochimica et Cosmochimica Acta* 32: 1279–1301.

Holser, W.T. and **Kaplan, I.R.** 1966. Isotope geochemistry of sedimentary sulfates. *Chemical Geology* 1: 93–135.

Imhoff, J.F. 1988. Halophilic phototrophic bacteria, p. 85–108 in Rodriguez-Valera, F. (editor), *Halophilic Bacteria*, Vol. I. CRC Press, Boca Raton.

Imhoff, J.F. and **Trüper, H.G.** 1977. *Ectothiorhodospira halochloris* sp. nov., a new extremely halophilic phototrophic bacterium containing bacteriochlorophyll b. *Archives of Microbiology* 114: 115–121.

Imhoff, J.F., Hashwa, F., and **Trüper, H.G.** 1978. Isolation of extremely halophilic phototrophic bacteria from the alkaline Wadi Natrun, Egypt. *Archiv für Hydrobiologie* 84: 381–388.

Imhoff, J.F., Sahl, H.G., Soliman, G.S.H., and **Trüper, H.G.** 1979. The Wadi Natrun: chemical composition and microbial mass developments in alkaline brines of eutrophic desert lakes. *Geomicrobiology Journal* 1: 219–234.

Imhoff, J.F., Tindall, B.J., Grant, W.D., and **Trüper, H.G.** 1981. *Ectothiorhodospira vacuolata* sp. nov., a new phototrophic bacterium from soda lakes. *Archives of Microbiology* 130: 238–242.

Iversen, N., Oremland, R.S., and **Klug, M.J.** 1987. Big Soda Lake (Nevada). 3. Pelagic methanogenesis and anaerobic methane oxidation. *Limnology and Oceanography* 32: 804–814.

Jannasch, H.W. 1957. Die bakterielle Rotfarbung der Salzseen des Wadi Natrun. *Archiv für Hydrobiologie* 53: 425–433.

Jaschof, H. and **Schwartz, W.** 1961. Untersuchungen über Lebensgemeinschaften halophiler Mikroorganismen. II. Über die Mikrobenassoziationen einer alkalischen Sole aus dem Hochtal von Mexiko. *Zeitschrift für Allgemeine Mikrobiologie* 1: 258–273.

Jones, B.F., Vandenburgh, A.S., Truesdell, A.H., and **Rettig, S.L.** 1969. Interstitial brines in playa sediments. *Chemical Geology* 4: 253–262.

Jones, B.F., Eugster, H.P., and **Rettig, S.L.** 1977. Hydrochemistry of the Lake Magadi basin, Kenya. *Geochimica et Cosmochimica Acta* 41: 53–72.

Kimmel, B.L., Gersberg, R.M., Paulson, L.J., Axler, R.P., and **Goldman, C.R.** 1978. Recent changes in the meromictic status of Big Soda Lake, Nevada. *Limnology and Oceanography* 23: 1021–1025.

Longinelli, A. and **Craig, H.** 1967. Oxygen-18 variations in sulfate ions in sea water and saline lakes. *Science* 156: 56–59.

Mason, D.T. 1967. Limnology of Mono Lake. *University of California Publications in Zoology* 83: 1–102.

Melack, J.M. 1981. Photosynthetic activity of phytoplankton in tropical African soda lakes. *Hydrobiologia* 81: 71–85.

Melack, J.M. 1983. Large, deep salt lakes: a comparative limnological analysis. *Hydrobiologia* 105: 223–230.

Morth, S. and **Tindall, B.J.** 1985. Variation of polar lipid composition within haloalkalophilic archaebacteria. *Systematic and Applied Microbiology* 6: 247–250.

Nehrkorn, A. and **Schwartz, W.** 1961. Untersuchungen über Lebensgemeinschaften halophiler Mikroorganismen. I. Mikroorganismen aus Salzeen der californischen Wustengebeite und aus einer Natriumchlorid-Sole. *Zeitschrift für Allgemeine Mikrobiologie* 1: 121–141.

Oremland, R.S., Marsh, L., and **Des Marais, D.J.** 1982. Methanogenesis in Big Soda Lake, Nevada: an alkaline, moderately hypersaline desert lake. *Applied and Environmental Microbiology* 43: 462–468.

Priscu, J.C., Axler, R.P., Carlton, R.G., Reuter, J.E., Arneson, P.A., and **Goldman, C.R.** 1982. Vertical profiles of primary productivity, biomass and physico-chemical properties in meromictic Big Soda Lake, Nevada, U.S.A. *Hydrobiologia* 96: 113–120.

Smith, G.I., and **Haines, D.V.** 1964. Character and distribution of nonclastic minerals in the Searles Lake evaporite deposit, California. *United States Geological Survey Bulletin* 1181-P, 58 pp.

Smith, R.L. and **Oremland, R.S.** 1983. Anaerobic oxalate degradation: widespread natural occurrence in aquatic sediments. *Applied and Environmental Microbiology* 46: 106–113.

Smith, R.L. and **Oremland, R.S.** 1987. Big Soda Lake (Nevada). 2. Pelagic sulfate reduction. *Limnology and Oceanography* 32: 794–803.

Talling, J.F., Wood, R.B., Prosser, M.V., and **Baxter, R.M.** 1973. The upper limit of photosynthetic productivity by phytoplankton: evidence from Ethiopian soda lakes. *Freshwater Biology* 3: 53–76.

Tew, R.W. 1980. Halotolerant *Ectothiorhodospira* survival in mirabilite: experiments with a model of chemical stratification by hydrate deposition in saline lakes. *Geomicrobiology Journal* 2: 13–20.

Tindall, B.J. 1985. Qualitative and quantitative distribution of diether lipids in haloalkalophilic archaebacteria. *Systematic and Applied Microbiology* 6: 243–246.

Tindall, B.J. 1988. Prokaryotic life in the alkaline, saline, athalassic environment, p.

31–67 in Rodriguez-Valera, F. (editor), *Halophilic Bacteria,* Vol. I. CRC Press, Boca Raton.

Tindall, B.J., Mills, A.A., and **Grant, W.D.** 1980. An alkalophilic red halophilic bacterium with a low magnesium requirement from a Kenyan soda lake. *Journal of General Microbiology* 116: 257–260.

Tindall, B.J., Ross, H.N.M., and **Grant, W.D.** 1984. *Natronobacterium,* gen. nov. and *Natronococcus* gen. nov., two new genera of haloalkalophilic archaebacteria. *Systematic and Applied Microbiology* 5: 41–57.

Weisser, J. and **Trüper, H.** 1985. Osmoregulation in a new haloalkalophilic bacillus from the Wadi Natrun (Egypt). *Systematic and Applied Microbiology* 6: 7–11.

Winkler, D.W. (editor). 1977. *An Ecological Study of Mono Lake, California.* Institute of Ecology Publication 12. University of California. Davis. 190 pp.

Zehr, J.P., Harvey, R.W., Oremland, R.S., Cloern, J.E., George, L.H., and **Lane, J.L.** 1987. Big Soda Lake (Nevada). 1. Pelagic bacterial heterotrophy and biomass. *Limnology and Oceanography* 32: 781–793.

Glossary

Anoxygenic photosynthesis Anoxygenic photosynthesis is the light-driven chemical reduction of CO_2 by electrons derived from a compound other than water, and thereby molecular oxygen is not a product. Typical reductants include H_2S and H_2. Example: $2\ H_2S + CO_2 \rightarrow (CH_2O) + H_2O + 2\ S^0$. Anoxygenic photosynthesis occurs in phototrophic bacteria and in some cyanobacteria. It never occurs in eucaryotic algae or plants.

Archaebacteria Archaebacteria include halobacteria, methanogenic bacteria, and certain sulfur-dependent bacteria. Archaebacteria have certain biochemical attributes that are distinct from those of eubacteria, including cell wall chemistry (they never produce peptidoglycans) and polar lipid composition (they produce glycerol diether-linked isoprenoids instead of ester-linked fatty acids). The distinctive structure of the 16 S fraction of archaebacterial ribosomal RNA has been used as a "molecular clock" to interpret the early divergence of archaebacteria from the branch of eubacterial procaryotic evolution.

Athalassic Athalassic refers to waters or salts of non-marine origin.

Autotrophic Autotrophic organisms obtain cell carbon from an inorganic source (CO_2). Energy for reducing CO_2 may be derived from light (photoautotrophy) or from the chemical oxidation of a reduced substance (chemoautotrophy).

Bitterns Bitterns are highly concentrated, marine-derived brines from which NaCl has already precipitated. Bitterns are rich in Mg^{2+}, Na^+, K^+, SO_4^{2-}, and Cl^-. If these brines are subject to further concentration, they will precipitate potash salts.

Chemoautotrophic Chemoautotrophic metabolism (found only among certain bacteria) involves the reduction of CO_2 for cell carbon using energy derived from the chemical oxidation of a reduced substance. Chemoautotrophs often live in the transition zone between aerobic and anaerobic conditions where reduced substrates and oxygen coexist. Chemoautotrophs include sulfur-oxidizing, nitrifying, iron-oxidizing, and hydrogen-oxidizing bacteria. Example: $H_2S + O_2 + H_2O + CO_2 \rightarrow (CH_2O) + H_2SO_4$. Because inorganic energy sources are used, these organisms are often called *lithotrophs.*

Compatible solute Compatible solutes are synthesized or taken up by cells to balance their intracellular osmotic pressure against that of the environment. Compatible solutes are deemed "compatible" because they do not perturb the internal functions of the cell. Most organisms accumulate organic compatible solutes when challenged by high salt. Halobacteria and certain halophilic eubacterial anaerobes accumulate inorganic salts.

Denitrification Denitrification involves anaerobic respiratory pathways found among many bacteria (facultative anaerobes) and certain fungi in which nitrate (NO_3^-) or other oxidized forms of inorganic nitrogen serve as terminal electron acceptors during the oxidation of organic carbon. These pathways are often induced when molecular oxygen is not available. Example: $(CH_2O) + 2\ NO_3^- \rightarrow CO_2 + H_2O + 2\ NO_2^-$.

Diagenesis Diagenesis describes the process of physical and chemical change that takes place within sediments after their deposition but before weathering or metamorphism.

Diapir A diapir is a dome or anticlinal fold of a rock formation in which a mobile, plastic core ruptures overlying rocks that are more brittle. Salt domes, formed by the upward "flow" of solid rock salt, are diapirs.

DOC Dissolved organic carbon. DOC is defined as the carbon released as CO_2 upon dry or chemical oxidation of a water or brine sample after acidification and sparging with N_2 or argon to remove inorganic CO_2.

EOC Extractable organic carbon. EOC is defined as the dissolved organic carbon that can be extracted from a water or brine sample by a chosen solvent.

Epigenetic Epigenetic processes are those that alter sedimentary structures or mineral content of rocks after they have been deposited.

Epilimnion The epilimnion of a stratified lake is the upper layer in which mixing occurs due to wind and convective currents. In a salinity-stratified lake, the epilimnion is the most dilute layer.

Eubacteria Eubacteria are a diverse group of procaryotes with certain molecular features in common, including peptidoglycan in cell walls (in most eubacteria), polar lipids containing fatty acid esters, and a high degree of similarity in nucleic acid sequences of evolutionarily conservative fractions of their ribosomal RNA.

Eucaryotes Eucaryotes include all organisms whose cells contain unit membrane-bound organelles (e.g., nuclei containing histone-associated DNA in chromosomes, centrioles, mitochondria, and chloroplasts) and in many cases, true flagella. The processes of vegetative (mitosis) and sexual (meiosis) cell division are concerted chains of events involving nuclear and centriolar components and are markedly different from cell division processes in procaryotes (all of which lack these components). Eucaryotes include protozoa, true algae, fungi, higher plants, and animals.

Eutrophic Eutrophic means "well nourished" and denotes high nutrient concentrations in a body of water which are often accompanied by high biological productivity.

Evaporite Evaporite minerals are those precipitated from solution as the result of the evaporation of water.

Extreme halophile Extremely halophilic organisms grow best in 2.5–5.2 M NaCl (range from about 15% salt to saturation). They may be inactivated or killed by lower salt concentrations.

Facultative anaerobe Facultative anaerobes are organisms that can grow under both aerobic and anaerobic conditions.

Fermentation Fermentation is the degradation of organic matter that does not involve molecular oxygen or other inorganic electron acceptors in a respiratory pathway. The products of fermentation usually include CO_2 in addition to reduced carbon compounds such as organic acids (e.g., lactate or acetate), ethanol, or volatile amines. Fermentation is sometimes called putrefaction when it is accompanied by the formation of malodorous amines.

Halobacteria Halobacteria (family Halobacteriaceae) are extreme halophiles of the archaebacterial kingdom. All grow by heterotrophy and some can obtain energy from light.

Halotolerant Halotolerant organisms can tolerate a wide range of salt concentrations. If they can thrive in salinities greater than 2.5 M NaCl, they may be considered extremely halotolerant.

Heterotrophic Heterotrophic organisms obtain their cell carbon from organic substrates, not from CO_2.

Holomixis Holomixis refers to a period of mixing in a lake when the lake water circulates completely from top to bottom.

Hypolimnion The hypolimnion is the lower (unmixed) layer of water in a stratified lake. It is usually characterized by a uniform density. In freshwater lakes it is characterized by uniform temperatures that are generally cooler than those of the upper layers. In salinity-stratified lakes, the metalimnion and hypolimnion may be subjected to extreme heating.

Liman Limans are narrow bays that cut deep inland. They form in previously existing river valleys as a result of coastal plain submergence or the transgression of seawater into the valleys. They may be open to the sea, partially blocked, or completely cut off from the sea. The term is used to describe these coastal features in the Black, Azov, and Chuckchi seas. A "ria", which is essentially a synonymous term, describes these coastal features in other parts of the world.

Meromictic Meromictic lakes are those that are permanently stratified and their waters do not circulate completely throughout the basin at any time of the year.

Metalimnion The metalimnion of a stratified lake is juxtaposed between the epilimnion and the hypolimnion. In freshwater lakes it is the layer in which the temperature gradient (the thermocline) is the greatest. In salinity-stratified lakes, the metalimnion corresponds to the pycnocline and sometimes to other steep chemical gradients such as those defined by O_2 concentration or Eh.

Methanogen Methanogens, which are obligately anaerobic archaebacteria, are the only organisms capable of producing methane as a major end-product of metabolism.

Microbial mat Microbial mats which receive light energy are typically laminated and composed of a surface layer of cyanobacteria (often with eukaryotic algae, especially diatoms in hypersaline environments), an underlying layer of phototrophic bacteria, sulfur-oxidizing bacteria at the O_2/H_2S interface, and bottom layers composed of organic matter at different stages of decomposition by fermenting and sulfate-reducing bacteria. Microbial mats have also been called algal mats, stromatolites, and potential stromatolites.

Mixolimnion The mixolimnion is the upper layer of a meromictic lake, characterized by low density, free circulation, and mixing by wind.

Moderate halophile Moderate halophiles are those that grow best in media containing 0.5–2.5 M NaCl (range from approximately seawater salinity to 15% salt).

Monimolimnion The monimolimnion of a meromictic lake is the dense, stagnant bottom layer that never mixes.

Monomictic Monomictic lakes are stratified lakes that completely turn over only once each year.

Natronobacteria Natronobacteria are halophiles and belong to the genera *Natronobacterium* and *Natronococcus*. They are the alkalophilic members of the family Halobacteriaceae.

Nitrate-reducing bacteria Nitrate-reducing bacteria are facultative anaerobes that respire using nitrate as the terminal electron acceptor instead of molecular oxygen (see DENITRIFICATION).

Nitrifying bacteria Nitrifying bacteria are autotrophs that obtain energy by oxidizing reduced forms of inorganic nitrogen (e.g., ammonia or nitrite).

Oligotrophic Oligotrophic conditions in an aqueous system denote low nutrient concentrations and characteristically low biomass.

Osmophilic Osmophilic organisms grow at high external solute concentrations. Halophilic bacteria are osmophiles. Many osmophilic fungi grow in high sugar rather than in high salt concentrations.

Osmoregulation Osmoregulation refers to an organism's capacity to regulate its internal osmotic pressure by regulating both cell water and solute concentrations. All procaryotes and wall-less algae such as *Dunaliella* are osmotic conformers, meaning they must maintain their internal osmotic pressure the same as that of their environment by selective synthesis or uptake (and conversely, breakdown or expulsion) of compatible solutes.

Oxygenic photosynthesis Oxygenic photosynthesis is the process in which CO_2 is reduced in a chlorophyll-mediated, light-driven reaction, electrons are derived from H_2O, and molecular O_2 is an end-product: $CO_2 + H_2O \rightarrow (CH_2O) + O_2$. Oxygenic photosynthesis occurs in cyanobacteria, eucaryotic algae, and plants.

Photoautotrophic Photoautotrophic organisms derive cell carbon by reducing CO_2 using energy from light.

Photoheterotrophic Photoheterotrophic organisms derive cell carbon from organic substrates and obtain energy from light.

Phototrophic Phototrophic organisms obtain energy from light. They may be further described as photoautotrophs or photoheterotrophs.

Playa A playa is a low, flat part of a basin or other undrained area in an arid region. A playa lake is a shallow temporary sheet of water covering a playa during the wet season.

POC Particulate organic matter. POC is separated from a water or brine sample by filtration on a glass-fiber filter, and the material retained on the filter is oxidized by wet or dry chemical techniques. POC is defined as the carbon removed as CO_2 by this technique.

Potash Potash refers to a variety of marine-derived salts of greater solubility than NaCl. They include potassium and/or magnesium salts containing chloride (e.g., sylvite and carnallite), sulfate (e.g., langbeinite), both chloride and sulfate (e.g., kainite), and salts containing calcium (e.g., polyhalite).

Procaryote Procaryotes include all bacteria (archaebacteria and eubacteria). They are structurally simpler than eucaryotes. They lack nuclei and other unit membrane-bound organelles. DNA and cell division lack the complexity of the eucaryotic processes of mitosis and meiosis. Most procaryotes are microscopic because they lack the capacity for cell differentiation and multicellularity of eucaryotes.

Psychrophilic Psychrophilic organisms are those that can live at low temperatures (generally less than 20° C).

Pycnocline The pycnocline of a stratified lake is the layer in which the density gradient is the greatest. In salinity-stratified lakes, the pycnocline is juxtaposed between the dilute mixolimnion above and the more saline hypolimnion below.

Sabkha Sabkhas are "salt flats" that include both coastal (supratidal) evaporitic environments as well as continental evaporitic environments associated with playa lakes. In sabkhas, the sediments are affected by capillary evaporation that results in the precipitation (and preservation) of evaporite salts. In sabkhas, hypersaline brines form within the sediments and occasionally on top of them.

Saltern A saltern is a commercially-operated system for producing NaCl (and sometimes other minerals) by the solar evaporation of brines. Salterns are also called salinas or saltworks.

Stromatolite Fossil and living stromatolites are organo-sedimentary structures that are usually laminated, and that were formed by the growth of microorganisms (often microbial mats) that trapped sediments. The sediments plus mats were cemented at the time of growth or diagenetically. The term "stromatolite" is preferably reserved for lithified structures while "microbial mat", "potential stromatolite", or other description is preferred for similar, uncemented structures.

Sulfate depletion Sulfate depletion refers to the relative loss of sulfate with respect to a conservative ion such as Cl^- in sediment porewaters due to the activity of sulfate-reducing bacteria.

Sulfate reduction Sulfate reduction to sulfide is a respiratory process found among a group of obligately anaerobic eubacteria that degrade organic acids and certain other low molecular weight organic compounds.

Syngenetic Syngenetic minerals are deposited at the same time as the surrounding rock.

TDS Total dissolved solids (determined gravimetrically).

Thalassic Thalassic refers to waters or salts of marine origin.

TOC Total organic carbon. TOC is the sum of DOC plus POC.

Volatile fatty acid Volatile fatty acids are simple and contain 2–6 carbon atoms (acetic, propionic, butyric, valeric, and hexanoic acids).

Water activity Water activity (a_w) of a solution at a given temperature is described by the quotient of the vapor pressure of the solution divided by the vapor pressure of pure water at the same temperature. This quotient also describes the relative humidity of the solution.

Index